THE BLACK HOLE AT THE CENTER OF THE MILKY WAY

Andreas Eckart • Rainer Schödel • Christian Straubmeier

University of Cologne, Germany

THE BLACK HOLE AT THE CENTER OF THE MILKY WAY

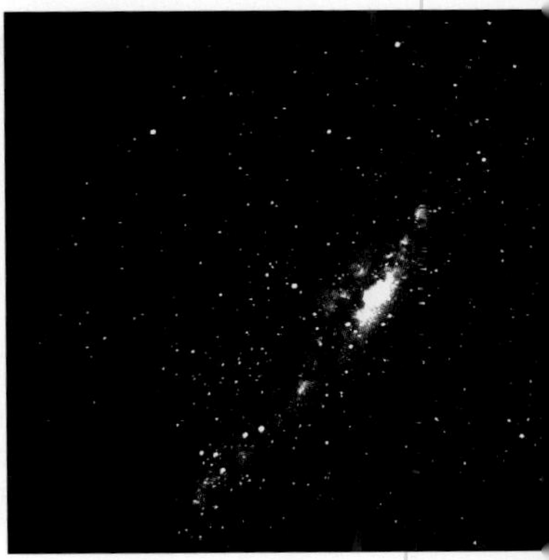

Imperial College Press

ICP

Published by

Imperial College Press
57 Shelton Street
Covent Garden
London WC2H 9HE

Distributed by

World Scientific Publishing Co. Pte. Ltd.
5 Toh Tuck Link, Singapore 596224
USA office: 27 Warren Street, Suite 401-402, Hackensack, NJ 07601
UK office: 57 Shelton Street, Covent Garden, London WC2H 9HE

British Library Cataloguing-in-Publication Data
A catalogue record for this book is available from the British Library.

THE BLACK HOLE AT THE CENTER OF THE MILKY WAY

ISBN 1-86094-567-8

Editor: Tjan Kwang Wei

Printed by FuIsland Offset Printing (S) Pte Ltd, Singapore

Preface

At a distance of only 8 kpc (26,000 light years) the Galactic Center is the closest nucleus of a galaxy, 100 to 1000 times closer than the nearest extragalactic systems. It is thus a unique laboratory in which physical processes that are relevant in general for the nuclei of all similar galaxies can be studied with the highest angular resolution possible. Research over the past 10 to 20 years has resulted in indisputable evidence for the presence of a 3-4 million solar mass Black Hole at the center of the Milky Way. In combination with spectroscopic measurements in extragalactic systems these extraordinary results suggest that there are nuclear black holes embedded in the bulges of all massive galaxies.

Due to the proximity of the Milky Way nucleus and thanks to the most recent instrumentation that allows observations at the diffraction limit of large telescopes we can study not only the evidence for the central black hole but also the detailed properties of the surrounding star cluster, such as the presence of a stellar cusp around the black hole, its mass function, its dynamics, and recent star formation processes.

The intention of this book is to provide the reader with a concise summary of some central aspects of the rapidly developing field of Galactic Center research. In chapters one and two we review a selection of fundamental instrumental techniques and observational results that are relevant for the Galactic Center. It covers the entire electromagnetic spectrum from decimeter radio-waves to high energy X-rays and γ-rays. A summary of the astrophysical interpretation is given in chapter three. Overall, we have concentrated more on aspects that deal with the nuclear stellar cluster in the central parsec and the compact radio source Sagittarius A* (Sgr A*; spoken "Sagittarius A star") at its center rather than the more extended components and large scale phenomena. For further details the reader is referred to the actual scientific articles and reviews in the corresponding sections. The reader may also consult general review articles on the topic by, e.g. Genzel & Townes (1987), Blitz (1992); Blitz *et al.* (1993), Genzel *et al.* (1994), Morris & Serabyn (1996), Mezger *et al.* (1996), Melia & Falcke (2001) and to the books and conference proceedings by Hehl *et al.* (1998), Falcke *et al.* (1999), Falcke & Hehl (2003), and Cotera *et al.* (2003). We would like to refer the reader to the web page of this book at *http://www.ph1.uni-koeln.de/~gcbook/*, where he can find additional electronic material and up-to-date information. Results of Galactic Center research are also continually published on the web site of the Galactic Center Newsletter (GCNEWS), *http://www.aoc.nrao.edu/~gcnews/*, a forum of all scientists involved in the field.

It also has to be mentioned that this compendium cannot be complete and is also biased. Most prominently it highlights observational techniques and especially image processing aspects. We have also put a high weight on a detailed presentation of the most recent work carried out in the near-infrared wavelength regime. All of these aspects had not yet been discussed in a unified manner in any available reference.

The book is aimed toward a number of audiences. It is a suitable basis for a single semester course at either the advanced undergraduate or graduate level. In fact it is based on a course given on that topic at the University of Cologne in summer 2003. The book has also been written for an audience with knowledge of elementary physics and calculus, who is interested in astrophysics and in the topic of the Galactic Center but who is not fully acquainted with the instrumental details and the terminology of most recent astrophysical research. Last, not least, we hope that this book will be a useful reference for scientists currently involved in Galactic Center research.

In general we suggest to use the book in the following way: the remaining part of the preface is thought as an overall summary and should be read first. For details the reader can then consult the chapters and sections of his interest. A list of abbreviations, a number of useful quantities, a bibliography comprising of more than 400 publications up to the editorial deadline in July 2004, and maps of the Galactic Center stellar cluster "for daily use" are included in this book.

The authors would also like to use this opportunity to thank all members of the MPE SHARP team that conducted the observations at the ESO NTT between 1991 and 2002. Especially we thank the ESO NTT team and the entire ESO staff that directly or indirectly helped to conduct these experiments. We are also grateful to the NAOS/CONICA team from MPIA/MPE, Meudon/Grenoble Observatories, ONERA, and ESO. In particular we are grateful to Reinhard Genzel for the extensive support and numerous scientific discussions. We are grateful to Mark Morris (UCLA) who made valuable comments to the manuscript. We also thank Sera Markoff (MIT) for her comments on some sections of this book. This work was supported in part by the Deutsche Forschungsgemeinschaft (DFG) via grant SFB 494.

Köln, Germany, July 2004

Andreas Eckart
Rainer Schödel
Christian Straubmeier

Towards the Center of the Milky Way

The gas and dust on the line of sight to the center make it impossible to observe the very center of the Milky Way at optical or UV wavelengths. It is only accessible in the radio, infrared, and X-ray domain. The optical extinction towards the Galactic Center (GC) amounts to almost 30 magnitudes (A_V=30^m; 1 magnitude corresponds to a factor of 2.5) or an attenuation of visible light by a factor of 10^{-12}. This corresponds in the near-infrared K-band at a wavelength of 2.2 μm to an extinction of only A_K=3.3^m.

Our home galaxy, the Milky Way, appears to be an ordinary, barred, spiral galaxy of type Sbc (Kormendy 2001). The central few 100 pc of the Milky Way is a region full of unique phenomena (Fig. 0.1). Due to its proximity, we can study the nucleus of the GC as an exemplary model for other galaxies of a similar type. In this preface, we will briefly review the phenomena at, and the properties of, the central few light years of the Milky Way. This includes its most peculiar object, the central supermassive black hole Sagittarius A* (Sgr A*). Figure 0.1 shows a representative sketch that gives a general overview of the central few parsecs of the Milky Way, and shows how various objects and phenomena are related there. We will tackle many topics which will be reviewed in the following chapters. For an easier readability of this short introduction we will not cite references

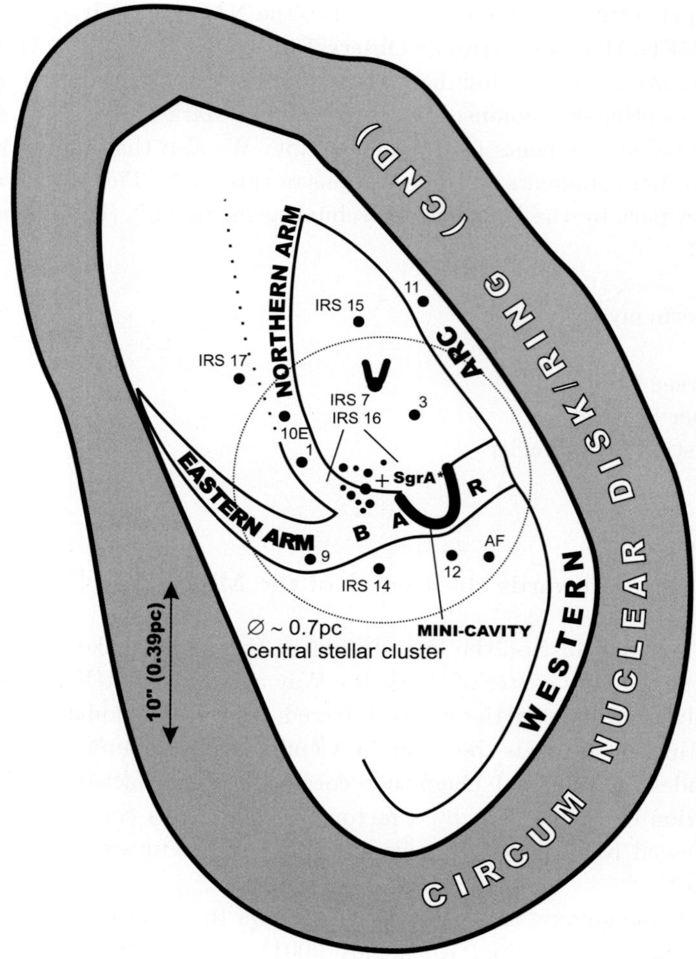

Fig. 0.1 Schematic picture of the different structures in the Galactic Center (see also Genzel *et al.* 1994). The "mini-spiral" arms of ionized gas – the eastern and northern arm as well as the "bar" – connect the central stellar cluster region with the circum nuclear gas ring. Although sketched as a continuous structure here, this ring is a rather clumpy association of molecular clouds. Some prominent stars are labeled with their IRS (InfraRed Source) number. The size of the central star cluster (two times its core radius of ~ 0.35 pc) is indicated by a dotted circle. The position of the compact radio source Sgr A* is indicated by a cross.

on the facts presented here. The reader will find all the relevant references in the corresponding sections of this book.

The Interstellar Medium at the Center

The central parsec is surrounded by an association of several clouds/filaments of dense ($10^4 - 10^7$ cm^{-3}) and warm (several hundred Kelvin) molecular gas, usually referred to as the circum nuclear disk (CND), see Fig. 0.1. The CND contains of the order 10^4 M$_\odot$ of gas and dust. The gas is generally interpreted to orbit the nucleus in a circular rotation pattern. The circum nuclear disk is probably fed by gas infall from dense molecular clouds at distances larger than 10 pc from the center. The CND is very clumpy (volume filling factor $\sim$ 10%) and extends no further than about 7 pc from the center. It has a rather sharp inner edge at a radius of 1.5 pc. Inside this radius there is a cavity, where only atomic and ionized, but hardly any molecular gas can be found. Several streamers of gas and dust appear to be on infalling trajectories from the inner edge of the CND. They are orbiting the center and interact with the intense winds emanating from the stars in the nuclear star cluster. The streamers of ionized gas form the so-called "mini-spiral", the intense thermal radio source Sgr A West. The total mass of the gas and dust in the mini-spiral does not exceed a few tens of solar masses. The mini spiral represents the brightest part of the thermal radio source Sgr A(West). The central 1.5 pc is comparatively devoid of interstellar matter and is referred to as the central cavity. Parts of the mini-spiral can be seen in the diffuse emission in the infrared L-band (3.8 μm) image of the central parsec shown in Fig. 0.4. More details on the interstellar medium in the GC can be found in the reviews by Genzel *et al.* (1994), Mezger *et al.* (1996), or Morris & Serabyn (1996).

The Nuclear Star Cluster

High-resolution near-infrared (NIR) adaptive optics observations at 8 m class telescopes allow the derivation of the distribution of stars in the GC that are brighter than $\sim 18^m$ in the K-band. This includes all red and most blue super giants, all red giants, and all main sequence stars down to about 2 M$_\odot$. Although several thousands of stars have been sampled by now in the central few arcseconds of the GC stellar cluster one has to keep in mind that all interpretations derived from these observations are still based on just around 1% of the total stellar content of the cluster.

On scales ≥ 0.5 pc, the stars at the center of the Milky Way are distributed according to a near-isothermal cluster. Stellar number counts indicate that the stellar volume density decreases with a power law of $\propto R^{-2}$ from projected radii of $100''$ down to about $10''$. Inside $\sim 10''$ ($1''$ corresponds to 0.039 pc at the distance of the GC) the number density of stars flattens out, indicating a core radius of the overall cluster of 0.34 ± 0.2 pc. A circle with twice the core radius indicates the scale of the nuclear cluster in Fig. 0.1. Within a few arcseconds (or about 0.1 pc) of the compact radio source Sgr A* there is a distinct peak in the stellar number density, exceeding the flat core of the isothermal cluster. The finding of this *cusp* is in agreement with theoretical predictions about the structure of a dense stellar system around supermassive black holes. Earlier, low resolution observations of the GC stellar cluster allowed examining the structure of the cluster only via the surface brightness distribution. The surface brightness peaks about two arcseconds east of Sgr A*, which was later on found to be caused by the presence of about a dozen very luminous stars that skewed the measurements. The number counts show, however, that the cluster peaks right on Sgr A*. Together with the finding of the cusp, this is in agreement with Sgr A* being a supermassive black hole that dominates the central light year by its gravity.

The most important components of the GC stellar cluster are indicated in Fig. 0.2. The position and names of the most prominent bright stars in the cluster are depicted in Appendix A, Fig. A.2. The brightest source in the central parsec is the red supergiant IRS 7, (spectral class MI, $T_{eff} \approx 3500$K) about $6''$ north of Sgr A*. It is a long period variable with an apparent average magnitude in K-band of ~ 6.5.

The luminosity in the central few arcseconds is dominated by about two dozen hot, massive, bright (K$\sim 9-12$), blue stars, which are characterized by strong HI/HeI emission lines. One of the so-called He-stars, IRS 16 SW, is most probably an eclipsing binary (period of 9.72 days) with a mass of $\leq 100\ M_\odot$. The He-stars have surface temperatures of $20000-30000$ K and are sources of intense stellar winds. They are mainly concentrated in the IRS 16 and IRS 13 clusters (marked in Fig. 0.2) and are probably witnesses of a star formation episode between 3 and 7 million years ago. The He-stars are not dynamically relaxed and display a coherent rotation pattern in counter direction to the overall galactic rotation. This could be interpreted as a remnant of the angular momentum of the original gas cloud from which these stars formed.

There is a second component dominating NIR images of the GC stellar

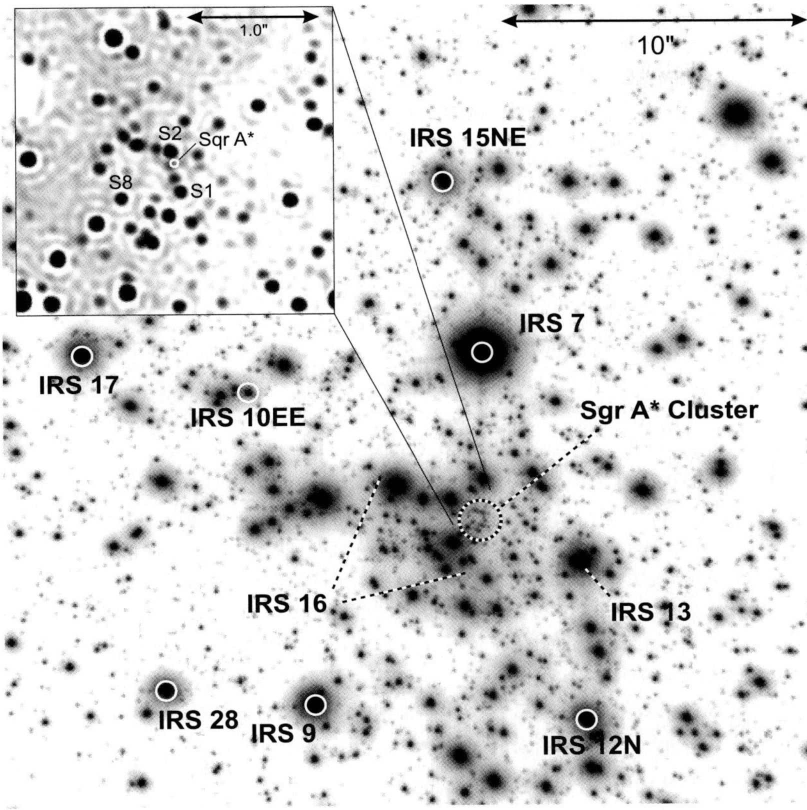

Fig. 0.2 This K-band (2.2 μm) image was obtained on May 10, 2003, with CONICA/NAOS at the VLT and covers the central $\sim 30 \times 30''$ ($\sim 1.2 \times 1.2$ pc) of the GC. The brightest source near the middle of the image is the late-type supergiant IRS 7 (K$\sim$ 6.5), which was used to lock the AO of NAOS. The IRS 16 and IRS 13 clusters of young, blue HI/HeI emission line stars are marked. Stars marked by circles are SiO maser stars that were used to establish the link between the NIR and the radio astrometry. A circle in the middle denotes the location of the cluster of stars in the immediate vicinity of Sgr A*, the so called "Sgr A* cluster". The inset in the left upper corner shows a region of $\sim 2 \times 2''$ ($\sim 0.08 \times 0.08$ pc) centered on this cluster. The inset image was obtained after applying a Lucy-Richardson deconvolution and subsequent restoration with a beam FWHM of ~ 60 mas. The location of Sgr A* is indicated by a white circle. The moderately bright (K$\sim 14 - 15$) sources S1, S2, and S8 near Sgr A* are labeled with their names. These stars and some other sources of similar brightness near Sgr A* are potential O/B type stars.

cluster. That is a group of intermediate bright (K$\sim$ 10 − 13), late-type stars. These stars have been identified as stars at the top of the asymptotic giant branch (AGB), which give testimony of another star formation episode about 100 million years ago. The strength of the CO absorption feature decreases in the central few arcseconds around Sgr A*. This may be due to an actual lack of late-type giants there because their envelopes are destroyed by collisions in the dense stellar cluster. Alternatively, the presence of luminous, blue stars (the "He-stars") in the central few arcseconds might dilute the CO absorption feature and/or there may be an increased presence of early-type stars in this region.

A specific concentration of K $\sim$ 14 − 15 stars can be found within about 1″ Sgr A* (see small inset in Fig. 0.2). Several of these stars exhibit proper motion velocities (i.e. velocities projected on the sky) in excess of 1000 kms^{-1} and even show signs of accelerated motion. Speckle imaging and adaptive optics spectroscopic observations showed that the so-called "S"-sources in the Sgr A* cluster are blue stars, possibly O or B main sequence stars. Absorption lines were found in recent spectroscopic observations of the brightest member of the Sgr A* cluster, S2 (alias S0-2), consistent with its classification as an $\sim$O9 star.

The presence of young stars in the very center of the nuclear cluster cannot be explained by standard theories of stellar evolution. Star formation in this region should be strongly inhibited by the tidal field of the black hole. Formation of O/B stars at greater distances, and their migration inwards via dynamical friction, appears hardly feasible within the short lifetime of massive stars. Various models are currently under discussion, such as stellar mergers, similar to blue stragglers in globular clusters, or the infall and dissolution of a massive young cluster, possibly aided by an intermediate mass black hole (IMBH). However, no convincing explanation could be found so far for the properties and origin of the He-stars and the O/B type stars in the Sgr A* cluster and their existence is currently one of most challenging problems in GC research.

A number of very cool objects with red, featureless spectra can be found mostly associated with the mini-spiral, such as IRS 21, IRS 1W, or IRS 10W. These sources are resolved in high-resolution imaging observations, i.e. they are not point sources. They have been interpreted as deeply dust enshrouded young stars, but more recent observations suggest that they might be rather massive stars, similar to the He-stars, that experience rapid mass loss. By plowing through the interstellar medium they create bow shocks and appear as extended sources.

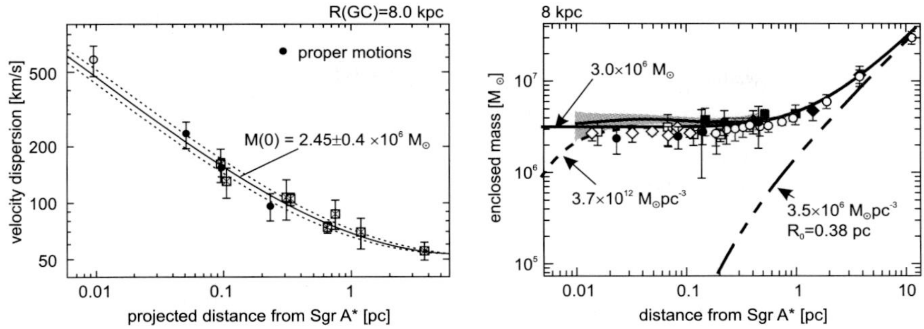

Fig. 0.3 Velocity dispersion and enclosed mass in the center of the Milky Way. Left hand plot: velocity dispersion plotted against the distance from Sgr A* as inferred by the evaluation of stellar proper motions in the nuclear cluster of the Milky Way. The Figure was taken from Genzel *et al.* (1997). For details, see their Figure 4. As expected in the potential of a point mass the velocity dispersion in the inner parsec follows Kepler's law and rises proportional $R^{-1/2}$. Here R is the distance to Sgr A*. Right hand plot: enclosed mass against distance from Sgr A* as inferred by gas and stellar dynamics. The various data points and error bars refer to different data sets and statistical methods. For details on this Figure, see Genzel *et al.* (2000), Figure 17. The long dash-short dash curve shows the mass of the near-isothermal, visible stellar cluster with a density of $\rho_0 = 3.5 \times 10^6$ $M_\odot pc^{-3}$ and a core radius of 0.38 pc, assuming a law of the form $\rho(r) = \rho_0[1 + (r/r_0)^2]^{-\alpha/2}$ ($\alpha \approx 2$ for an isothermal cluster). The straight line shows a model of this stellar cluster plus a point mass of 3.0×10^6 $M_\odot$. The dashed line is a model of the cluster plus a hypothetical, very dense cluster of central density 3.7×10^{12} $M_\odot pc^{-3}$ and core radius 0.0058 pc. A Plummer model has been assumed for this cluster ($\rho(r)$ defined as above, but with $\alpha = 5$).

Since stars are not subject to large scale forces produced by interstellar magnetic fields or winds, but only to gravity, they are ideal tracer particles for determining the shape of the gravitational potential in the GC. High-resolution NIR speckle imaging observations, followed by adaptive optics imaging at 8-10 m class telescopes that allow measuring stellar positions to within a few milli-arcseconds, have opened the way for measuring stellar proper motion velocities and accelerations in the nuclear cluster. In agreement with spectroscopic observations of line-of-sight velocities, these experiments showed that the velocity dispersion increases with distance R to Sgr A* as $R^{-1/2}$ as expected for particles on Keplerian orbits in the potential field of a point mass (see left hand plot of Fig. 0.3). Measurements from gas and stellar dynamics showed that the enclosed dark mass is constant from distances of 1 pc down to 0.01 pc from Sgr A* (see right hand plot Fig. 0.3). The observed accelerations of the innermost stars allowed pinpointing the location of the dark mass.

Recently, the orbits of a number of stars in the S-cluster could be measured. These Keplerian orbits added the so far strongest piece of evidence for the compactness of the dark mass in the central parsec. The mass density inferred from the orbits excludes clusters of dark astrophysical objects or agglomerations of exotic particles, such as a ball of heavy, degenerate neutrinos, as explanations for the central dark mass. The only plausible explanation left is that the dark mass is present in the form of a supermassive black hole that manifests itself as the compact radio, infrared, and X-ray source Sgr A*. The next generation of telescopes and interferometers operating in the near-infrared will possibly discover stars orbiting Sgr A* with periods as short as about a year. The observation of stellar orbits near the Milky Way's black hole will then provide a unique tool for testing predictions of special and general relativity.

Sagittarius A*

Right at the center of the stellar cluster one can find one of the most mysterious objects of the central region of the Milky Way, the manifestation of the central supermassive black hole, Sgr A*. It was originally identified as a compact radio source. It has a radio luminosity of $\sim 2\times 10^{34}$erg s^{-1}, corresponding to a flux density of 1.1 Jy at λ=2 cm, and a size of less than 2.4×10^{13}cm or roughly about the diameter of the Earth's orbit around the sun. Only in the past years was it possible to identify Sgr A* also in the infrared and X-ray domains.

The spectrum of Sgr A* is only well known at radio to sub-millimeter wavelengths, where it is flat to inverted, i.e. increasing flux density with decreasing wavelength. There is a specific excess of emission at sub-millimeter wavelengths; the so-called "sub-millimeter bump". At these wavelengths the electromagnetic radiation must come from plasma within a few Schwarzschild radii of the event horizon of the black hole.

There is a sharp drop-off of the emission toward the infrared domain and there had been no unambiguous detection of Sgr A* at infrared wavelengths until early 2003, when a faint and highly variable infrared source was found at the position of Sgr A*. In the infrared and X-ray domain, Sgr A* shows bursts of emission, generally called "flares", when its luminosity increases within minutes by factors of a few in the near-infrared and of a few up to 100 in X-rays. There are strong indications that activity in these two wavelength domains is related. The short variability time scales involved in the flares indicate that the emission must come from within about ten

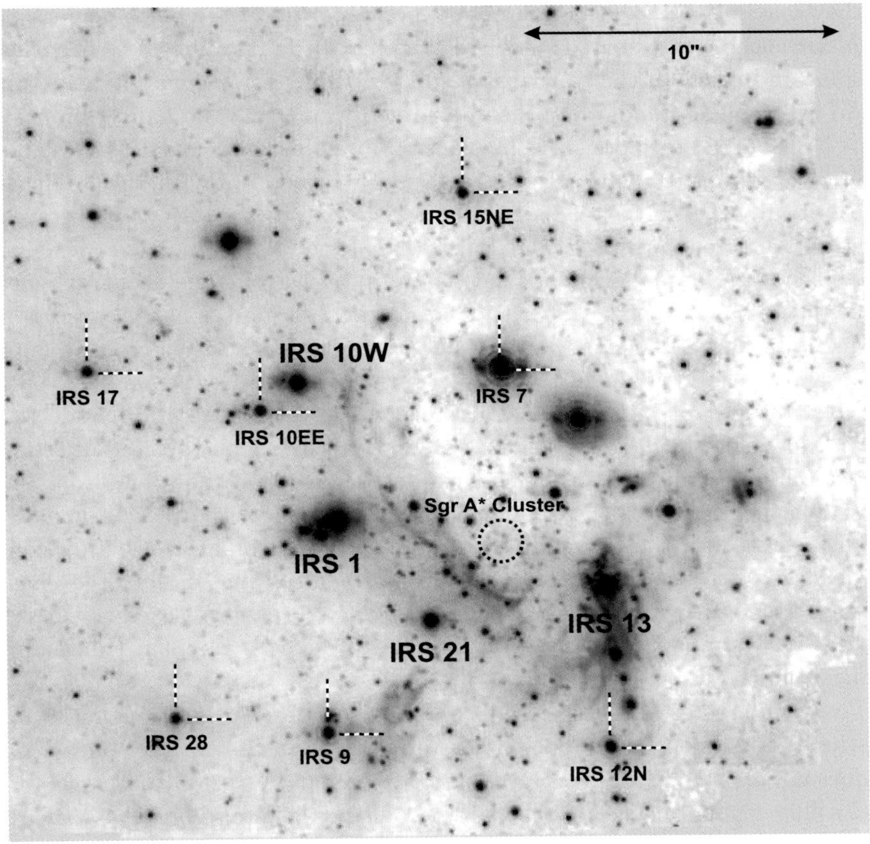

Fig. 0.4 This L-band (3.8 μm) image was obtained on May 10, 2003, with CONICA/NAOS at the VLT. The bright source near the middle of the image is the late-type supergiant IRS 7 (K$\sim$ 6.5), which was used to lock the AO of NAOS. The circle in the middle marks the location of the Sgr A* stellar cluster. Sources marked with upright, bold letters denote cool, deeply embedded, dust enshrouded sources, which are found to be associated with the gas and dust of the mini-spiral. Stars marked by circles and labeled with italic letters are SiO maser stars that were used to establish the link between the NIR and the radio astrometry (see chapter 2.7). The diffuse NE-SW emission feature near the middle of the image is part of the northern arm of the mini-spiral. The "bar" feature of the mini-spiral is also indicated in the image: it is the diffuse, broad emission feature running NW-SE across the IRS 13 complex.

Schwarzschild radii of the black hole.

Since the X-ray and near-infrared flares probe the region in the immediate vicinity of the event horizon and show rapid temporal substructures, they may provide information about the properties of the black hole itself. A currently hotly debated subject is whether the flares allow measuring the spin (and mass) of the supermassive black hole. Near the end of chapter 3 we will provide some of the theoretical background for these discussions.

Although there appears to be plenty of interstellar gas and dust available for accretion at the GC, Sgr A* radiates far below its Eddington luminosity at all wavelengths. This is only in part due to the very low observed accretion rate, which is probably less than 10^{-7} $M_\odot$ yr^{-1}. The conversion efficiency of thermal energy of the plasma into electromagnetic energy must be much lower than the 10% assumed in standard models of thin accretion disks. In fact, Sgr A* is the most under-luminous black hole accessible to observations. Therefore, observations of the Milky Way's central black hole have considerably stimulated theoretical research into accretion models of very low efficiency. These so-called Radiatively Inefficient Accretion Flows (RIAF) are connected to phenomena such as a massive outflow from the region near Sgr A* (i.e. that only a tiny fraction of the plasma is actually accreted) or the advection of thermal energy beyond the event horizon, which is generally related to a more radial infall of plasma into the black hole. Models involving a jet appear to be very successful in describing the emission from Sgr A* and are consistent with the most recent size measurements of the radio source.

At millimeter and longer radio wavelengths observations of Sgr A* are strongly hindered by interstellar scattering that smears out the source and does not allow imaging of its intrinsic structure. In the near future, however, possibly within this decade, it might become possible to resolve Sgr A* with very long baseline interferometry at sub-millimeter wavelengths. The most exciting prospect in this undertaking will be the possibility of imaging the shadow cast by the event horizon through gravitational deflection of the light in its vicinity. This imaging of the immediate environment of the event horizon will provide the ultimate proof of the black hole nature of Sgr A*.

Andreas Eckart, born 1957 in Schwelm/Westf. (Germany), studied physics and astronomy at the Westfälische Wilhelms Universität Münster, Germany. For his diploma and PhD thesis he worked at the Max-Planck-Institut für Radioastronomie in Bonn (Germany) studying the radio properties of quasars via very long baseline interferometry. After receiving his doctorate in 1984 from the University of Münster he became a scholarship holder of the Alexander von Humboldt-Stiftung at Steward Observatory in Tucson, Arizona (USA), where he worked for two years in the speckle interferometry group. In 1986 he joined the Max-Planck-Institut für extraterrestrische Physik in Garching (Germany) as a research staff scientist in the submillimeter-infrared group, working on the Galactic Center and the infrared and millimeter properties of extragalactic nuclei. After his habilitation at the Ludwig-Maximilians-Universität in Munich, Germany, he is now active as a professor for experimental physics at the Albertus Magnus University of Cologne (Germany) since January 2000. For his scientific work he recieved the "Otto Hahn Medaille" awarded by the Max Planck society in 1984, and the "Manne Siegbahn Medaille" awarded by the Manne Siegbahn Laboratory (Stockholm University, Sweden) in 2004.

Rainer Schödel, born 1971 in the Bavarian town of Burghausen (Germany), spent a year as an exchange student in Prague, Oklahoma (USA), at the age of 18. While studying geophysics at the Ludwig-Maximilians-Universität in Munich, Germany, he took the opportunity to study physics for a year in Sevilla (Spain), which was funded by the EU Erasmus program. He prepared his master thesis in the space plasma group of the Max-Planck-Institut für extraterrestrische Physik (MPE) in Garching, Germany, and obtained his diploma as a geophysicist in 2000. Also at MPE, he wrote his PhD thesis on the Galactic Center in the submillimeter-infrared group. In 2003 Rainer Schödel accepted a postdoc position at the Albertus Magnus University of Cologne (Germany).

Christian Straubmeier, born 1970 in Munich (Germany), studied physics at the Technische Universität München. For his diploma and PhD thesis he worked at the Max-Planck-Institut für extraterrestrische Physik in Garching (Germany) developing the detector system OPTIMA, a single-photon counting optical highspeed photometer, which he used to study the optical emission of pulsars, cataclysmic variable objects, and black hole candidates. After finishing his thesis he joined the astrophysical submillimeter-infrared group of Andreas Eckart at the Albertus Magnus University of Cologne (Germany) as a research staff scientist in December 2000. Concentrating his astrophysical studies at the Galactic Center of the Milky Way and nearby quasar host galaxies, he spends most of his time leading the near- and mid-infrared instrumentation group, which (at the moment of writing this book) contributes to the development of astronomical detector systems for the Large Binocular Telescope and the James Webb Space Telescope.

Contents

Chapter 1

Observational Techniques

Key to our understanding of the physics at the center of the Milky Way and in particular the central compact radio source Sagittarius A* and its infrared and X-ray counterparts is our capability to gather observational data. These observations have to be carried out over a wide range of wavelengths and spectral resolutions in order to identify and analyze the relevant physical emission mechanisms. In order to distinguish between the contributions from different sources and to isolate Sgr A* from the surrounding stars, gas, and dust, it is necessary to achieve the highest possible angular resolution at each observing wavelength. In this chapter we describe how these measurements are done. We outline how the radiation is detected, what the technical requirements of the used instrumentation are, and how high angular resolution imaging and spectroscopy are performed. When explaining the general aspects in each spectral domain we put special emphasis on the case of the Galactic Center. This information may be particularly useful for understanding the observational results and the astrophysical interpretation put forward in chapters 2 and 3.

1.1 The GC Across the Electromagnetic Spectrum

Most of the information that is required to understand the physics of an astronomical target is obtained through electromagnetic radiation. The physics and technology involved in the detection of this radiation are quite different for the individual wavelength bands. In the following sections we cover some of the fundamental principles that apply to observations from the radio to the γ-ray domain.

1.1.1 *Radio Wavelengths*

Electromagnetic radiation from the longest wavelengths all the way into the sub-millimeter domain is usually detected with radio heterodyne receivers, that convert the received radiation to a lower, intermediate frequency using a stable local oscillator source. The detection takes place at the intermediate frequency. Bolometer systems allow a direct detection with large fractional bandwidth and become important in the mm to sub-mm regime (see following section). At the focus of a large parabolic mirror with diameter D the angular resolution A of the receiving systems at a wavelength λ is given via $A \sim \frac{\lambda}{D}$. For the currently largest single dish telescopes operating at the shortest radio wavelengths the angular resolution is limited to about 10 arcseconds.

For the Galactic Center such a resolution is sufficient to map the large scale distribution of the radio emission. A substantially higher angular resolution is needed if the fine structure of that emission or the direct environment of a compact source has to be investigated. This can be achieved by combining the output signals of an array of several individual antennas to an interferometer, where the highest angular resolution is determined by the longest baseline B as seen in its projection from the target source in the sky, rather than the diameter D of the telescopes which then defines the field of view (FOV) of the interferometer. Technical and physical concepts of radio interferometry are given in a number of review articles and text books (e.g. Fomalont & Wright 1974; Meeks 1976; Thompson *et al.* 1986). For the Galactic Center the technique of radio interferometry is indispensable for the investigation of structure on angular scales of a few arcseconds and below. Since this method is currently in the process of also becoming feasible in the infrared domain (first experiments on the Galactic Center are currently being conducted) we will now give a concise summary of the relevant facts.

The output of an interferometer is called the visibility function $V(u, v)$. Here u and v are are the orthogonal components of the two dimensional projection of a baseline for each telescope pair as seen from the target. The complex visibility function $V(u, v)$ is linked to the intensity distribution $I(x, y)$ on the sky via a Fourier relation:

$$V(u,v) = \int\int e^{-2\pi i(ux+vy)} I(x,y)dxdy \quad . \qquad (1.1)$$

Mathematically the integral can be thought of as being carried out from

$-\infty$ to $+\infty$. However, due to the finite response of the primary beam and a limited sampling rate of the data, it is in practice only carried out over the solid angle of Ω_B of the primary beam or an even smaller region. Since $I(x, y)$ is a real function, V is hermitian and the following relation to its complex conjugate V^* holds:

$$V(-u, -v) = V^*(u, v) \quad . \tag{1.2}$$

The intensity distribution can be obtained via the inverse Fourier transform

$$I(x, y) = \int\int e^{2\pi i(ux+vy)} V(u, v) du dv \quad . \tag{1.3}$$

This means that at each position (u, v) the complex visibility function V represents a Fourier component of the source structure given by $I(x, y)$. Here the quantities u and v are spatial frequencies that are proportional to the angular resolution achieved by the projected baseline(s) of the interferometer. The integrals need to be carried out over the range covered by the u and v values. Unfortunately the coverage of visibility data in the u, v-plane is sparse since only a limited number of telescopes can participate in the measurement. For M discrete measurements of V this can be expressed using the weights w_j with j running from 1 to M as:

$$V(u_j, v_j) = V(u, v) w_j \quad , \tag{1.4}$$

with w_j=1 for $u = u_j$, $v = v_j$, and w_j=0 else. The measured intensity distribution is therefore given by

$$I'(x, y) = \sum_{J=1}^{M} V(u_j, v_j) e^{2\pi i(u_j x + v_j y)} \quad . \tag{1.5}$$

Since V is the product of two functions, the measured quantity $I'(x, y)$, also called the "dirty map", can be written as a convolution between the real intensity distribution $I(x, y)$ and the so called "dirty beam" DB:

$$I'(x, y) = I(x, y) \odot DB \quad . \tag{1.6}$$

Here $\odot$ denotes the convolution operator. The dirty beam is the interferometer response to a point source that results in a constant output signal for each telescope pair defining one interferometer baseline. Introducing a function $W(u_j^2 + v_j^2)$ to weight the contribution from the longest baselines, the dirty beam is given by

$$DB(x, y) = \sum_{J=1}^{M} W(u_j^2 + v_j^2) e^{2\pi i(u_j x + v_j y)} \quad . \tag{1.7}$$

$DB(x, y)$ consists of a central component and side lobes representing a more complicated response to a point source resulting from the insufficient coverage of the u, v-plane. The so called "clean beam" is in most cases constructed via a Gaussian fit to the central component of the dirty beam. Correspondingly the measured distribution $I'(x, y)$ is called the "dirty map". The desired true intensity distribution $I(x, y)$ can be obtained by deconvolving $I'(x, y)$ with $DB(x, y)$. Here $DB(x, y)$ can be determined from the u, v-plane coverage. The deconvolution is mostly done via an iterative subtraction of point source response functions $DB(x, y)$ and is then called "cleaning".

The resulting list of "clean components" is restored to a "clean map" by convolution with a "clean beam". There are "clean algorithms" of various complexity optimized to deal with large maps and extended structures for which the implicit assumption of a point source representation is not a good approximation. The quality of the map $I(x, y)$ also depends critically on the calibration of the complex amplitudes and phases of the visibility function, the phase calibration is especially problematic. The visibility phases carry, in addition to information on the source structure, information on the source position. For compact radio interferometer arrays the calibration can be tied to a common local oscillator signal. Very long baseline interferometry includes trans- and intercontinental baselines or even baselines to satellite antennas. Here the local oscillator signal is replaced by independent exact clocks and so called "closure quantities" between a minimum of three stations must be used for mapping (see the review articles mentioned before for details).

For the Galactic Center, radio interferometry is an essential observational method to study the emission on all scales. Deep and sensitive maps in the decimeter to short centimeter wavelength range are obtained with compact arrays like the Very Large Array (VLA). With more than 27 antennas the u, v-plane coverage is excellent and ideally suited to map for instance details of the gas streamers in the central parsec. The achievable dynamic range (intensity range between the brightest and weakest

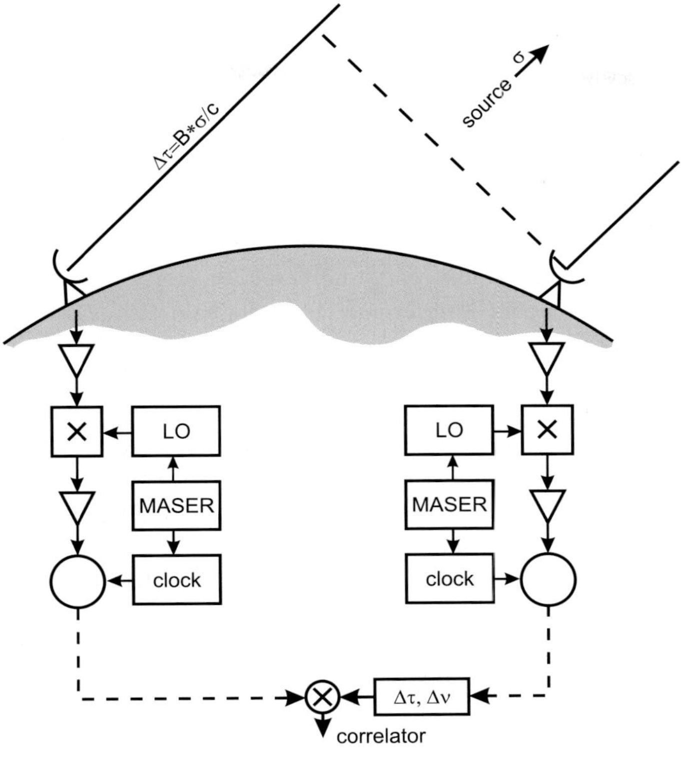

Fig. 1.1 Simplified schematic representation of a single baseline VLBI experiment. Each station, both of which define a baseline vector B, is equipped with an independent frequency and time standard (local interferometers share a common standard). The vector σ points towards the source. The amplified and digitized data stream is written to a storage medium and brought to a common location. Here the delay $\Delta\tau$ and Doppler shift correction $\Delta\nu$ is applied before the correlation. The resulting data stream (which is complex, since the correlation is also performed with a $\pi/2$ shift in one arm of the interferometer) contains the visibility information, i.e. Fourier amplitude and phase information of the source structure.

features that can be reliably detected) in the maps is, however, limited by the presence of the strong point source Sgr A* associated with the central massive black hole. At the longest wavelengths (e.g. 90 cm) the calibration is complicated additionally by the disturbing influence of the ionosphere on wave propagation. For radio interferometers operating in the (sub)mm-wavelength range the calibration must cope with similar problems due to the humid atmosphere.

Highest angular resolutions can be obtained using intercontinental baselines or even baselines that include satellite dishes. This Very Long Baseline Interferometry (VLBI; see Fig.1.1) involves radio and millimeter telescopes all over the world. The highest angular resolution for the Galactic Center can currently be achieved at wavelengths of 3 and 7 mm. With a FWHM (full width at half maximum intensity) of the beam of $\sim$100 microarcseconds the linear resolution at a distance of 8 kpc is 0.8 astronomical units, corresponding to about 14 times the Schwarzschild radius of a 3 million solar mass black hole or 124 million kilometers.

Most of the compact interferometers and single dish VLBI radio telescopes are located on the northern hemisphere. For the Galactic Center this means that the u, v-plane suffers correspondingly and that the final beam shape is elliptical with a lower angular resolution in north–south direction. Since the Galactic Center is located at a declination of about -29° this also means that for observations from the northern hemisphere the source will always be at low elevations above the horizon. For the longest and shortest wavelengths phase and amplitude calibration due to disturbing influences of the ionosphere and atmosphere become especially demanding.

Millimeter arrays that are actively involved in observations of the Galactic Center are currently the Berkeley-Illinois-Maryland Association (BIMA) and the Owens Valley mm-array. Both arrays will form the new Common ARray for Millimeter Astronomy (CARMA) within the next few years. The Sub-Millimeter Array (SMA) will become of increasing importance for observing the Galactic Center. The SMA consists of eight 6 m elements that are reconfigurable to achieve baselines from 8 to 508 m (Moran 1998). The array is located on Mauna Kea in "Millimeter Valley", at an elevation of 4,080 m. It covers all bands from 180 to 900 GHz. The efficiency of the antennas is high: excluding the outer most 10 cm of the 3 m radius dishes to avoid edge effects, the average surface rms is 13 μm and for the inner 2.75 m radius that value is 12 μm. Zhao *et al.* (2003) report results from observations of Sgr A* at short-/submillimeter wavelengths made with the partially finished SMA on Mauna Kea. These observations resulted in the detection of three flares from Sgr A*.

In the mm and sub-mm domains further progress in dynamic range and sensitivity, as well as u, v-plane coverage and achievable angular resolution will be possible in the near future with the Atacama Large Millimeter Array (ALMA). Limited operation will start in 2006-2007. Located in the Atacama desert in northern Chile at an altitude of 5000 m an interferometer of 64 antennas with 12 m diameter each will allow for a typical resolution of 100 mas at 300 GHz in a 10" FOV (Field of View). The frequency range will be at least 100 (possibly 30) to 1000 GHz. Since the Galactic Center will culminate close to zenith at that location, ALMA will be an ideal instrument for this source. At longer radio wavelengths the Square Kilometer Array (SKA — at a not yet determined location) is planned and may lead to even higher dynamic range maps of the Galactic Center. The SKA will operate over a possible frequency range of 0.15 to $\sim$20 GHz allowing for an angular resolution of 10 mas at 20 GHz and a FOV of the order of 1 degree at 1.4 GHz. Both ALMA and SKA could be used as phased arrays in mm- and cm-VLBI networks, which would improve high angular resolution radio interferometry for the Galactic Center considerably.

1.1.2 *Far-Infrared Wavelengths*

The far infrared (FIR) is probably one of the technically most challenging wavelength domains, but the development of detectors, cryogenic instrumentation, and the telescopes themselves make rapid progress. Between about 30 and 400 μm the Earth's atmosphere is opaque for radiation (Fig.1.2). FIR measurements therefore have to be carried out from high sites with low humidity, from airborne telescopes, or satellites. However, a major drawback — basically resulting from these technical restrictions — is the limited angular resolution in this spectral domain caused by the modest diameters of suitable primary telescope mirrors (e.g. 2.5 m in the case of the airborne observatory SOFIA).

For the Galactic Center this spectral range is of special interest since it allows one to trace the larger scale (typically in the ten to a few ten arcseconds range) distribution of warm and cold dust as well as the atomic neutral and partially ionized interstellar medium at and towards the center of the Milky Way. The primary temperature range of the gas and dust that is covered between 10 μm wavelength and the sub-mm domain stretches from about 300 K to a few Kelvin. The broad band spectrum of the central source Sgr A* peaks in the sub-mm/FIR domain and then falls sharply off towards the shorter wavelength infrared. Flux densities — or their limits — of Sgr A* are essential to study and distinguish between different emission processes.

In the near sub-millimeter wavelength range (a few 100 to 1000 μm) radio heterodyne techniques are applicable for high spectral resolution work. Schottky and SIS (Superconductor-Insulator-Superconductor) mixers are in use. This section of the FIR is also the range within which most bolome-

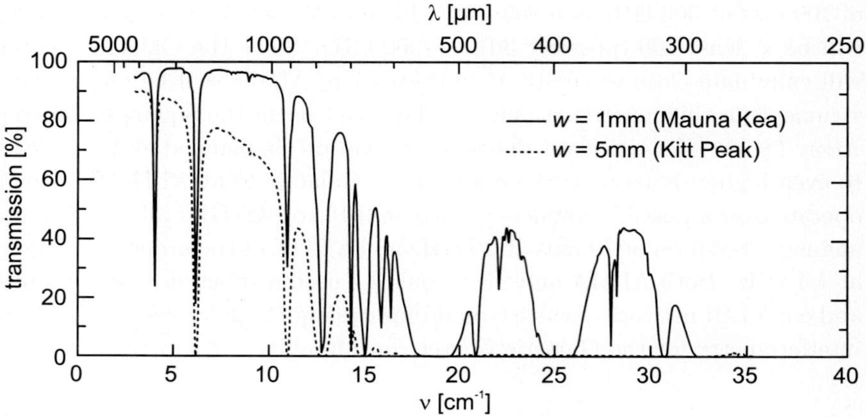

Fig. 1.2 The atmospheric transmission spectrum between 0.25 and 3 mm wavelength at the Mauna Kea and Kitt Peak sites (Griffin *et al.* 1986; Cox 1999). "w" is the precipitable water vapor at the sites. The graph demonstrates that — with the exception of two windows — the atmosphere is opaque shortward of 0.6 mm wavelength.

ters are operated. These devices absorb photons and convert them into heat and a corresponding variation in their electrical resistance.

The energy response of semi-conductors can be optimized for the deep FIR domain (a few 100 to about 30 μm) using the addition of impurities as well as the application of mechanical stress. This extends the operation of these extrinsic photo-conductors from the near- and mid-infrared to just beyond 200 μm wavelengths.

Bolometers and extrinsic photo-conductors can be used as broad band detectors or can be combined with Fabry-Perot interferometers to achieve high spectral resolution. In cases of high reflectivity and a low absorption coefficient of the detector material (e.g. in the case of Ge:Ga) each detector is located in an integrating cavity to maximize the quantum efficiency.

In the following paragraphs, we will give a brief overview of FIR observations of the Galactic Center.

Heterodyne measurements in the near sub-millimeter domain:
In the near sub-mm domain radio heterodyne techniques can be used to perform high frequency resolution measurements of selected lines. More recently such observations were performed during the austral winter seasons of 2001 and 2002 at the Antarctic Sub-millimeter Telescope and Remote Observatory (AST/RO). This station is located at 2847 m altitude at the Amundsen-Scott South Pole Station. Due to its very low water vapor, high

atmospheric stability, and a thin troposphere the site is exceptionally well suited for sub-millimeter observations. AST/RO is a 1.7 m diameter, offset Gregorian telescope capable of observing at wavelengths between 200 μm and 1.3 mm (e.g. Stark *et al.* 1997, 2001). Simultaneous 461-492 and 807 GHz observations were performed using a dual-channel SIS waveguide receiver (Walker *et al.* 1992; Honingh *et al.* 1997) in order to study the distribution of warm molecular and atomic gas traced by the 461 GHz CO(4-3), 807 GHz CO(7-6), and 492 GHz [CI] lines in the inner 3 degrees of the Galactic Center area.

Bolometric measurements in the near sub-millimeter domain:
Near sub-mm bolometric measurements of the Galactic Center are possible from the ground. Chini *et al.* (1986) used the ^{3}He cooled MPIFR (Max Planck Institute für Radioastronomie) bolometer at a wavelength of 1.3 mm at the 3 m diameter IRTF (Infrared Telescope Facility) on Mauna Kea, Hawaii. These measurements resulted in a map of the large scale ($\sim$20'$\times$10') distribution of the cold dust. Zylka *et al.* (1995) used a single element ^{3}He cooled Ge:In:Sb bolometer at 800, 600, and 450 μm (UKT14 receiver Duncan *et al.* 1990) at the James Clerk Maxwell Telescope (JCMT) to determine the flux densities of Sgr A* and to discriminate its contribution from the cold dust distribution in the central parsec.

The SCUBA array was used to map the distribution of warm dust in the Galactic Center region (e.g. Pierce-Price *et al.* 2000). SCUBA (Holland *et al.* 1999) is a sub-millimeter continuum bolometer camera with two arrays. It is operated at the JCMT. One is optimized for 450 μm wavelength and has 91 pixels. A second array with 37 pixels is optimized for 850 μm. Both arrays can be used simultaneously. The angular resolution is 8" FWHM at 450 μm and 15" at 850 μm. It is usually operated in combination with a secondary mirror chopping between the source and the sky at a rate of a few Hertz.

SPIFI (Benford et al. 2003) is an imaging Fabry-Perot interferometer designed at Cornell University for use in the 350 and 450 μm telluric windows available on Mauna Kea as well as the 200, 350 and 450 μm windows available to the AST/RO telescope at the South Pole. The system employs a 5 $\times$ 5 array of silicon bolometers operated at a temperature of 60 mK using an adiabatic demagnetization refrigerator. SPIFI has been used to study the line emission of the molecular (the CO(7-6) feature at 371.651 GHz) and neutral atomic (the $^3P_2$-$^3P_1$ [CI] line at 370.415 GHz) gas phase in the central few parsecs — especially the circum nuclear ring (Stacey *et al.* 2004).

Staguhn *et al.* (2003) presented the first preliminary sub-millimeter continuum images of the Galactic Center region obtained with the new Caltech Sub-millimeter Observatory facility camera SHARC II (Dowell *et al.* 2003). The instrument allows observations at 350 μm wavelength with unprecedented sensitivity and instantaneous spatial coverage. SHARC II is a 12$\times$32 high filling factor array of doped silicon bolometers each 1 mm$\times$1 mm in size.

Continuum measurements in the deep FIR:

In this wavelength domain measurements are only possible from satellites or high altitude airplanes (or — mostly on larger angular scales — from balloons, which are not covered here). Werner *et al.* (1988) presented the first detection of linear polarization of the far-infrared (100 μm) emission from the about 3 parsec diameter dust ring surrounding the Galactic Center. The observations were carried out using the University of Chicago single-beam far-infrared polarimeter on board of the NASA Kuiper Airborne Observatory (KAO). Morris *et al.* (1992) used the array polarimeter STOKES on board of the KAO to measure the polarization of the 100 μm continuum emission at 14 positions in the dense, warm molecular cloud associated with the arched filaments near the Galactic Center.

Dust distribution, composition, and energetics of dust at the Galactic Center were studied via broad band continuum measurements in the 16 to 45 μm wavelength domain using instrumentation flown on the KAO (e.g., Rieke *et al.* 1978; Chan 1995; Chan *et al.* 1997; Telesco *et al.* 1995; Latvakoski *et al.* 1996, 1999). Far-infrared $\sim$30 μm KAO continuum observations of the Galactic Center filaments, as well as the process of dust destruction in that region, are also discussed in Erickson *et al.* (1995) and Stolovy *et al.* (1995).

Spectroscopy in the deep FIR:

Deep FIR spectral line measurements towards the Galactic Center are well suited to investigate the interstellar medium in that region. Some of the brightest and most important cooling lines of the diffuse ISM are found in the deep FIR, e.g. the [CII] 158 μm and [OI] 63 μm fine structure lines. Measurements in this spectral domain reveal structure, temperature and density of the more extended atomic neutral and partially ionized gas component. Early spectrally resolved FIR measurements were obtained by Genzel *et al.* (1990) using the Mark II UC Berkeley cryogenic Fabry-Perot spectrometer (Lugten 1987) on the Kuiper Airborne Observatory (KAO). These measurements were continued with the MPE UCB FIR Imaging Fabry-Perot Interferometer using a 5$\times$5 stressed Ge:Ga array (Poglitsch *et al.* 1991; Jackson *et al.* 1993). In addition the [OIII] 52 and 88 μm, [NIII] 57 μm, and [OI] 63 μm lines in the Radio Arc (Thermal Arches) regions of the Galactic Center (Timmermann *et al.* 1996) were mapped.

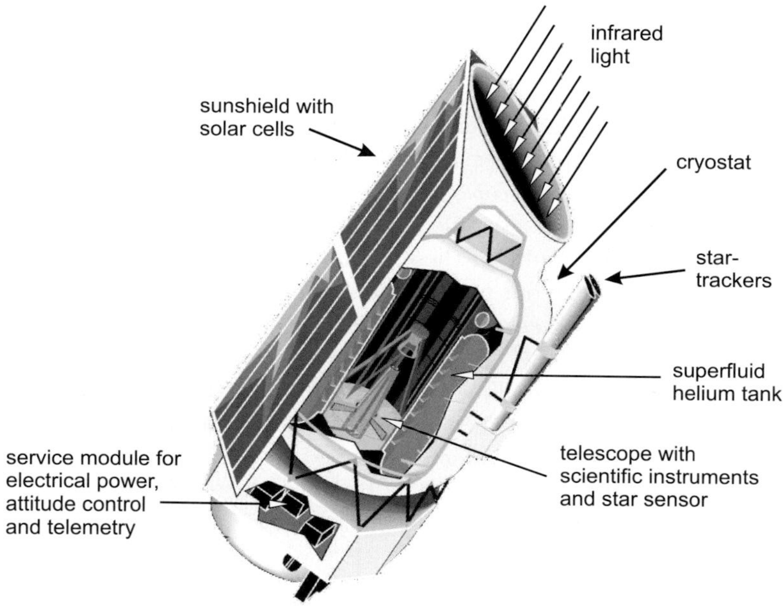

Fig. 1.3 Schematic drawing of the Infrared Space Observatory (ISO), launched in November, 1995, by the European Space Agency (ESA).

Further progress in terms of sensitivity, dynamic range, and angular resolution can be expected from future instrumental developments. As an example, the MPE built Far-Infrared Field-Imaging Line Spectrometer (FIFI LS; e.g., Looney *et al.* 2003) including for the first time at these wavelengths an integral field spectrometer, will be flown on board of the joint NASA/DLR airborne observatory *SOFIA*. It will be observing in the spectral ranges between 40-105 μm and 105-210 μm. The integral field unit consists of a reflective image slicer system that slices the 5$\times$5 pixels field-of-view into a 25$\times$1 pixels pseudo-long slit. The system therefore permits taking data cubes i.e. combined positional and spectral (R = $\lambda/\Delta\lambda$ = 1400 - 6500) information.

Space projects:

Comprehensive information about the FIR spectrum towards the Galactic

Center was obtained using the spectrometers on-board of ESA's *Infrared Space Observatory (ISO)*, launched in November 1995 (see Fig.1.3). The measurements covered a wavelength range between 2.38 μm and 45.2 μm (SWS; Short Wavelength Spectrometer de Graauw *et al.* 1996a) and 60 μm and 168 μm (LWS; Long Wavelength Spectrometer Clegg *et al.* 1996; Swinyard *et al.* 1996). The observations gave information on the composition and extinction of the gas, dust, and ices within a few arcseconds diameter region towards the Galactic Center (Lutz *et al.* 1996; Lis & Menten 1998; Chiar & et al. 2001; Moneti *et al.* 2001; Rodríguez-Fernández *et al.* 2001).

The *Spitzer Space Telescope*, launched by NASA in 2003, carries out a galactic plane survey and a detailed, sensitive investigation of many nearby galactic nuclei. The results of these ongoing measurements will certainly have a deep influence on our understanding of the center of the Milky Way. The *Herschel Space Observatory* with its FIR cameras and spectrometers will allow for a broad range of opportunities to measure the continuum and line emission in the sub-millimeter and far-infrared domain. The Herschel Space Observatory is the 4th cornerstone mission within the ESA HORIZON 2000 programme. The launch is planned for 2007. Three major instruments will be flown on Herschel: The Photoconductor Array Camera and Spectrometer (PACS) for medium resolution spectroscopy (R=1-2$\times10^3$) in the wavelength range 60-210 μm, SPIRE (Spectral and Potometric Imaging Receiver) a camera and low resolution spectrometer (R=10^{2-3}) for λ>200 μm, and HIFI, the Heterodyne Instrument for the Far-Infrared for high resolution spectroscopy (R$\sim10^7$) at wavelengths 127 μm<λ<625 μm. The goal of Herschel's mission will be detailed studies of the physics and kinematics of the interstellar medium in the gas layer within the central 50 parsecs, including outstanding gaseous components like the circum nuclear disk (CND), the arches, and radio filaments (see chapter 2).

1.1.3 *Near- and Mid-Infrared Wavelengths*

For the Galactic Center near-infrared imaging and spectroscopy of the wavelength range between 1 μm and 30 μm is of special importance. In the wavelength domain from 1 to about 5 μm sub-arcsecond angular resolution observations are possible, the extinction in that domain is more than ten times lower than in the optical, and at the same time it is the perfect window through which the bulk of the stars within the central stellar cluster can be observed. At wavelengths longward of 3 μm the emission of warm/hot dust becomes increasingly important. Observations of the motions and populations of the stars in the Galactic Center allow conclusions about the mass distribution and star formation history in that area. In order to achieve this goal infrared focal plane detector arrays with a fine pixel scale are required in order to obtain diffraction limited images using large apertures or interferometers. Here we give only a very short summary of the camera and detector requirements and a few aspects that are important for the Galactic Center.

Camera systems in that wavelength domain have to use either reflective optics (mirrors) or transmissive optics (lenses and windows) made out of materials which are highly transmissive in that spectral range (e.g. BaF_2 0.15–15 μm; KBr 0.23–25 μm; NaCl 0.21–26 μm; ZnSe 0.5–22 μm; ZnS 1.0–14 μm to name a few common ones; or special glasses). Optics and detectors have to be cooled to liquid nitrogen (77 K) or liquid helium temperature (4 K) in camera dewars, and the thermal heat load through the shielding and entrance window(s) must be minimized in order to keep the background low on the detector. Comprehensive information on these topics is given by Rieke (1994) and Glass (1999). Due to the variability of an increasingly brighter sky background at wavelengths longer than about 3 μm position chopping is required — preferentially with a fast telescope secondary mirror.

Between wavelengths of about 1 and 10 μm photodiode technology can be applied using HgCdTe, InSb, PtSi or other materials. With the application of anti-reflection coating, quantum efficiencies in the range of 90% and a read noise of about 10 to several 10 electrons can be achieved. Between 4 and 40 μm array detectors based on extrinsic silicon photoconductors or silicon blocked impurity band (BIB) detectors can be used. These devices have typical quantum efficiencies of 30-80% and read noise values of the order 50 electrons.

Very sensitive and increasingly large format near-infrared arrays have been built using Indium bump bonded hybrid technology (Fig. 1.4; see also reviews by Norton 1991; Scribner *et al.* 1991; Rieke 1994; Glass 1999; Amico *et al.* 2004). In these devices a monolithic infrared active layer made out of HgCdTe or InSb is contacted via small Indium bumps to a multiplexer unit. This unit is based on MOSFETs that can collect charge on their gate capacitors. The entire area over which the charge is collected defines a pixel. Pixel sizes are of the order of 40 μm diameter and the overall filling factor of the array with pixels approaches 100%. The pixels are made on a wafer of appropriate semiconductor substrates (multiplexer unit) and bonded with small Indium contacts to one side of the infrared detector

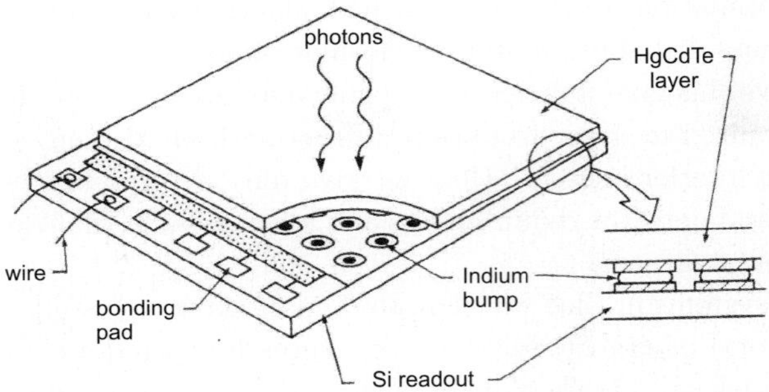

Fig. 1.4 The hybrid structure used for the Rockwell Hawaii 2 and ALADDIN arrays. The IR sensitive HgCdTe layer is bonded to the multiplexer unit via Indium bumps.

material. Therefore this material has to be illuminated from the back. The detector substrate must be thin enough ($\sim$10 μm) to let the created photo charge carriers migrate from where they have been produced to the gate capacitors as inputs for the readout electronics. Electronic switches can activate individual columns and rows such that individual pixels can be addressed. This random access mode allows any combination of sub-units of the array which will result in higher readout compared to the overall array. These hybrid arrays have been used for most of the telescope camera systems with which the Galactic Center has been observed to date.

The Galactic Center has been observed with the Hubble Space Telescope (HST) on a regular basis. The NICMOS (Near-Infrared Camera and Multi-Object Spectrometer) instrument was inserted into the HST in February 1996. It is equipped with three cameras with plate scales of 0.045"/pixel, 0.75"/pixel, and 0.2"/pixel. The cameras can be operated simultaneously, and a selection of filters, polarizers, and 3 slitless grisms for low resolution spectroscopy were in use. Especially in the 1-2 μm region, where HST's background is extremely low, the high sensitivity and very stable point spread function as well as the accessibility of the wavelength ranges between the atmospheric windows are a strength of NICMOS. These properties allowed for a variety of studies including sensitive searches for variable sources and accurate colors across the 1 to 2.5 μm region (e.g. Stolovy *et al.* 1999; Rieke 2003). Furthermore, its 0.2" pixel scale was very well suited for sensitive mapping of extended emission lines. High spatial resolution measurements of shocked molecular hydrogen lines have been carried out (Yusef-Zadeh *et al.* 2001). The atomic hydrogen Paα-line has

Fig. 1.5 Images of the Keck and VLT sites. The ESO VLT (top) is located in Chile on Paranal in the northern part of the Atacama desert near Antofagasta. The site harbors four 8.4 m diameter telescopes (see *http://www.eso.org*). The Keck telescope (bottom) is located on Mauna Kea, Hawaii, consisting of two 10 m diameter telescopes (see *http://www2.keck.hawaii.edu*).

been used in combination with information from other wavelength domains to derive a high angular resolution extinction map for the central 3 pc of the Galactic Center (Scoville *et al.* 2003).

The angular resolution that can be achieved from the ground or from satellites is an important topic. The HST primary mirror has a diameter of 2.4 m. At a wavelength of 2.1 μm this results in a diffraction limited angular resolution of about 0.2". For investigations at higher angular resolutions, large ground-based telescopes like the ESO Very Large Telescopes or the Keck telescope (see Fig.1.5 and following sections) equipped with adaptive optics are currently best suited. From the ground the angular resolution of the observations is either determined by the atmosphere (in direct long exposures) or the diffraction limit of the primary telescope mirror (for speckle interferometric and adaptive optics observations). At a wavelength of 2.1 μm one obtains a resolution of about 45 milli-arcseconds (mas) with a 10 m class telescope. In order to achieve even higher angular resolutions telescopes have to be combined to near-infrared interferometers (see section 1.6.1).

1.1.4 *Optical Wavelengths*

The overall extinction toward the Galactic Center is large. Approximately 27 magnitudes of visual extinction correspond to a factor of 1.6×10^{11} by which the intensity of yellow, optical light at a wavelength around 550 nm that originates at the center is reduced. However, the situation improves dramatically towards near infrared wavelengths (see preceeding section), where the most sensitive observations can be done using charge coupled devices (CCD). These are arrays based on intrinsic silicon detectors in which the charges collected on a single pixel are electronically conveyed from pixel to pixel to a common readout amplifier outside the photosensitive array. Biretta *et al.* (1982) obtained direct CCD images and spectra at the 5 m Hale telescope. With two-color CCD observations of the Galactic Center region at 800 nm and 920 nm they confirmed the discovery of two very red optical objects by Grindlay & Liller (1978) that are located at an angular distance of only a few arcseconds from the non-thermal compact radio source Sgr A*. They refer to them as sources A and B (labeled "star 1" and "star 2" in Fig. 1.6). With optical spectra covering the wavelength range between 700 nm and 1000 nm, Biretta *et al.* (1983) could exclude that these objects are HII regions or compact star clusters at the Galactic Center. They concluded that these sources are most likely reddened foreground stars located a few kiloparsecs from the Galactic Center.

In the following years the Galactic Center area was observed at 1 μm both from the ground (Henry *et al.* 1984; Rosa *et al.* 1992) as well as using the HST (Liu *et al.* 1993). In addition to the foreground sources A and B

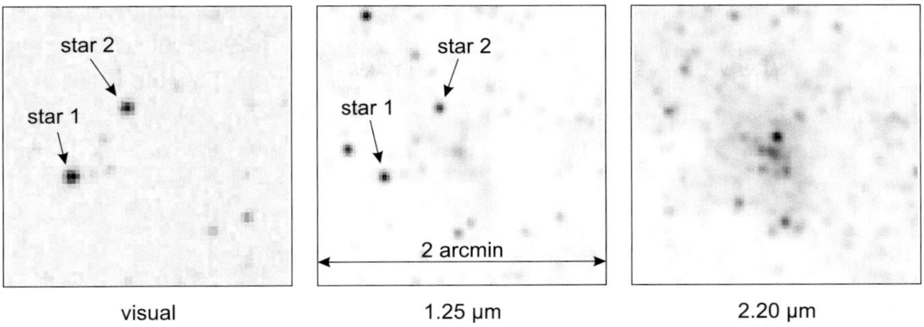

Fig. 1.6 Images of the central 2 arcminutes of the Galactic Center showing the effect of 27 magnitudes of visual extinction (source: Digitized Sky Survey). They show two stars (Biretta *et al.* 1982) that are suitable for optical wavefront sensing. Star 2 (USNO 0600-28579500) located 30″ NNE of the very center has an R-band magnitude of 13.2 (e.g. Ghez *et al.* 2003b).

(referred to as IRR1 and IRR2 by Henry *et al.* 1984), weak emission of quite a number of sources actually located within the central star cluster has been observed. Among them are IRS1, IRS21, IRS12, IRS7, as well as several of the extremely luminous blue stars in the IRS 16 complex. However, despite applying deconvolution algorithms (Rosa *et al.* 1992) a positive identification of the non-thermal radio source Sgr A* was not possible in this spectral domain.

Some of the stars that are detectable at visible wavelengths have become important for high angular resolution imaging as they are close to the position of Sgr A* and can therefore be used as adaptive optics reference stars.

1.1.5 *X-Ray Energies*

The X-ray emission from astrophysical objects can only be investigated above the Earth's atmosphere with X-ray telescopes on board of satellites orbiting the Earth. Some of the earliest X-ray observations of the Galactic Center region were done with the ROSAT satellite observatory (e.g. Predehl & Truemper 1994; Predehl 1995, and references therein). The most recent observations were done with telescopes on board of the NASA satellite Chandra and the European ESA satellite XMM-Newton (Fig.1.7). The telescopes of the satellite observatories consist basically of highly-nested, grazing-incidence mirrors (Wolter mirrors; Fig.1.8) which focus the X-ray radiation onto the image plane and the science instruments which detect and record the radiation. The exposures towards the Galactic Center are dominated by the diffuse emission of the Sgr A East region, which is thermal in origin and rich in emission lines. Collected by the mirror apertures the average combined count rates from that region are of the order of several events per second.

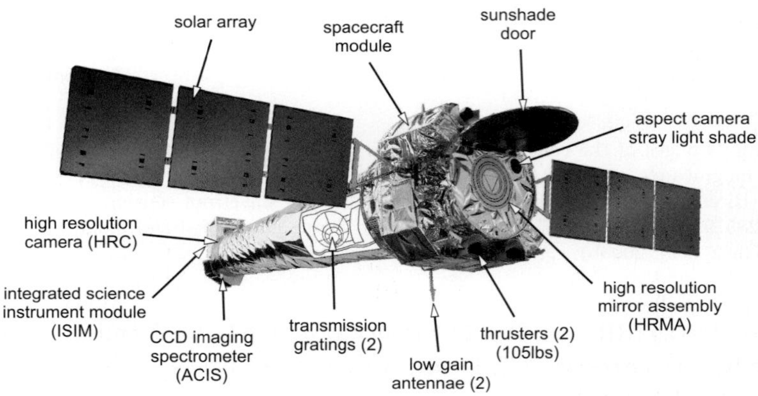

Fig. 1.7 The NASA satellite Chandra as an example of an X-ray observatory.

NASA's X-ray observatory Chandra was launched and deployed by Space Shuttle Columbia in July 1999. Chandra is on a very elliptical orbit around the Earth. The most distant point from Earth (apogee) is located at more than about one third of the way to the moon. The closest point to Earth on its orbit (perigee) is at a distance of about 16,000 kilometers. A complete orbit lasts for 64 hours and 18 minutes. This way the spacecraft orbit is optimized for maximum integration times above the belts of charged particles that surround the Earth. Uninterrupted observations as long as 55 hours (almost 180 kilo-seconds) are possible. The collecting area of Chandra's mirrors is about 400 cm^2 at 1 keV with an angular resolution of about 0.3 arcseconds.

Observations of the Galactic Center (e.g. Baganoff *et al.* 2001, 2003; Muno *et al.* 2003; Eckart *et al.* 2004a) were mostly carried out using two specialized CCD camera systems comprised of an imaging array (ACIS-I) and an element spectroscopy array (ACIS-S). Except for one, all of the detectors are front-side-illuminated CCDs operated at a focal plane temperature of -110° C. Data frames are produced after an integration time of 3.2 seconds. A typical total integration time over which the data is collected is of the order of 50 ks. Each detected photon on the CCD chips results in a characteristic response covering several pixels. The Galactic Center is a weak X-ray source. Here the individual photon detection events are recorded as their pulse-height amplitudes of a 5×5 pixel region centered on each of the events. Some of the data analysis including the rejection of bad frames or unwanted cosmic ray events is done already on board of the spacecraft.

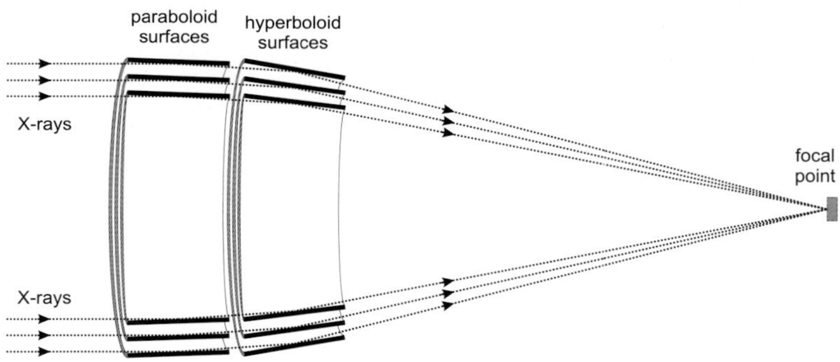

Fig. 1.8 A schematic view of a section of nested grazing-incidence Wolter mirrors as used in X-ray telescopes like Chandra and XMM-Newton.

The X-ray spectroscopy Multi-Mirror Mission (XMM-Newton) is one of the cornerstone projects in the ESA long-term program HORIZON 2000 for space science. The satellite observatory was launched on December 10, 1999, with an Ariane 5. The collecting area of the XMM mirrors is about 4300 cm^2 at 1.5 keV with an angular resolution of about $\sim$5 arcseconds. The primary scientific objective of XMM is to perform high throughput spectroscopy of cosmic X-ray sources over a broad band of energies ranging from 0.1 keV to 10 keV. The XMM observatory includes three modules of type Wolter I (Fig.1.8), which are coupled to reflection grating spectrometers and X-ray charge-coupled device (CCD) cameras with energy resolving powers ranging from 10 up to 1000 as well as one small optical/UV telescope.

The European Photon Imaging Camera (EPIC) consortium has provided the focal plane instruments for the three X-ray mirror systems on XMM-Newton (Strüder et al. 2001, Turner et al. 2001). The EPIC focal plane imaging spectrometers use passively cooled CCDs to record the images and spectra of celestial X-ray sources focused by the three mirrors. There is one camera at the focus of each mirror; two of the cameras contain seven Metal Oxide Semiconductor (MOS) CCD cameras, while the third uses twelve PN CCDs, defining a circular field-of-view of 30' diameter in each case. Each EPIC instrument is equipped with a filter wheel that carries three kinds of X-ray transparent light blocking filters, as well as a fully closed, and a fully open position. Data processing on-board removes signal tracks caused on the detector by cosmic rays and generates X-ray event files for further data analysis. The instruments were calibrated on the ground using laboratory X-ray sources and in orbit using a variety of celestial X-ray targets. The current calibration is better than 10% over the entire energy range of 0.2 to 10 keV.

Since 2001 XMM-Newton repeatedly observed the Galactic Center detecting the onset of a flare (Goldwurm *et al.* 2003), and in October 2002, a powerful X-ray flare of Sgr A* which was about 160 times stronger than the quiescent level (Porquet *et al.* 2003).

1.1.6 γ-*Ray Energies*

In order to explore the nature of the central source Sgr A* it is of great interest to determine its flux density at the highest accessible energies. This can be achieved by γ-ray observatories orbiting the Earth. However, these energies are beyond the limit where mirrors can be used to focus the light onto the detector plane. In the domain of γ-rays, coded mask imaging is used — as briefly outlined below. As a result the current observations lack angular resolution in comparison to what is achieved in the X-ray or infrared/optical wavelength domains.

The EGRET instrument on the Compton Gamma-Ray Observatory has observed the Galactic Center (GC) region several times (Mayer-Hasselwander *et al.* 1998). A strong excess of emission is observed, peaking at energies >500 MeV towards the central position. However, a larger range of sources (including the very center) contributes to this emission.

The most recent results on the Galactic Center (Bélanger *et al.* 2004) have been obtained with the International Gamma-Ray Astrophysics Laboratory, INTEGRAL. INTEGRAL (Winkler *et al.* 2003) is a European Space Agency observatory that began its mission in October 2002 carrying four instruments. These consist of two main ones, IBIS (Ubertini *et al.* 2003) and SPI, the Spectrometer on INTEGRAL (Vedrenne *et al.* 2003), and two monitor instruments, JEM-X (Lund *et al.* 2003) and OMC (Mas-Hesse *et al.* 2003).

The IBIS coded mask instrument is characterized by a wide field of view (FOV) of $29^o \times 29^o$ ($9^o \times 9^o$ fully coded). From the event list for a given pointing, subsets of events are selected according to energy bins. Each subset is used to build a detector image or shadowgram generated by the source shining through a Tungsten alloy mask coded with a fixed pattern that has an area filling factor of $\sim$50%. Convolution of the shadowgram with the known decoding array gives rise to a sky image containing the main peak of all sources in the FOV and their secondary lobes. Source identification and subtraction of secondary lobes results in the final reconstructed sky image. The point spread function (PSF) has a beam width of 12' FWHM (full width at half maximum intensity). Fluxes are derived using INTEGRAL observations of the Crab Nebula. The system is sensitive over the energy range between 15 keV and 8 MeV. This response is achieved via two detector layers. One of these is a soft gamma-ray instrument with the upper CdTe layer sensitive between 15 keV and 1 MeV with peak sensitivity between 15 and 200 keV (ISGRI; Lebrun *et al.* 2003). The second, bottom CsI layer is sensitive between 200 keV and 8 MeV (PICsIT).

The Galactic Center was observed by INTEGRAL in 2003 between February 28 and May 1. The combined data cover a total integration time of about 1100 ks. Mosaicked images were constructed using data obtained with ISGRI in the energy ranges 20-40 and 40-100 keV.

The resulting final images give an unprecedented view of the high-energy sources of this region in hard X-rays and gamma-rays with an angular resolution of 12' (FWHM). Bélanger *et al.* (2004) report on the discovery of a source, IGR J1745.6-2901, coincident with the Sgr A* to within 0.9 arcminutes. Located at α_{2000}=$17^h45^m38.5^s$, δ_{2000}=-29°01'15", the source is visible up to about 100 keV with a 20-100 keV luminosity of $(2.89 \pm 0.41) \times 10^{35}$ ergs s^{-1} (assuming a distance of 8 kpc). The new INTEGRAL source cannot be associated unambiguously with the Galactic Center. However, this is the first time that significant hard X-ray emission from within the inner 10' of the Galaxy has been reported. Therefore, a flux density contribution from the galactic supermassive black hole itself cannot be excluded.

1.2 NIR Imaging Through the Atmosphere

Following basic physics the angular resolution A that can be achieved with a telescope is given by the ratio between the observing wavelength λ and the telescope diameter D. In this case the resolving power is given only by the diffraction of light at the edges of the mirror or imaging optics. Therefore these conditions are called "diffraction limited".

$$A = 1.22\frac{\lambda}{D} \quad ; \quad A_{FWHM} = 1.02\frac{\lambda}{D} \quad . \tag{1.8}$$

Here A is given in radian and λ and D in the same units, i.e. meters. The quantity A is the angular distance between the maximum and the first minimum of the Airy diffraction pattern of a circular aperture. The corresponding full width of the main diffraction component at its half maximum value (FWHM) is given as well. The resolving power A can only be reached if nothing disturbs the light propagation between the source and the telescope or the image plane where the detector is located. For optical and near-infrared wavelengths these disturbances can be produced by turbulent air. Therefore — if no other precautions are taken — ideal conditions in which diffraction limited operation of a single telescope is possible are only given in space or in rare situations in which the disturbing influence of the air turbulence are small or even negligible compared to the resolving power A. This situation is approached if the telescopes have diameters of less than a few meters or the observing wavelength is larger than a few micrometers. In general the presence of the atmosphere will result in NIR images in which point sources are blurred and have apparent angular diameters of the order one arcsecond.

Telescopes in space, like the Hubble Space Telescope (HST) or the future James Webb Space Telescope (JWST), can deliver diffraction limited images. However, the diameter of the telescope and its optical quality both limit the resolution that can be achieved and it is quite expensive to bring large aperture telescopes into space. On the ground, existing telescopes with diameters of up to 10 m and even larger telescopes that are under construction or planned can be used for high angular resolution measurements. Here, techniques like speckle interferometry or adaptive optics can retrieve the diffraction limited information. Detailed descriptions of the physical processes involved in imaging through the atmosphere are given in several articles (e.g. Dainty & Scaddan 1975; Mariotti *et al.* 1983; Roddier 1989; Christou 1991). In this section we give a brief summary of the essential processes and introduce expressions that will be used afterwards.

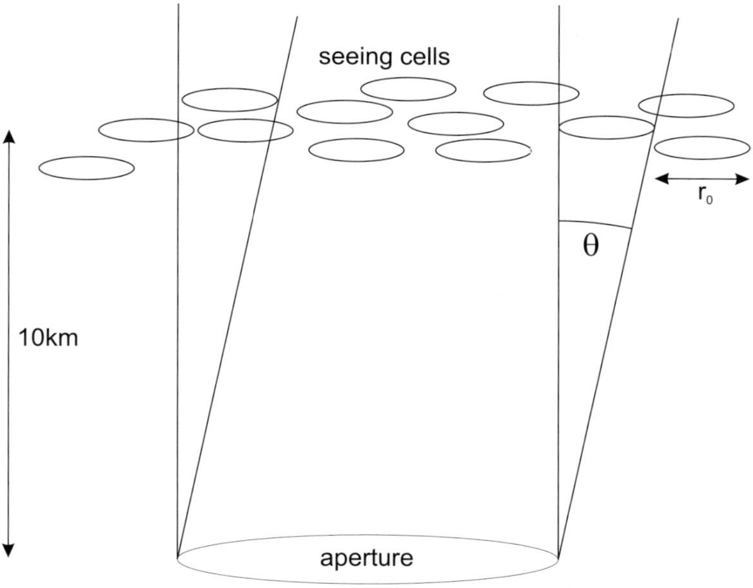

Fig. 1.9 The case of a single turbulent layer at a hight of 10 km above the telescope aperture. A distribution of seeing cells, each with a typical diameter r_o is moved across the telescope pupil with the wind speed v_{wind}. The angle θ is given by the approximate angular diameter of a seeing cell as seen from the telescope. It encloses the isoplanatic patch — the region on the sky over which the distribution of seeing cells over the aperture — and therefore the atmospheric point spread function — does not change significantly.

The atmospheric turbulence above the telescope distorts the otherwise planar wavefronts of astronomical sources and is responsible for a point spread function that varies rapidly with time. This phenomenon is called seeing. Over small entities of the turbulent layer, sections of the phase front experience predominantly a linear phase change only. These so-called seeing cells are characterized by the Fried diameter (Fried 1966), r_0. It has a value of the order 10 cm in the optical and approximately 20 cm to 50 cm in the near-infrared domain.

For small angular distances in the sky the light of neighboring stars is distorted by almost the same set of seeing cells, resulting in almost identical atmospheric point spread functions for stars within a sufficiently small area. This area is called the isoplanatic patch and has a diameter of approximately 20" to 30" in the near-infrared, depending on the seeing and the height of the turbulent layer(s) (Fig.1.9). Image degradation at high spatial frequencies is caused by a turbulent atmospheric layer at a typical hight of 10 km. Sometimes, several layers at different heights have to be taken into account. However, local (telescope dome) or low(er) level layers (e.g. the so called "ground layer") are usually of importance as well. The angular diameter of the isoplanatic patch, i.e. the isoplanatic angle, corresponds approximately to the angular diameter of a seeing cell at the given height of the turbulent layer (assuming a single layer; Fig.1.9). Across the isoplanatic patch the arrangement of the seeing cells is approximately the same, which leads to a stable (only slowly degrading) PSF in that area.

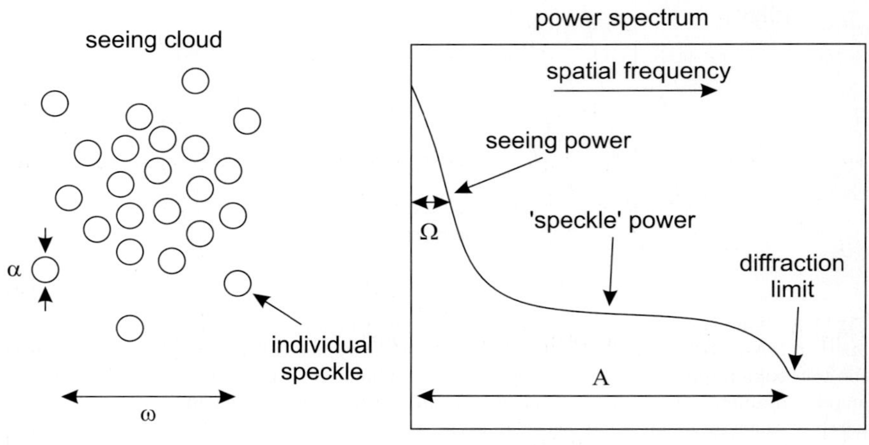

Fig. 1.10 As a simple model, the seeing cloud with angular diameter $\omega \sim \lambda / r_o$ can be thought of as being composed of a number of distorted diffraction limited images with angular diameter $\alpha \sim \lambda / D$. This is reflected in the distribution of image power. Here A and Ω are inversely proportional to the telescope resolution and the seeing width. The image power drops rapidly to zero beyond the diffraction limit.

In short exposures of a bright source individual spots are visible, which are essentially diffraction limited images. These are the result of a constructive interference of an unfilled aperture composed of subsets of different seeing cells. These cells refract portions of the light across the telescope pupil into common directions. The resulting speckles can therefore be regarded as distorted images of a point source at the diffraction limit of the telescope (Fig.1.10). In order to record such speckle images (or seeing clouds) the exposure time must be of the order of the coherence time of the atmosphere which turns out to be around one hundred milliseconds in the near-infrared wavelength range. Recording a large number of these short exposure images allows one to later compensate for the degrading influences of the turbulent atmosphere and to reconstruct images at the diffraction limit of the telescope (see section 1.3). Alternatively the wavefront degradations can be corrected with an adaptive optics system that delivers diffraction limited images in real time (see section 1.4).

In the case of a long exposure image through the telescope optics and the atmosphere, the process of image formation can be described in the following way. An image I(x,y) of an object O(x,y) taken in the focal plane (with coordinates x and y) through the combination of the telescope and the turbulent atmosphere can be written as the convolution of O(x,y) with the combined telescope and atmospheric point spread function P(x,y):

$$I(x, y) = O(x, y) \odot P(x, y). \tag{1.9}$$

Here $\odot$ denotes the convolution operator. In Fourier space with the spatial frequency coordinates u and v this convolution can be written as:

$$i(u, v) = o(u, v)p(u, v). \tag{1.10}$$

Here i(u,v), o(u,v), and p(u,v) are the Fourier transforms of I(x,y), O(x,y), and P(x,y).

The image power as a function of spatial frequencies u and v is transferred in an incoherent two-dimensional imaging system For conventional long exposures the power spectrum after taking the time average of all images in a given ensemble can be written as (Dainty 1975; Christou 1991):

$$| < i(u, v) > |^2 = o^2(u, v)| < T(u, v) > |^2. \tag{1.11}$$

The long exposure transfer function can then be written as:

$$< T(u, v) >= T_s(u, v)T_0(u, v). \tag{1.12}$$

Here $T_0(u, v)$ is the telescope transfer function and $T_s(u, v)$ is the seeing (atmospheric) transfer function.

Direct long exposure imaging experiments of the Galactic Center in the near-infrared carried out under best sub-arcsecond seeing conditions can result in excellent, deep images (e.g. Forrest *et al.* 1986; Depoy & Sharp 1991a; Herbst *et al.* 1993; Close *et al.* 1992). Due to the high stellar surface density, these experiments, however, were not able to fully resolve the stellar complexes within the central stellar cluster of the Milky Way. Lunar occultation measurements (Simons *et al.* 1990; Simon *et al.* 1990; Adams *et al.* 1988) achieved that with effective angular resolutions of about 10 mas. However, they suffer severely from the limited integration times due to the short time scales of the occultation and re-appearance of the sources. Lunar occultation measurements are currently only useful for the brightest sources.

1.3 Speckle Imaging

Detailed descriptions of (infrared) speckle interferometry and imaging can be found in various reviews (e.g. Dainty 1975; Mariotti 1989; Roddier 1989; Christou 1991). Although most of the more recent high resolution imaging is done via adaptive optics we consider it essential to outline some of the fundamental aspects of speckle interferometry in the context of the Galactic Center observations. There are two main reasons for this 1) Currently a significant amount of the available high resolution imaging data has been obtained via speckle imaging. 2) Even if an adaptive optics system is used there are cases in which some of the basic observational and image reduction techniques outlined in the following sections become important, leading to so called "hybrid modes" (see section 1.4).

Here, we only give a brief summary in order to introduce expressions that are linked to this technique and that are used in the following description of data processing and interpretation. We will give a short summary of the data reduction used to reconstruct diffraction limited images and spectra from speckle data.

Rather than proceeding as in the case of the long exposure (Eq. 1.12) one can alternatively calculate the power spectrum for each short exposure image of the ensemble before averaging:

$$< |i(u,v)|^2 >= o^2(u,v) < |T(u,v)|^2 > . \tag{1.13}$$

To evaluate these expressions further and to decide which of the imaging techniques allows one to preserve information at the diffraction limit of the telescope, one must have a closer look at the properties of the transfer function T(u,v). T(u,v) is the (peak) normalized autocorrelation function of the pupil function H(ξ,η) which describes the field distribution of the received wavefront across the pupil (Dainty 1975). Here ξ and η are distances in the pupil plane. They are related to the spatial frequencies u and v via the observing wavelength λ and the focal length f:

$$\xi = \lambda f u$$
$$\eta = \lambda f v.$$

This field distribution H(ξ,η) contains a contribution from the pupil function H_0(ξ,η) of the telescope and the complex amplitude A(ξ,η) due to the light from the observed source that was propagated through the atmosphere:

$$H(\xi, \eta) = A(\xi, \eta) H_0(\xi, \eta). \quad (1.14)$$

In this case we find that in the long exposure transfer function

$$< T(u, v) >= T_s(u, v) T_0(u, v). \quad (1.15)$$

The quantity $T_0(u, v)$ is the telescope transfer function which can be expressed as the the autocorrelation of the pupil function H_0(ξ,η). $T_s(u, v)$ is the seeing transfer function and equals the ensemble average autocorrelation of the complex amplitude A(ξ,η). A Kolmogorov power law spectrum for the atmospheric turbulence results in (e.g. Fried 1966; Mariotti 1989; Perrier 1989):

$$T_s(w) \approx exp(-3.44 \frac{\lambda |w|^{5/3}}{r_0}). \quad (1.16)$$

Here w^2=u^2 + v^2, and the Fried size (i.e. diameter) of the turbulent cells is

$$r_o \propto \lambda^{6/5} \times (cos(\gamma))^{3/5}, \quad (1.17)$$

with γ being the zenith angle (Fried 1966, Perrier 1989). With the angular diameter of the seeing cloud

$$\omega \approx \frac{\lambda}{r_0} \tag{1.18}$$

one finds that it depends to the -1/5th power on the observing wavelength λ:

$$\omega \propto \lambda^{-1/5}. \tag{1.19}$$

This implies that the seeing in the infrared domain is in general better than in the optical. If, for example, the typical angular diameter of a speckle cloud at a wavelength of λ=0.5 μm is ~1", it will be of the order of only 0.7" in the near-infrared at a wavelength of λ=2.2 μm. This is a major advantage both for direct imaging as well as for speckle imaging, since the number of speckles and the coherence time (i.e. the speckle noise or the atmospheric image degradation of a point source in a single short exposure) is much smaller in the infrared.

As shown above, the speckle interferometry transfer function is equal to the averaged squared modulus of T(u,v), that is $< |T(u,v)|^2 >$. The evaluation of this quantity infers a fourth order moment of the complex amplitude A(ξ,η) (Dainty 1975). Assuming a complex Gaussian process one can show for these moments that

$$< A_1^* A_2 A_3 A_4^* > = < A_1^* A_2 >< A_3 A_4^* > + < A_1^* A_3 >< A_2 A_4^* > . \tag{1.20}$$

The quantities A_1 through A_4 correspond to complex amplitudes at four different coordinates in the pupil plane.

Using this relationship the speckle transfer function can finally be written as (Dainty 1975; Korff *et al.* 1972):

$$< |T(u,v)|^2 > \propto | < T(u,v) > |^2 > + \frac{1}{N_{sp}} T_0(u,v). \tag{1.21}$$

Here N_{sp} is the number of speckles per seeing cloud which is proportional to the square of the telescope diameter D over the Fried cell size r_o:

$$N_{sp} \propto (\frac{D}{r_o})^2. \tag{1.22}$$

This shows that the speckle interferometry transfer function contains a component up to the diffraction limit of the telescope that is weighted inversely proportional to the number of speckles per seeing cloud. The diffraction limited information can be used for imaging. Three different methods that have been applied to Galactic Center imaging data are described in the following sections.

The overall conversion of the raw speckle data into the final diffraction-limited images is done in two steps. First the individual speckle frames have to be calibrated, and secondly an appropriate speckle reconstruction algorithm has to be applied. Currently all near-infrared high angular resolution imaging is done in a read-noise limited mode. In contrast to optical speckle work, no individual photon addresses are recorded but full short exposure images are taken. Depending on the wavelength, the correction for the sky-background and the read noise bias can be done by subtraction of a mean sky exposure of the same integration time as the individual images on the target. The data for that exposure is taken before, after and in most cases interleaved with the target data. After the subtraction, all individual images are divided by a flat field that corrects for the pixel dependent sensitivity. In addition, a bad pixel correction and removal of other possible artifacts from the individual frames has to be carried out.

After these steps the resulting individual specklegrams are ready for being processed by the different speckle data reduction algorithms explained below. Most of them can be carried out on-line — as implemented in the SHARP camera system (see below) — or more elaborately, off-line. For the SHARP system the corresponding off-line data reduction package in which all commonly used speckle algorithms are implemented was written in 1989/90 and was then further improved through the following years (Eckart & Duhoux 1990; Eckart *et al.* 1994; Cruzalebes *et al.* 1996). It includes algorithms that preferentially work in the image plane, such as direct imaging combined with seeing based image selection and the simple shift-and-add algorithm (SSA; Christou 1991), as well as those that work in the Fourier plane, such as the Knox-Thompson method (Knox & Thompson 1974; Knox 1976) and triple-correlation (Lohmann *et al.* 1983) algorithms. All these algorithms have been applied to the Galactic Center data and have lead to comparable results. A combination of a Knox-Thompson and SSA map was published in Eckart *et al.* (1992), and an image obtained using the triple-correlation algorithm was presented in Eckart *et al.* (1994). By far the best results were obtained using the combination of the SSA and a Lucy deconvolution algorithm (Lucy 1974). A detailed description of this combination of imaging algorithms used for the Galactic Center SHARP (Fig.1.11) data is given in Eckart *et al.* (1994).

In the following sections we summarize the basics of the different speckle

Fig. 1.11 Thomas Ott filling the cryogenic tanks of the SHARP dewar with liquid nitrogen. The cylindrical camera cryostat is mounted to the NTT rotator flange behind which the NTT mirror itself is located.

imaging data reduction algorithms used in the case of the Galactic Center.

1.3.1 *The Knox-Thompson Method*

Since each short exposure contains speckles and therefore information at the diffraction limit of the telescope, the Fourier amplitudes a(u,v) of the observed object can be obtained by calculating an ensemble average PS(u,v) of the power spectra of all available m=1, ... ,M short exposures:

$$\begin{aligned} a(u,v) &= PS(u,v)^{1/2} \\ &= [\frac{1}{M}\sum_{m=1}^{M} i(u,v)i^*(u,v)]^{1/2}. \end{aligned} \tag{1.23}$$

Here the superscript * denotes the complex conjugate. However, calculating the power spectrum results in the loss of all phase information. A way to preserve the phase information and to have power transferred at high spatial frequencies is to calculate a cross spectrum in which the conjugate complex is shifted by one resolution element Δu or Δv, respectively, in the Fourier plane — thus preserving the phase difference information between neighboring resolution elements. This technique was first proposed by Knox & Thompson (1974) (see also Knox 1976). For a shift by Δu the cross spectrum has the following form:

$$\begin{aligned} DPX(u,v,\Delta u) &= i(u,v)i^*(u+\Delta u,v) \\ &= a_{u,v}exp(-j\phi)\ \ a_{u+\Delta u,v}exp(-j(\phi+\Delta\phi)) \\ &= a_{u,v}a_{u+\Delta u,v}exp(-j\Delta\phi) \end{aligned} \tag{1.24}$$

with $j=(-1)^{1/2}$ and $a_{u,v}=\mid i(u,v)\mid$. Integration of the complex phase differences along the u- and v-direction starting at the lowest spatial frequency possible then allows one to recover the complex phase front that can be combined with the Fourier amplitudes for final calibration and imaging. A resulting image for the Galactic Center is shown in Eckart *et al.* (1992).

1.3.2 *The Bispectrum Analysis*

Using triple products in Fourier space the complex phase front can be retrieved in an even more general way (Lohmann *et al.* 1983). The complex intensity can be written as

$$i(p) = \mid i(p) \mid exp(j\phi(u)). \tag{1.25}$$

In the one-dimensional case the triple product can be calculated in the following way:

$$\begin{aligned} i^{(3)}_{p,q} &= i(p) \times i(q) \times i^*(p+q) \\ &= \mid i^{(3)}_{p,q} \mid exp(j\beta_{p,q}) \end{aligned} \tag{1.26}$$

with the triple amplitude $\mid i^{(3)}_{p,q} \mid$ and the complex triple phase

$$exp(j\beta_{p,q}) = exp(j(\phi(p)+\phi(q)-\phi(p+q))). \tag{1.27}$$

The much larger number of triple phases compared to phase differences results in a higher redundancy and the phase front can be retrieved with a higher signal to noise ratio. The multitude of all triple products is called bispectrum and has twice the dimension of the original image because two displacement vectors p and q are used. When calculating the full bispectrum for two-dimensional images this results in a substantial computational effort. Using estimates of two phases at the lowest spatial frequencies, e.g. extracted from a long exposure, the actual phases can be calculated recursively from the triple phases making use of the phase closure relation:

$$exp(i\phi(p+q)) = exp(i(\phi(p) + \phi(q) - \beta_{p,q}). \tag{1.28}$$

As in the case of the Knox-Thompson algorithm the recovered complex phase front can be combined with the Fourier amplitudes for final calibration and imaging (see Eckart *et al.* 1994). For a more detailed description of the actual phase retrieval algorithm that was used for the Galactic Center (see also Cruzalebes *et al.* 1996).

1.3.3 *Seeing Calibration of Fourier Data*

To obtain the true Fourier amplitudes and phases of the source, the power spectra, cross-spectra and bispectra have to be calibrated by division with the same quantities calculated for an unresolved reference star. If the telescope aberrations, the detector properties, as well as the statistical properties of the atmosphere were the same for the ensembles of images taken on the target object and the reference star, this division will result in Fourier amplitudes and phases calibrated for seeing and instrumental influences. An image of the object can then be obtained after an inverse Fourier transformation into the image domain.

For weak sources, noise contributions have to be properly taken into account as well. Christou (1991) describes the noise bias correction in speckle image reductions. Major noise sources for short exposures used in speckle interferometry are the detector read noise in the near-infrared and photon noise in the optical. In practice this calibration by Fourier division turns out to be very delicate and difficult to control. Even if the reference object is within the same field of view as the target object, the conditions mentioned before are violated to a certain degree. In particular since the reference source has to be extracted as only a portion out of the original image, the calibration by Fourier division cannot fully account for the spatial frequency structure of the detector array. In addition, small number division problems (or the effect of noise) may not result in meaningful quotients at high spatial frequencies at which the transfer functions usually exhibit values that may be many orders of magnitude smaller than at the lowest spatial frequencies. In these cases the quotient has to be multiplied with an apodization filter that smoothly roles off towards high spatial frequencies in order to suppress the influence of the data at the corresponding spatial frequencies. *All these problems usually result in a limited dynamic range combined with a loss in spatial resolution.*

For the Galactic Center the results of the Knox-Thompson method and bispectrum analysis are quite comparable (Cruzalebes *et al.* 1996). However, as described in the following section, the results obtained via the simple shift-and-add method give images with higher dynamic ranges and more stable seeing calibration.

1.3.4 *The Shift-and-Add Algorithm*

The Knox-Thompson method and the bispectrum analysis both work in the Fourier plane. This is not really necessary because there exists a very simple and efficient algorithm to produce diffraction limited images that can be applied in the image plane. It is the shift-and-add algorithm that was first used by Bates & Fright (1982). The 12.8"×12.8" field of view of the SHARP speckle camera as well as the 5.3"×5.3" field of view of the NIRC camera (see section 1.3.6) is always smaller than the approximately 20" to 30" diameter isoplanatic patch size in the near-infrared. This means that for all the stars within the field of view of the camera, the position of the brightest speckle is at the same position within each speckle cloud. A basic assumption of the shift-and-add algorithm (Bates & Fright 1982; Christou 1991) is that individual speckles can be regarded as distorted images of the object as already indicated in earlier sections. This assumption is well justified if there are relatively few, well separated speckles (as in the NIR in good seeing on a 4 m telescope) and for a source structure consisting of a collection of point sources, as in the Galactic Center nuclear star cluster. Co-adding all short exposures, after shifting the positions of the brightest speckle in the seeing cloud of a bright reference object to a common location, results in an image containing substantial power at the diffraction limit of the telescope.

The speckle image processing can formally be described in the following way: each of the M two-dimensional images $I_m(x,y)$ of a series of short exposures can be written as the object O(x,y) convolved with the combined telescope and atmospheric point spread function $P_m(x,y)$ plus an additive noise component $N_m(x,y)$.

$$I_m(x,y) = O(x,y) \odot P_m(x,y) + N_m(x,y) \tag{1.29}$$

Here $\odot$ denotes the convolution operator. In the simple shift-and-add algorithm one determines the location of the brightest speckle in the seeing disk of a reference star in each of the images $I_m(x,y)$ and shifts them to the same position before summing them.

$$S(x,y) = \frac{1}{M}\sum_{m=1}^{M} I_m(x+x_m, y+y_m). \tag{1.30}$$

Here x_m and y_m are the corresponding shift vector components. One obtains a shift-and-add image $S(x,y)$ that can be written as the object convolved with a shift-and-add point spread function $P_s(x,y)$ plus a modified noise contribution $N_s(x,y)$.

$$S(x,y) = \frac{1}{M}(O(x,y) \odot \sum_{m=1}^{M} P_m(x+x_m, y+y_m) + \sum_{m=1}^{M} N_m(x+x_m, y+y_m)) \tag{1.31}$$

or

$$S(x,y) = \frac{1}{M}(O(x,y) \odot P_s(x,y) + N_s(x,y)). \tag{1.32}$$

For each star in the field of view (smaller than the isoplanatic patch size) this point spread function results in a diffraction limited image of the object on top of a residual seeing background built up by the contributions of all fainter speckles. In contrast to the long exposure (a simple co-addition of individual frames) the raw shift-and-add image contains information at the diffraction limit of the telescope. In the near-infrared and especially for the Galactic Center observations the shift-and-add algorithm has the following advantages over the Knox-Thompson method and the bispectrum analysis.

- No Fourier transforms have to be involved resulting in a fast processing of a large amount of data.
- No linear deconvolution is necessary, therefore no small number division problems at high spatial frequencies and no loss in spatial resolution due to apodization in Fourier space occurs.

- Most importantly: local artifacts in the image plane are not spread all over the Fourier plane resulting in problems for the seeing and source power calibration. Such artifacts are detector imperfections or — in a crowded field like the Galactic Center area — point spread functions of sources at the very edges of the detector that are affected by the limited size of the field of view.
- The algorithm is very efficient in the near-infrared due to the small number of speckles — less than 5 to 10 for a 4 m-class telescope at a wavelength of 2.2 μm — under good seeing conditions. This results directly in an image that contains of the order of 10% to 30% of the total image power at the diffraction limit of the telescope. During most of the Galactic Center observations the short-exposure seeing cloud diameter (e.g. the second order moment) varied between 0.3" and 0.5" FWHM and the bright reference sources (IRS 7 or IRS 16NE) were dominated by a single, bright speckle.
- For most of the observations the comparatively isolated bright super giant IRS 7 (or IRS 16NE for the southern parts of the stellar cluster) was in the same field of view as the Sgr A* region and could therefore be used as a very reliable calibrator source to correct for the shape of the shift-and-add point spread function.

The point spread function of a shift-and-add image is composed of a diffraction-limited core containing typically 10% (up to 30%) of the light on top of a broad seeing halo. Airy rings (mainly the diffraction pattern of the primary mirror) around the cores of bright stars indicate that the diffraction limit has truly been reached (see e.g. Fig. 1.12). The algorithm described above is also referred to as the simple shift-and-add (SSA) algorithm since it only makes use of the brightest speckle in the seeing cloud. A shift-and-add algorithm that uses all bright speckles in a seeing cloud weighted accordingly to their brightness is called weighted shift-and-add (WSA) algorithm and is described in Christou (1991). It is therefore suitable for shorter wavelengths or larger telescope diameters. However, the algorithm involves Fourier transformations and a linear deconvolution and does not share all of the advantages the SSA algorithm has over the Knox-Thompson method and the bispectrum analysis.

The quality of SSA images can be improved considerably by selecting only the frames of highest quality, i.e. with the best S/N-ratio. On smaller telescopes (such as the ESO NTT) the best speckle exposures resemble diffraction limited images. Schödel *et al.* (2003) could reach Strehl ratios

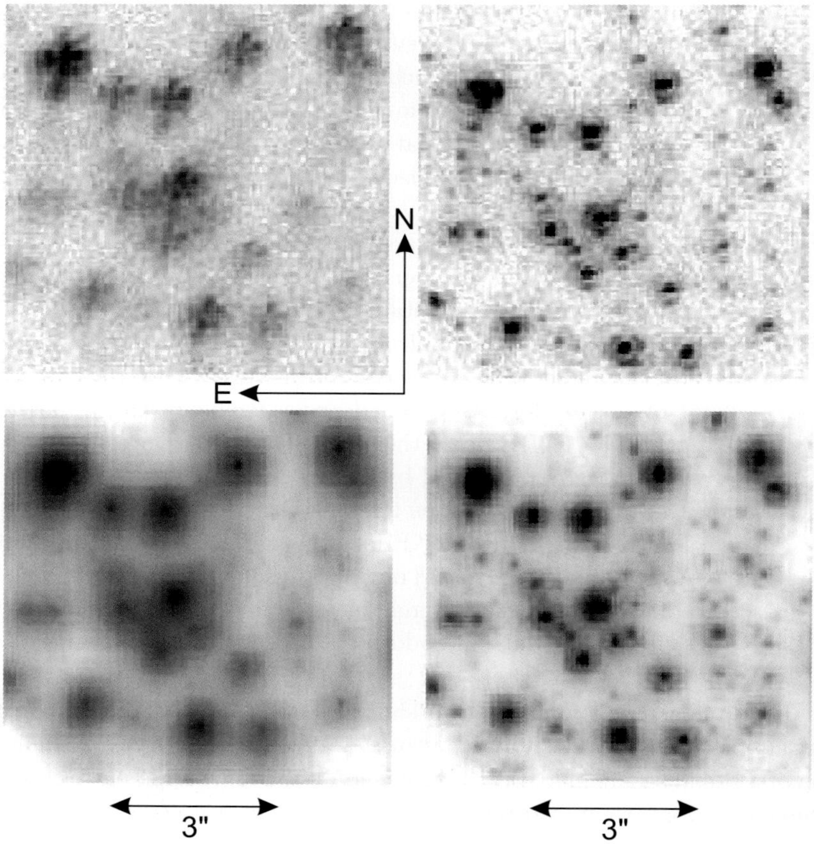

Fig. 1.12 Top left: speckle frame of the IRS 16 cluster in the Galactic Center obtained under moderate to bad atmospheric conditions. The S/N is low, faint sources are not visible and the brighter sources are broken up into numerous speckles. Top right: speckle frame observed under excellent atmospheric conditions. The S/N is high and the image is basically a diffraction limited short exposure with the first Airy rings clearly visible around the brighter sources. The visible stars have K-magnitudes between about 9 and 14. Bottom left: SSA image of an average speckle imaging data cube; Strehl ratio about 15%. Bottom right: SSA image of a cube of about 100 high quality speckle frames; Strehl ratio of about 35%. Numerous faint sources, which cannot be seen in the bottom left image, can be distinguished in this high-quality image. All speckle images were obtained with SHARP at the NTT telescope.

of up to 30% by rigorous frame selection (see section 1.4 for a brief definition of the Strehl ratio). The difference in quality between average SSA images and SSA images resulting from such a rigorous selection of the best speckle frames can be seen in Figure 1.12.

The shift-and-add images still have to be corrected for the SSA point spread function. For a description of this seeing calibration step we refer to section 1.5, where various deconvolution techniques will be described.

1.3.5 *Speckle Spectroscopy*

Although the combination of adaptive optics and dedicated spectrographs now permits spectroscopy combined with high angular resolution, the method of speckle spectroscopy still has some relevance for the observation of faint objects. The method presented here is based on slitless spectroscopy and can best be employed for obtaining simultaneous low resolution spectra of multiple faint sources for which otherwise high resolution spectroscopy would not have been possible or too cumbersome to collect using a slit. Since that method can also be employed over a large field-of-view it may also have advantages over low spectral resolution integral field spectroscopy in not too crowded fields. Here, the term "large field-of-view" refers to large fields for diffraction limited imaging i.e.
indexFOV significant fractions of an arcminute or more.

The stars in the central arcsecond close to the position of the radio point source Sgr A* (in the following referred to as the Sgr A* stellar cluster) are essentially all located within one single seeing disk. In addition, they are very close to the bright members of the IRS 16 complex that are located 1.5″ to 2″ to the east. The members of this complex are 100 times brighter than the stars close to Sgr A*. In order to conduct spectroscopic observations of these stars, the SHARP speckle camera was equipped with an objective prism (Genzel *et al.* 1997). The additional device consisted of two optically contacted prisms. One was made from BaF_2 and the other from Schott IRG 9 glass. Mounted in front of the SHARP camera the two prisms had an on axis net-dispersion that gave K-band spectra of 1.0″ length in the image plane. This dispersion was chosen on purpose in order to minimize confusion problems in the highly crowded area surrounding the compact radio source Sgr A*. The spectral range was limited by the K-band filter (1.95–2.40 μm) inside the camera. In this configuration low resolution ($\lambda/\Delta\lambda$=35) K-band spectra of 4 objects within the central Sgr A* stellar cluster could be obtained. This data were taken during NTT observing runs in April and June, 1996. Several thousands of spectrally dispersed speckle interferograms were then recorded with exposure times of 0.5 seconds. The region described above was observed alternately through the full K-band

and through a He I 2.09 μm narrow band ($\lambda/\Delta\lambda$=150) filter. The raw images were sky subtracted, a flat field was applied and dead pixels were corrected for by interpolation. The spectrally dispersed data could be flat-fielded since the relative spectral response of individual pixels at the chosen spectral resolution was confirmed to be equal to within $\leq$1%.

Since speckle spectroscopy is a rarely used observational technique we include an outline of the theory of image formation and the applied data reduction at this point. Speckle spectroscopic interferograms IS(x,y) taken in the NIR K-band can be thought of as the usual short exposure image I(x,y) of a seeing cloud – i.e. the image of the object O(x,y) convolved with the atmospheric point spread function P(x,y) – convolved with a product of the source K-band spectrum K(x) and the combined K-band atmospheric and instrumental spectral response function SP_K(x)

$$IS(x,y) = I(x,y) \odot (K(x) \times SP_K(x)) \tag{1.33}$$

or

$$IS(x,y) = [O(x,y) \odot P(x,y)] \odot (K(x) \times SP_K(x)). \tag{1.34}$$

Here $\odot$ and $\times$ denote the convolution and multiplication operators respectively. For point-like objects (stars), $O(x,y) \odot P(x,y)$ can be represented by an Airy pattern. The image of the star at the K-band diffraction limit of the NTT telescope has a FWHM of the order of 0.13". On the spectrally dispersed and diffraction limited image the K-band spectrum was about 1" long. The maximum spectral resolving power was about R=35. In order to extract the spectral information one has to obtain an independent measure of the atmospheric point spread function. This was achieved by interleaving the described observations with observations of the same field (containing the structural calibrator source IRS 7) through the combination of the objective prism and the R=150 He I 2.09 μm narrow-band filter. With SP_{NB}(x) we denote the corresponding combined atmospheric and narrow band instrumental response function. The spectral resolution of the narrow band filter was higher than what one could obtain with the speckle spectroscopic observations described above. Therefore, the narrow-band filter spectral response function SP_{NB}(x), as well as its product with the source spectrum K(x) can be approximated by a delta function δ(x-x_{NB}) where x_{NB} corresponds to the center wavelength of the NB filter. Individual reference speckle interferograms IR(x,y) of the IRS 7 reference source REF(x,y) can then be written as

$$IR(x,y) = I(x,y) \odot (K(x) \times SP_{NB}(x)), \tag{1.35}$$

$$IR(x,y) = [REF(x,y) \odot P(x,y)] \odot (K(x) \times SP_{NB}(x)) \tag{1.36}$$

or

$$IR(x,y) = [REF(x,y) \odot P(x,y)] \odot \delta(x - x_{NB}) = REF(x,y) \odot P(x,y). \tag{1.37}$$

This way one can record images of the structural calibrator through the objective prism at the same telescope focus. Alternated observations through the full K-band and the narrow band filter allowed one to sample the statistical variations of the atmospheric PSF and to seeing correct the dispersed speckle data.

The SSA algorithm could be used to reconstruct spectra of individual sources from the dispersed speckle frames. The highest spatial and spectral resolution SSA images were obtained using a matched filter approach. This was done by searching through individual short exposures and extracting an image of a single bright dispersed IRS 7 speckle sufficiently isolated from the overall seeing cloud and using it as a matched filter. We calculated the cross-correlation between this spectrum and all the individual images of the full speckle interferograms. The peak in the cross-correlated image of IRS 7 then denoted the corresponding location of the brightest dispersed speckle in each image. The difference between this location and the long exposure centroid of the IRS 7 images was used as a shift vector in the SSA process.

The resulting broad-band spectrally dispersed SSA images were then deconvolved with the PSF obtained from the narrow band SSA images on IRS 7 using the Lucy algorithm. The results of the speckle spectroscopy experiment are discussed in chapters 2 and 3.

1.3.6 *Speckle Instrumentation*

In this section we present some of the technical details of speckle camera systems that have carried out observations of the Galactic Center on a regular basis. These are the SHARP camera (Hofmann *et al.* 1993, 1995) at the ESO NTT and NIRC (Matthews & Soifer 1994; Matthews *et al.* 1996) at the Keck telescope. These cameras have been equipped with sensitive near-infrared array detectors. Before the 1990s, sensitive NIR chips were not easily available and initial speckle observations of the Galactic Center had been carried out using one dimensional slit-scan speckle techniques (e.g. Beckwith *et al.* 1986).

1.3.6.1 *SHARP*

The goal of the SHARP (System for High Angular Resolution Pictures) experiment was to study the structure of the central stellar cluster at diffraction limited resolution (see Fig.1.11). Given the distance of 8 kpc and the high extinction of $A_V \approx 25 - 30^m$ this required diffraction limited imaging in the near-infrared with a cooled, sensitive, high speed camera at a 4 m-class telescope and special speckle interferometric data reduction software. All observations with the SHARP camera were done at the ESO (European Southern Observatory) New Technology telescope (NTT) in La Silla, Chile. The camera was mounted at one of the Nasmyth foci with a focal ratio of f/11. The camera used liquid nitrogen as a coolant and was built around a 256^2 pixel NICMOS 3 array from Rockwell. This is a HgCdTe based focal plane array sensitive in the 1 to 2.5 μm wavelength range. The read noise for a single read process was about 30 electrons. The camera was equipped with standard near-infrared broad and narrow band filters. During the initial observing runs a circular variable filter wheel (CVF) was used instead of the narrow band filters. The array is read with a maximum frame rate of 8 Hz resulting in a data rate of about 1 Mbyte/s. The camera optics has been designed such that the diffraction limited beam of the NTT at 2.2 μm (~130 milli-arcseconds; mas) could be sampled with 3 pixels of 50 mas each. The array size and pixel scale resulted in a total field of view of 12.8"×12.8". At the distance of the Galactic Center this corresponds to about 0.51 pc ×0.51 pc. Therefore the central part of the stellar cluster as well as the bright star IRS 7 could be included in the same field of view.

The camera had on-line data analysis capabilities. The analog data stream from the four quadrants of the NICMOS 3 array was digitized and sent via a fiber link into the memories of 4 digital signal processors (DSP). It was there where the images were calculated by subtracting the value read of each pixel obtained right after the reset from the read value after the chosen integration time. The data was then stored on disk and the image was available for further on-line processing. After correcting for bad pixels, subtracting a sky image and multiplying with an inverse flat field, the DSPs performed an on-line simple shift-and-add algorithm. After typically 1000 short exposure images, the intermediate result was displayed and written to disk as well. This allowed a data quality check at the telescope during the observations and could also be used for further real time image analysis at the observatory.

Since its first installation at the ESO NTT in August 1991 until the final observing run in June 2002 the whole camera system was shipped to the NTT and operated there once or twice per year. Further details on the camera design and electronics are given in Hofmann *et al.* (1993, 1995).

1.3.6.2 *NIRC*

At the Keck Observatory the Galactic Center observations were carried out with the Near Infrared Camera (NIRC) (Matthews & Soifer 1994; Matthews *et al.* 1996). NIRC is an instrument designed to produce both infrared images and low resolution spectra from 1 to 5 μm. NIRC not only produces direct images but also allows for coronographic imaging (bright sources can then be covered by a finger in the light path) and has low resolution (R $\sim$ 60-120) grisms available for spectroscopy.

For NIR speckle operation the f/25 telescope beam from the secondary mirror is first transformed via an image converter into an f/180 beam. Then the light encounters NIRC's tertiary mirror which is still outside of the camera cryostat. This tertiary mirror can be tilted in order to align the pupil onto the instrument's pupil mask. After the tertiary, the dewar window is encountered, followed by the focal plane which contains two movable slides. These slides provide a choice of slits and masks for spectroscopic use. After a field mirror, a folding mirror, and a cold, circular pupil stop the light path reaches two filter wheels with 20 positions each. These wheels provide a choice of infrared bandpass filters that allow one to select a suitable wavelength and bandwidth for the observations. The Galactic Center data (e.g. Ghez *et al.* 1998) was taken with a 256$\times$256 pixel InSb detector produced by Hughes-SBRC. The detector has a quantum efficiency of 89.5% at 1.7 μm wavelength. That means almost 90% of all photons that reach the chip at this point are converted into an electrical signal. The detector is attached to a cold block that provides cooling through the liquid helium dewar. A small heater attached to the block also provides thermal regulation. The pixels have a diameter of 30 μm. Projected onto the sky for the NIR speckle operation that corresponds to a pixel scale of 20.6 milli-arcseconds per pixel and results in a total field of view of 5.3 arcsec. The detector read-out noise of the imaging chip is 75 electrons in the best timing pattern. Output from the NIRC detector is sent to the readout electronics rack. A four-channel readout is provided, along with coadders, filter banks, and user-selectable gains. The analog-to-digital conversion is done by 12-bit or 16-bit ADCs.

After the standard data processing (sky subtraction, flat field and bad pixel correction) a few additional steps were carried out in the case of the NIRC Galactic Center observations at the Keck telescope (see e.g. Ghez *et al.* 1998). The imaging distortion and artifacts of the camera were accounted for. Each pixel was expanded into 2×2 pixels with equal flux in order to allow for a higher precision by shifting the specklegrams by fractions of a pixel before adding them. For the speckle observations the image rotator was turned off so that the field orientation on the detector changed throughout the night. This procedure minimizes the systematic effects of possible flat fielding imperfections and image distortion, and maximizes the sky coverage in the speckle image reconstruction. In order to compensate for the field rotation the specklegrams are rotated such that they have a common orientation. Due to field rotation and differences in the centering of individual frames, the fields of view of the final shift-and-add images are larger than the original frames and cover a 6"×6" region. Following these data processing steps the NIRC specklegrams of the Galactic Center are free of camera artifacts and are dominated by the short exposure structure of the four brightest stars of the central star cluster.

1.4 Single Telescope Adaptive Optics

On single telescopes diffraction limited images can be obtained in real time by using adaptive optics systems (e.g. Tyson 1998). In the following we describe how these systems work and how they can be used for high angular resolution imaging and spectroscopy. We also give a closer description of some of the adaptive optics systems that have been used to observe the Galactic Center.

1.4.1 *Adaptive Optics Imaging*

Adaptive optics allows for a real time correction of the disturbing influences on the shape of the wavefront entering the telescope aperture. The initial wavefront of the astronomical object can be regarded as being planar just outside the Earth's atmosphere. It then travels about 20 km through the turbulent atmosphere. Across the diameter of a large telescope, it accumulates phase errors of a few micrometers. In most systems the low frequency component of the seeing — the image motion — is corrected separately by a tip-tilt sensor and a corresponding correction mirror. The high frequency part of the fluctuations is taken over by the adaptive optics system and requires corrections on time scales of the order of a few 10 to a few 100 Hertz. The correcting loop must therefore run at a multiple of that frequency. The wavefront shape is measured by a wavefront sensor. It has to be corrected to a small fraction of the observing wavelength ($\lambda/20$ to $\lambda/50$) about every millisecond. In a conventional AO system light with wavelengths short-ward of 1 μm is sent to the wavefront sensor while light long-ward of 1 μm is sent to the science instrument (Fig. 1.13).

In the terminology of the previous sections, the adaptive optics system measures and corrects the transfer function $T(u, v)$. It corrects possible static abberations contained in the telescope transfer function $T_0(u, v)$ and high frequency atmospheric abberations described by the seeing transfer function $T_s(u, v)$. As shown in section 1.3, seeing conditions are better in the NIR than in the visible domain, where the requirements on an AO system (atmospheric coherence time and Fried diameter of seeing cells) are much higher. In observational astrophysics, current AO systems are preferentially used to correct images at wavelengths longer than about 1 μm.

There are two frequently used types of wavefront sensors: Shack-Hartmann and curvature wavefront sensors.

In a *Shack-Hartmann* device the telescope pupil is subdivided into multiple sub-pupils each resulting in an image of a light source (star or calibration laser) on a detector. From the resulting image one can measure the slope of the wavefront at the position of the sub-pupils from the displacement of the images of the reference star with respect to a reference position (Fig. 1.14). That position can be calibrated using an internal light source (e.g. a laser). *Curvature wavefront sensors* make wavefront measurements using pairs of extra-focal images. The differences between them are proportional to the curvature of the wavefront as a function of position in the pupil. With both methods, wavefront sensing is done on a guide star, or even on the observed object itself if it is sufficiently bright and compact (or if it has sufficiently sharp light gradients). The measurement can be performed in the visible domain for observations in the infrared, or in the infrared itself if the guide star is too faint in the visible and/or closer to the target than a visible reference source. However, currently only the NAOS AO system at the ESO VLT unit telescope 4 is equipped with an infrared wavefront sensor, giving the combined NAOS/CONICA system unique capabilities.

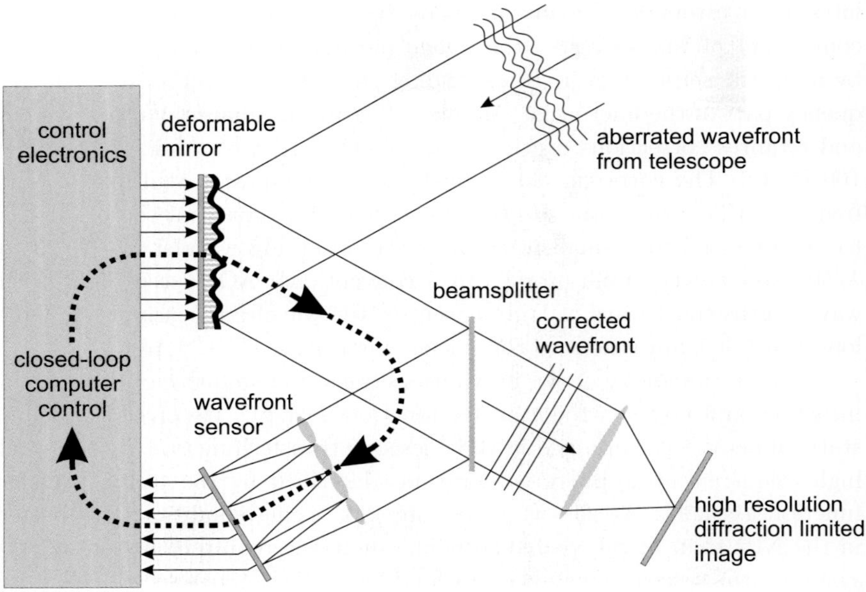

Fig. 1.13 Schematic view of a typical adaptive optics loop. The wavefront is reflected off a deformable mirror (DM) and sent to a beam splitter. There the optical light is sent to the wavefront sensor (WFS) and the infrared light to the science camera. The wavefront sensor signal is used by a closed-loop control computer to drive the DM. The wavefront straightened by the DM produces an image that contains a significant amount of power at the diffraction limit of the telescope.

Correction of the wavefront in adaptive optics systems is done with a deformable mirror which is located close to a pupil image of a conjugated atmospheric layer and has a diameter of the order of 10 to 20 cm. This mirror with an arrangement of actuators matched to the wavefront sensor is shaped into a conjugate of the measured aberrations to correct the wavefront. Correction of the wavefront in adaptive optics systems using a Shack-Hartmann wavefront sensor is usually made with a deformable mirror supported by individual piezoelectric actuators. For 8 m class telescopes and a wavefront correction in the near-infrared, one requires a few hundred actuators. In a system that senses curvature the correction is usually accomplished with a bimorph adaptive mirror, made of two bonded piezoelectric plates.

The control loop between the wavefront sensor and the deformable mirror is run by a fast real-time computer. Within only 1 millisecond or less

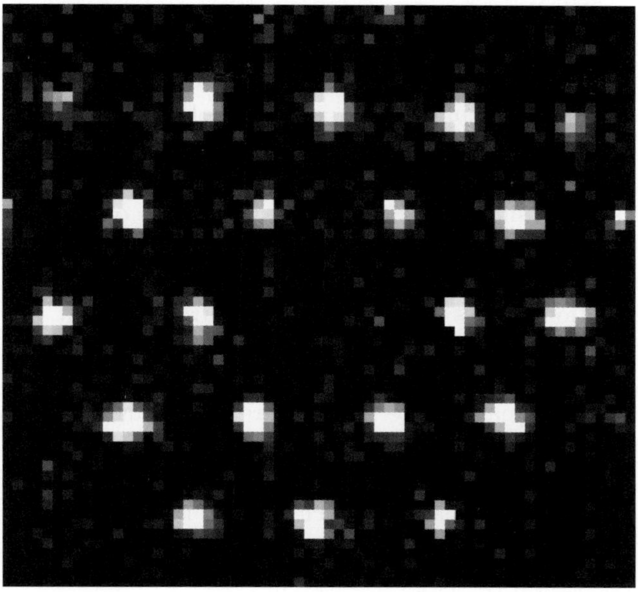

Fig. 1.14 Typical image of a star produced by a Shack-Hartmann wavefront sensor through an array of lenslets across the pupil. Each lenslet produces an image of the star. The varying positional offset of the individual stellar images from a reference position is a measure of the varying gradients across the wavefront within the pupil at the position of the corresponding lenslet. Using this information a fast computer calculates the signal to drive the deformable mirror in the control loop. This image was generated by the ALFA adaptive optics system (a common MPE/MPIA project) on the Calar Alto 3.6 m telescope (see Eckart *et al.* 2000, and references therein).

the system uses the wavefront sensor measurements to calculate the commands to set the deformable mirror. This is the reason why bright stars are needed as reference sources (with magnitudes brighter than about 17 in the visible and brighter than about 10 in the NIR domain). The correction schemes include zonal or modal control methods. In zonal control, each zone or segment of the mirror is controlled independently by wavefront signals that are measured for the sub-aperture corresponding to that zone. In modal control, the wavefront is expressed as the linear combination of basic wavefront modes that fit the measured phase front. Modal control systems are very flexible and allow limitation to a certain spectrum of basic modes or an a priori choice of certain offsets in order to compensate for fixed aberrations.

The actual delivered performance of the system depends strongly on several factors: the guide-star characteristics, the intrinsic spatial and temporal properties of the atmosphere, telescope wind shake, and so forth. The goal of course is to obtain a stable correction of the image that allows for a long integration time. If the correction is less stable (or if the observed source is too bright to allow for long exposures) one can obtain series of AO assisted images — similar to Speckle interferometric observing techniques — but with integration times in the range of a few seconds. After further post-processing and image selection such "hybrid modes" will result in high angular resolution images at or close to the diffraction limit.

The atmospheric restrictions for AO operation in the infrared are less severe than in the optical. Adaptive optics operation is strongly affected by the size of the isoplanatic angle. In the infrared it is usually of the order of 20" and shrinks to less than 5" in the optical. However, even at a wavelength of 2.2 μm the sky coverage, i.e. the probability of finding a suitably bright reference star in the isoplanatic patch around the chosen target, is only of the order of 0.5 to 1%. The sky coverage can be extended by using either artificial laser guide stars projected high above the telescope (e.g. Eckart *et al.* 2000; Bonaccini *et al.* 2003; McLean & Sprayberry 2003a) or by employing multi-conjugate wavefront systems that use multiple reference stars to measure the atmospheric properties towards the target and allow to correct a field-of-view that is much larger than the isoplanatic patch (Marchetti *et al.* 2003; Ragazzoni *et al.* 2003). Both methods are currently being developed and have not yet been applied to the Galactic Center.

Adaptive optics imaging produces a point spread function (PSF) in which a substantial fraction of the light is gathered in a diffraction limited core. The remainder is contained in a "seeing foot" which has a width similar to or larger than the seeing cloud corrected for image motion (tip-tilt) (Fig. 1.15). A frequently used measure of the quality of the adaptive optics correction is the so-called Strehl ratio. It is defined as the ratio between the peak intensity in the AO-corrected (but through instrumental and external effects still degenerated) and the hypothetical fully diffraction limited image of a point source, e.g. an Airy pattern. A direct seeing limited long time exposure (i.e. adaptive optics switched off), has a Strehl ratio of the order of 1 to 2%. Under optimum conditions (excellent input seeing of about $0.5''$ or less, a bright reference star, and a fully operative adaptive optics system) Strehl ratios of about 50–60% can routinely be achieved at wavelengths of 2 μm and of 70–90% at wavelengths of 4 μm.

For observations of the Galactic Center one is in the fortunate situation

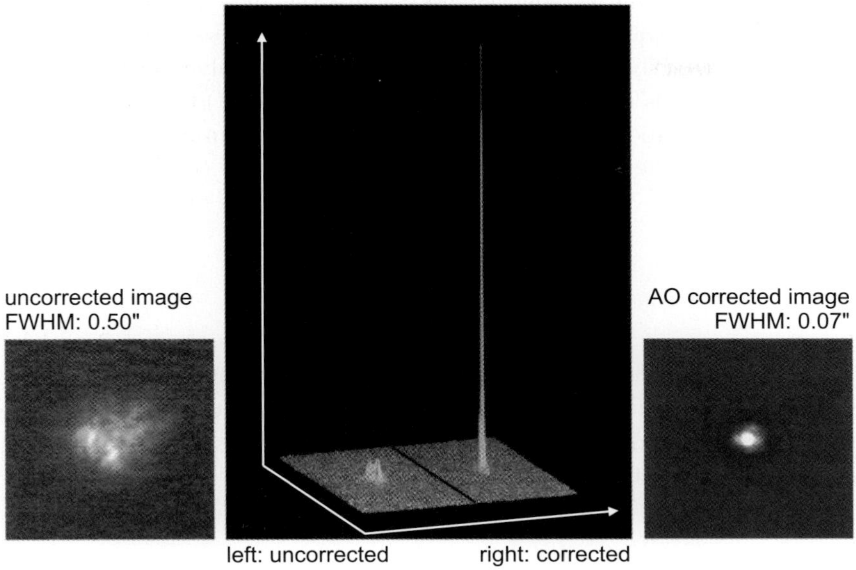

Fig. 1.15 A demonstration of how adaptive optics improves the Strehl ratio of a seeing limited image.

to have infrared and optical reference stars in the vicinity of the target, i.e. the position of the radio source Sagittarius A* at the core of the central stellar cluster of the Milky Way. There are two foreground stars (Biretta, Lo, & Young 1982) that are just suitable for optical wavefront sensing. Star 2 (USNO 0600-28579500), located 30″ NNE of the very center, has a V-band magnitude of ∼14 (Biretta *et al.* 1982) and an R-band magnitude of 13.2 (e.g., Ghez *et al.* 2003a). For the Galactic Center currently the best conditions for adaptive optics measurements can be achieved if an infrared wavefront sensor is available. In this case the adaptive optics correction loop can be closed on the K∼ 6.5 mag supergiant IRS 7 (e.g. Clénet *et al.* 2001; Schödel *et al.* 2002) which is located ∼ 5.5″ north of Sgr A*.

As an example of how the atmospheric phase front across the telescope aperture can be calculated from the wavefront sensor signal we briefly outline the situation for the case of a Shack-Hartmann wavefront sensor and the more frequently used modal control. The observables here are the displacements of the stellar images at various positions (x_m, y_m) on the Shack-Hartmann wavefront sensor image, with m running from $m = 1$ to

$M/2$ (there are M measured gradients), across the aperture, which are proportional to the wavefront slopes. The phase that we seek to reconstruct can be conveniently expressed in terms of Zernicke polynomials $Z_k(x,y)$. These polynomials correspond to individual (k=1 to K) orthogonal phase front aberration modes like defocus, coma, astigmatism and higher order modes. In total the phase front can be written as:

$$\phi = \sum_{k=1}^{K} a_k Z_k(x,y) \quad . \tag{1.38}$$

Here the coefficients a_k measure the power contained in that mode. Since there are M total measurements, half of which account for the phase gradients in the x-direction and half of which for those in the y-direction, a set of linear equations is defined:

$$\left.\frac{\partial\phi}{\partial x}\right|_m = \sum_{k=1}^{K} a_k \left.\frac{\partial Z_k(x,y)}{\partial x}\right|_m \quad . \tag{1.39}$$

This set can be written as a matrix equation $s = [B]\ a$, where the vector s describes the slopes and vector a the modal coefficients:

$$s = \begin{pmatrix} \left.\frac{\partial\phi}{dx}\right|_1 \\ \left.\frac{\partial\phi}{dx}\right|_2 \\ \cdots \\ \left.\frac{\partial\phi}{dx}\right|_{\frac{M}{2}} \\ \left.\frac{\partial\phi}{dy}\right|_1 \\ \left.\frac{\partial\phi}{dy}\right|_2 \\ \cdots \\ \left.\frac{\partial\phi}{dy}\right|_{\frac{M}{2}} \end{pmatrix} \qquad a = \begin{pmatrix} a_1 \\ a_2 \\ \cdots \\ a_{\frac{M}{2}} \\ a_1 \\ a_2 \\ \cdots \\ a_{\frac{M}{2}} \end{pmatrix} . \tag{1.40}$$

The matrix B contains the derivatives of the Zernicke polynomials which are here taken as the polynomial basis functions. Other sets of basis functions on the pupil annulus that can be used in this context are Karhunen-Loeve functions. The basis elements need to be linearly independent but not necessarily orthogonal,

$$[B] = \begin{pmatrix} \mu_{1,1} & \mu_{2,1} & ..... & \mu_{K,1} \\ \mu_{1,2} & \mu_{2,2} & ..... & \mu_{K,2} \\ .... & & & .... \\ \mu_{1,\frac{M}{2}} & \mu_{2,\frac{M}{2}} & ..... & \mu_{K,\frac{M}{2}} \\ \xi_{1,1} & \xi_{2,1} & ..... & \xi_{K,1} \\ \xi_{1,2} & \xi_{2,2} & ..... & \xi_{K,2} \\ .... & & & .... \\ \xi_{1,\frac{M}{2}} & \xi_{2,\frac{M}{2}} & ..... & \xi_{K,\frac{M}{2}} \end{pmatrix} \tag{1.41}$$

with

$$\mu_{k,m} = \frac{\partial Z_k(x,y)}{\partial x} \Big|_m \qquad \textit{and} \qquad \xi_{k,m} = \frac{\partial Z_k(x,y)}{\partial y} \Big|_m \quad . \tag{1.42}$$

The solution of this equation is found via the pseudo inverse of matrix [B] and results in the vector of modal coefficients a :

$$a = [(B^T B)^{-1} B^T] s. \tag{1.43}$$

Knowing the elements of vector a the phase front can be reconstructed. In order to correct for it and and straighten out the phase across the aperture it is required to drive the deformable mirror with an appropriate signal that impresses exactly one half of the phase front amplitude onto it. If g are the gradients that result from the actuator commands this can be written in the form of a matrix equation: $g = [G]\ a$. This assumes that one knows the dynamic response of the mirror if driven by an individual mode of a given amplitude coefficient. That response has to be calibrated. One way of doing this is to expose the mirror to a constant modal injection using an artificial light source. In this way the elements of a response matrix G can be measured.

1.4.2 *Adaptive Optics Spectroscopy*

It is most advantageous to combine high angular resolution with high spectral resolution. For the Galactic Center this allows one to measure spectra of the high velocity stars closest to Sgr A*. Using NIRSPEC (McLean *et al.* 1998, 2000) behind the adaptive optics (AO) system (Wizinowich *et al.* 2000b) on the W.M. Keck II 10 m telescope one can obtain diffraction limited K-band (2.0 to 2.4 μm) spectra and images of stars in the Sgr A* cluster. Using an optical wavefront sensor, near-diffraction-limited spatial resolution can be achieved with adaptive optics using the R = 13.2 mag natural guide star located 30" northeast from Sgr A*. Spectra were obtained using a 2 pixel wide slit (3.96"$\times$0.036") in low-resolution mode, resulting in a moderate spectral resolution of R=$\lambda/\Delta\lambda$=2600 (e.g. Gezari *et al.* 2002).

For NAOS/CONICA (see below; Lenzen *et al.* 1998; Rousset *et al.* 1998; Brandner *et al.* 2002) at the VLT UT4 a set of slits and associated masks for spectroscopy are available. The slits have a width of 86 mas and 172 mas and a length of 40 arcseconds. The spectral resolutions range from R=$\lambda/\Delta\lambda$=400-1400.

For NIRSPEC and NACO the spectral dispersion is performed via a set of objectives, sorting filters, and grisms. The grisms are a combination of a grating applied to one surface of a prism. The sorting filters select a diffraction order of the grating which is responsible for the required dispersion and the prism corrects the beam tilt due to the selected grating order. This is a compact setup that allows dispersion preserving the overall optical axis.

Of high value are the observations of the Galactic Center that have been carried out using near-infrared integral field spectrometers. These devices use a slicing mirror device that transforms an entire two-dimensional image into the equivalent of a long-slit. After spectral dispersion onto the detector chip a three-dimensional data cube can be reconstructed, i.e. one obtains an image of the object with spectral information for every pixel in the image (the 3rd dimension in the data cube). This technique is extremely advantageous for spectroscopy of crowded fields such as the GC stellar cluster. The 3D camera (Weitzel *et al.* 1996; Hofmann *et al.* 1995) was the first of these systems to be used in the NIR on the Galactic Center. Recently the near-infrared integral field spectrometer SPIFFI (Tecza *et al.* 1998; Eisenhauer *et al.* 2000) has come into operation at the ESO VLT. Data on the Galactic Center in a seeing limited mode have already been taken in early 2003 (see Eisenhauer *et al.* 2003b). Starting in 2004 SPIFFI will operate in combination with adaptive optics at the VLT (SINFONI; e.g. Eisenhauer *et al.* 2003a).

1.4.3 *Adaptive Optics Instrumentation*

There are quite a number of operating astronomical adaptive optics systems. In this section we restrict the discussion to some that have often been used to observe the Galactic Center. We will give a short summary of the technical aspects of the adaptive optics systems with which the large body of recent diffraction limited, or close to diffraction limited near-infrared images have been obtained. This information may be helpful to assess and appreciate the high quality of the data sets. It gives important complementary information that may in some cases be required to better understand the content or presentation of some of the observations described in chapter 2. These descriptions also represent a record of technical developments that started at 4 m-class telescopes and ended in highly developed and efficiently working adaptive optics systems operated at 8 m-class telescopes.

1.4.3.1 *NAOS/CONICA*

NAOS/CONICA ("NACO") is the AO system and near-infrared camera at the ESO VLT unit telescope 4 (UT4; Yepun). The system was installed in spring/summer 2002 (Lenzen *et al.* 1998; Rousset *et al.* 1998; Brandner *et al.* 2002). CONICA is the infrared camera and spectrometer attached to the NAOS AO system. NAOS is installed at one of the UT4 Nasmyth foci. It picks up an f/15 beam, corrects for atmospheric turbulence and provides an f/15 beam to CONICA again (Fig. 1.16). Having passed the Nasmyth focal plane, the beam is led to a first collimating parabola. Then it is reflected successively onto the tip-tilt and the deformable mirror. The following dichroic mirror separates the optical train into the imaging path and the wavefront sensing path. NAOS offers five different dichroic beam splitters in order to adapt to the flux and the spectral characteristics of the guide star as well as to the desired observing wavelength. In the imaging path the light is refocused onto the entrance focal plane of CONICA, which is located behind the entrance window in the cold cryostat. An atmospheric dispersion compensator (ADC) can be positioned between NAOS and CONICA in case of observations at a high zenith angle. CONICA is equipped with a 1024x1024 pixel InSb array detector sensitive from 1 to 5 μm.

The wavefront sensing path consists of a field selector (Spanoudakis *et al.* 2000) and two wavefront sensors. They are located between the dichroic mirror and the WFS input focus. The two wavefront sensors, one in the visible and one in the near-infrared spectral range, enhance the sky

Fig. 1.16 The ESO adaptive optics imaging system NAOS/CONICA mounted to a flanch (to the left) of UT4.

coverage of the instrument because bright NIR sources can be used for wavefront sensing as well as visible guide stars. The field selector chooses the guide star in a 2 arcmin field-of-view and allows differential object tracking, pre-calibrated flexure compensation and counter-chopping. In combination with the deformable mirror it is also able to correct for a certain amount of defocus, as needed when the prisms of the atmospheric dispersion compensator are shifted into the beam.

The high angular resolution camera CONICA is equipped with an ALADDIN array (produced by the Santa Barbara Research Center) covering the 1-5 μm spectral region. Splitting the wavelength region into two parts (1 to 2.5 μm and 2.0 to 5 μm) allows one to keep the light path achromatic. Therefore the available pixel scales are realized by seven cameras. To each pixel scale a camera is associated with the short wavelength region and another one with the long wavelength region. The only exception is the camera with the highest magnification. There is no long wavelength counterpart needed. A variety of different observing modes is provided by the analyzing optics: coronography, low resolution long slit spectroscopy, imag-

Table 1.1 *CONICA NIR broad band filters (values taken from the ESO NAOS-CONICA user manual).*

Filter Name	Central Wavelength [μm]	Width [μm]
H	1.66	0.33
Ks	2.18	0.35
L'	3.80	0.62

ing spectroscopy by a tunable cold Fabry-Perot spectrometer, polarimetry by wire-grids or Wollaston prisms, and about 40 broad- and narrow-band filters can be chosen.

The GC stellar cluster was observed several times during the commissioning and science verification of the NACO AO system and near-infrared camera in spring/summer 2002. In 2003, several epochs of regular H, K_S, and L'-band GC imaging observations were obtained with NACO (see Table 1.1 for the properties of the filters used). For simplicity, we will use the expressions "K-band" and "L-band" in this book because the differences in transmission to Ks and L' are negligible for our purpose. For the observations in the K- and H-bands, the loop of the AO was closed with the infrared wavefront sensor on the K$\sim$ 6.5 mag supergiant IRS 7, which is located $\sim 5.5''$ north of Sgr A*. Because of instrument restrictions, AO correction for the L-band observations in the 2002 epoch could only be carried out with the visible wavefront sensor, locking on a faint $V \sim 14$ star about $30''$ north-east of Sgr A*. Since 2003, L-band observations of the GC stellar cluster use the NIR wavefront sensor routinely, as well. NACO observations of the Galactic Center are typically made in blocks of several tens of short exposures images, with typical integration times of $\sim$ 15s for H- and K_S-band and 0.2s for L'-band observations. The performance of the NAOS AO system depends strongly on the momentary atmospheric turbulence, especially on coherence time, but Strehl ratios as high as 50% in H-band, 50–70% in K band and about 90% in L-band can be routinely achieved when observing the Galactic Center under good conditions. The unique NIR wavefront sensor of NAOS in combination with the bright reference star IRS 7 close to Sgr A* make NAOS/CONICA at the ESO VLT currently the best suited instrument for observations of the GC stellar cluster, especially since the GC passes close to zenith at the ESO Paranal site and therefore allows for long integration times at small airmasses.

1.4.3.2 *NIRC2*

The Near-Infrared Camera 2 (NIRC2) is a facility near-infrared instrument (e.g. McLean & Sprayberry 2003b) designed to take full advantage of the adaptive optics on the Keck II telescope (Wizinowich *et al.* 2000b). It is equipped with a Shack-Hartmann wavefront sensor which is locked on the R $= 13.2^m$ natural guide star (USNO 0600-28579500) for observations of the Galactic Center. This star is located about 30" away from Sgr A*. NIRC2 is in operation since the middle of 2001. NIRC2 is positioned behind the AO bench on the Left Nasmyth Platform of Keck II. The instrument operates over a wavelength range of 1 to 5 μm, providing three selectable cameras to cover the expected range in image sizes. Two filter wheels with 18 positions each provide a variety of filters and/or grisms, while a focal plane mechanism provides slits and occulting masks for coronography. A dedicated slide carries larger grisms for spectroscopy. Six selectable pupil masks are available to reduce background noise sources; four of these rotate in concert with the telescope pupil and one serves specifically for spectroscopy. The detector is a 1024$\times$1024 Aladdin-3 InSb array with four-quadrant readout into 32 channels. NIRC2 was used for the first measurement of an absorption line in the star S0-2 that is on an orbit around the supermassive black hole (Ghez *et al.* 2003b).

1.4.3.3 *PUEO*

The adaptive optics system PUEO was developed for the Canada-France-Hawaii Telescope (CFHT), based on F.Roddier's curvature concept (Roddier 1991; Arsenault *et al.* 1994). It is a facility instrument mounted at the f/8 Cassegrain focus of the CFH 3.6 m telescope on top of Mauna Kea, Hawaii. The instrument is equipped with a 19 element optical curvature wavefront sensor (Graves & McKenna 1991; Arsenault *et al.* 1994; Graves *et al.* 1994), and a 19 electrode bimorph deformable curvature mirror, all under control of a Real Time Computer (see Rigaut *et al.* 1998). The system was commissioned at the CFHT during three runs in the first semester of 1996. PUEO has only few mainly reflecting optical parts (Richardson 1994). Off-axis parabolic mirrors allow for a compact instrument with small optical components resulting in reduced flexures. Davidge *et al.* (1997) carried out JHK observations of the Galactic Center covering a mosaicked field of approximately 12"$\times$12". The data was combined with L-band data by Simons & Becklin (1996). Main goal of the project was the investigation of the stellar content of the central $\sim$0.7 parsec diameter star cluster surrounding Sgr A*. The data was reduced using point spread fitting routines installed in standard data reduction programs. Further observations of the GC stellar cluster using PUEO at the CFHT were performed in 1998 (see Paumard *et al.* 2001).

1.4.3.4 *ADONIS*

For several years the adaptive optics system ADONIS (Beuzit *et al.* 1997; Bonaccini *et al.* 1997; Hubin 1997) was available at the ESO 3.6 m telescope at La Silla, Chile. In this system, atmospheric wavefront distortions are measured with a Shack-Hartmann sensor at visible wavelengths and are corrected by a deformable mirror with 52 piezo actuators. This mirror is driven by a closed control loop with a correction bandwidth of up to 17 Hz. The natural guide star within the near-infrared isoplanatic patch must be brighter than about m_V=13. The AO system was also used in conjunction with a prototype of the VLT infrared wavefront sensor.

For recent AO measurements with ADONIS Clénet *et al.* (2001) used both the optical and the RASOIR prototype infrared wavefront sensor. For optical sensing the R=13.8^m star 16" north and 15" east and for infrared sensing the bright supergiant IRS7 located about 5.5" north of Sgr A* were used. For their measurements in the 3-5 micrometer range Clénet *et al.* (2001) used the COMIC camera (Bonaccini *et al.* 1996), at the f/45 Cassegrain focus, which provides an image scale of 0.1"/pixel, resulting in a field of view of 12.8“×12.8“. The COMIC detector is a 128×128 HgCdTe photovoltaic focal plane array. At 77 K, the average dark current is 2000 electrons per second and the total capacity is about 6×10^6 electrons. The readout noise is about 1000 electrons rms allowing to observe under conditions of background limiting performances (BLIP) for individual exposure times larger than 500, 260 and 150 ms for the L-, L'- and M-bands respectively (Lacombe *et al.* 1998). In the L-, L'-, and M-bands, one can observe in a chopping mode, alternating object and sky images using a field selection mirror. For the Galactic Center observations an offset position 20" to the south and 10" to the east was chosen.

1.4.3.5 *HOKUPA'A*

Hokupa'a is a natural guide star (NGS), curvature-sensing adaptive optics system (AOS) built by the University of Hawaii (Graves *et al.* 2000, 1998). It was originally built for 4 m-class telescopes but is successfully used to improve the image quality at the 8 m-class Gemini North telescope. The corrected output beam of Hokupa'a is fed at f/26 to a near-infrared camera (QUIRC). With QUIRC operating in the 1-2.5 μm wavelength bands, the system provides a fixed plate-scale of 20 milli-arcseconds per pixel. This results in diffraction limited images at H- and K-band. Hokupa'a is based on a 36-element curvature wavefront sensor (WFS) in combination with a bimorph mirror. The AO system was constructed to deliver images with near diffraction-limited resolutions at 2.2 microns with seeing of $1''$ or better. Under favorable seeing conditions, a FWHM of about twice the diffraction limit can be achieved in the J-band.

The Gemini North observatory Galactic Center Demonstration Science Data Set with imaging observations from the year 2000 is publicly available. This data set provides NIR observations of the Galactic Center stellar cluster with the Gemini North telescope, the QUIRC near-infrared camera and the Hokupa'a AO system. Since the Hokupa'a AO system was designed for a 4 m-class telescope, it could only partially provide the correction needed for the 8 m-class Gemini telescope. Moreover, the visible guiding star used for the observations is located at $\sim 30''$ from Sgr A*, so the Strehl ratio of the image is rather low. However, the central sources around Sgr A* are clearly resolved at an estimated resolution of 100 mas.

1.4.3.6 *FASTTRAC*

FASTTRAC is a rapid guiding unit that was used several times to observe the Galactic Center. It was operated at the Steward Observatory Bok (2.3 m) telescope in the early 90's. The FASTTRAC instrument (Close & McCarthy 1994) incorporates a tip-tilt Cassegrain secondary mirror to rapidly ($\leq$100 Hz) stabilize image motion during long exposures. It utilizes a high-speed InSb camera onto which a guide star — IRS7 in the case of the Galactic Center — is imaged via a small pick off probe mirror. The infrared science camera is based on a 256$\times$256 NICMOS3 array with a pixel scale of 0.21" and an overall field of view of 54"$\times$54". Given its speed and separate control loop it can be regarded as a low-order adaptive optics system. It was mainly used to search for an infrared counterpart of Sgr A* and investigate its variability (Close *et al.* 1995) and to examine the population of hot stars in the central cluster (Tamblyn *et al.* 1996).

1.5 Cleaning and Deconvolution

When an image of some object is produced with any kind of optical system, this image will suffer a certain degradation because of the general limitations of the instrument and because of the influence of numerous external factors. For example, in the absence of an atmosphere or other degrading factors, the image of a point source, such as a very distant star, by an aberration-free telescope with a circular aperture will always be the familiar Airy pattern. This pattern arises because the limited size of the telescope will truncate the information on the spatial frequencies of the object. A perfect point image could only be produced with an infinitely large aperture. When using ground-based telescopes, the most important additional source of image degradation is the turbulence in the Earth's atmosphere. The consequence of the combined influences of the instrument, the atmosphere, and other external factors is that the image of a point source will be characterized by a specific function, the so-called point spread function (PSF). This function will generally vary with time, e.g. due to temporal variations associated with the instrument or due to the limited atmospheric coherence time, as well as with the position in the imaging plane, e.g. due to the limited size of the isoplanatic angle.

Generally, the observed signal, $g(x, y)$, can be expressed by the convolution of the observed object, $f(x, y)$, with a function describing the PSF, $h(x, y)$, plus an additional term, $c(x, y)$, that includes the remaining degradation effects, such as anisoplanaticity, detector read-out noise and other non-linear terms:

$$g(x, y) = f(x, y) \odot h(x, y) + c(x, y). \tag{1.44}$$

The symbol $\odot$ denotes the convolution operator.

There are two ways how one can improve the imaging process. On the one hand, there are technical achievements, such as improving the imaging quality of the telescopes, reducing detector noise, or suppressing the influence of the atmosphere by using techniques such as adaptive optics or even launching a telescope into space. On the other hand, in spite of all possible improvements, there will always remain PSF residuals and the Airy pattern, of course, which will for example hamper the detection of faint sources near bright stars or complicate the photometry of sources with overlapping PSFs. Therefore it is always desirable to improve astronomical images further by reducing the degrading influence of the PSF with an appropriate deconvolution algorithm. This "cleaning" of astronomical images can be roughly illustrated by stating that the PSFs of point sources are replaced with Gaussian PSFs that have a FWHM comparable to the diffraction limit of the respective telescope used.

Convolution corresponds to a multiplication in Fourier space. Thus, in the case that the image is free of noise and free of non-linear effects such as anisoplanaticity, one would obtain the Fourier transform of the original object distribution by a simple division in Fourier space:

$$F(u,v) = \frac{G(u,v)}{H(u,v)}. \tag{1.45}$$

Here, $G(u,v)$, $F(u,v)$, and $H(u,v)$ denote the Fourier transforms of the observed signal, $g(x,y)$, of the object, $f(x,y)$, and of the PSF, $h(x,y)$. u and v are the spatial frequencies corresponding to the x and y coordinates. We have ignored here that $H(u,v)$ will have zero values at high frequencies because a telescope cannot be infinitely large and therefore information at high spatial frequencies will be lost. In practice, one will also encounter additional problems: the term $c(x,y)$ — with its Fourier transform $C(u,v)$ — that describes the image degradation through noise or non-linear effects, can never be neglected. Also, the shape of the PSF will never be known with arbitrary accuracy. It is usually approximated analytically or extracted from individual point sources in the observed image.

Because of these limitations, deconvolution is always an ill-posed problem that allows for more than one solution. The existing deconvolution algorithms intend to overcome the difficulties by various approaches in order to find the most probable object distribution. In the following sections we describe some algorithms that were used in the case of the Galactic Center.

1.5.1 *Linear Deconvolution*

The linear deconvolution technique departs from the straightforward Fourier space devision suggested by Eq. (1.45). The linear deconvolution tries to circumvent the problems that the Fourier transform of the PSF is not known to arbitrarily high frequencies and that the noise term becomes dominant at high frequencies by constructing an appropriate filter, a so-called Wiener filter, that suppresses high spatial frequencies. In Fourier space, the reconstruction of the object can then be described as

$$O(u,v) = \frac{[G(u,v) + C(u,v)] \cdot \Phi(u,v)}{H(u,v)}, \tag{1.46}$$

where Φ designates the filter function and C is the Fourier transform of the noise term. The filter is chosen such that the square of the deviations between the reconstructed object $O(u,v)$ and the "true" object $F(u,v)$ is minimized, i.e.

$$\int_{-\infty}^{+\infty} |O(u,v) - F(u,v)|^2 d(u,v) = Minimum. \tag{1.47}$$

Using Eqs. (1.45) and (1.46) we obtain

$$\int_{-\infty}^{+\infty} \left|\frac{[G(u,v) + C(u,v)] \cdot \Phi(u,v)}{H(u,v)} - \frac{G(u,v)}{H(u,v)}\right|^2 = Minimum. \tag{1.48}$$

In order to find the filter $\Phi(u,v)$ that minimizes this function, we have to derive this equation with respect to $\Phi(u,v)$ and set the result equal to zero. We point out that because the signal and the noise are uncorrelated $\int_{-\infty}^{+\infty}(G(u,v) \cdot C(u,v))d(u,v) = 0$.

We finally obtain

$$\Phi(u,v) = \frac{|G(u,v)|^2}{|G(u,v)|^2 + |C(u,v)|^2}. \tag{1.49}$$

In practice, we can estimate the structure of $\Phi(u,v)$ by replacing the spectral energy distribution of the observed signal, $|G(u,v)|^2$, with the spectral energy distribution of the observed PSF, $|H(u,v)|^2$. The spectral energy distribution of the noise, $|C(u,v)|^2$, can be replaced in a good approximation by the Fourier transformation of a delta function, $|\tilde{\delta}(u,v)|^2$. Taking into account that the power spectrum of a function is the product of its Fourier transform with its complex conjugate, we obtain from Eq. (1.46)

$$F(u,v) \approx O(u,v) \approx \frac{G(u,v) \cdot \overline{H}(u,v)}{|H(u,v)|^2 + |\tilde{\delta}(u,v)|^2}, \tag{1.50}$$

where $\overline{H}(u,v)$ denotes the complex conjugate of $H(u,v)$. The noise term $C(u,v)$ in Eq. (1.46) was neglected when deriving Eq. (1.50).

Application of Wiener filtering leads to a characteristic "ringing" around point sources in the reconstructed image. There are two main reasons for ringing that is not linked to diffraction: a) extraction of the PSF in image space and b) apodization of the PSF in Fourier space. The influence of a characteristic pattern due to the PSF extraction can be reduced by avoiding to cut out the PSF with a function that has steep flanks. The cause for the ringing due to apodization can be described in a simplified way as follows: since in practice we cannot know the PSF $H(u,v)$ up to infinitely high frequencies, we have to cut off $H(u, v)$ at some upper frequencies u and v. In the most simple case, this cutting off is done by multiplication with a boxcar function. The Fourier transform of such a boxcar function is a function of the form $\sin(x)/x$ in the image domain, and is a major source of ringing.

1.5.2 *Lucy-Richardson Deconvolution*

The Lucy-Richardson (LR) deconvolution algorithm is an iterative method based on the scheme for the rectification of observed probability distributions (Richardson 1992; Lucy 1974). The LR method is based on the iteration of three steps. In the first step, the current estimate of the object distribution, $O_k(x, y)$ is convolved with the PSF, $H(x, y)$.

$$\Psi_k(x, y) = O_k(x, y) \odot H(x, y) \tag{1.51}$$

Subsequently, the obtained image, $\Psi_k(x, y)$, is compared with the observed image, $G(x, y)$.

$$R(x, y) = \frac{G(x, y)}{\Psi_k(x, y)} \odot H(x, y) \tag{1.52}$$

Here, the convolution with the PSF acts as a low-pass filter that reduces the influence of high frequencies on the result because they are affected much stronger by noise. In the last step, the current object estimate, $O_k(x, y)$, is multiplied with the correction function, $R(x, y)$ in order to obtain the new estimate

$$O_{k+1}(x, y) = R(x, y) \cdot O_k(x, y). \tag{1.53}$$

Because of Eq. (1.52), high spatial frequencies will be suppressed, thus avoiding the amplification of noise peaks. As a consequence, however, details of the image that are related to high frequencies (e.g. close double stars) will only be adequately resolved after a sufficient number of iterations has been carried out (in our experience on the order of a few 10^4). The disadvantages of LR deconvolution are the high demand for computing time, its tendency to resolve the diffuse, noisy background into distinct point sources when running a large number of iterations, and the effect that in the case of close sources that differ by several magnitudes the fainter one will be "sucked" up by the brighter one. This leads to characteristic dark areas around bright stars. The latter problem can be minimized effectively by a very accurate determination of the PSF.

1.5.3 *Estimating the PSF*

For most deconvolution algorithms such as linear or LR deconvolution one needs to find an estimate for the image PSF before the deconvolution algorithm can be started. Such a PSF estimate can be obtained from a PSF reference source, such as an isolated star that was observed close in time to the observation of the scientific target and under similar atmospheric conditions. Alternatively, one can try to determine the PSF directly from the observed field. In the case of Galactic Center observations, for example, there is a bright (K$\sim$6.5) reference star just $5.5''$ north of Sgr A*. It is more than 5 magnitudes brighter than any other source located within $\sim 2''$ of it and thus is very well suited as a PSF reference. The PSF can also be estimated by taking the average image of several fairly isolated stars in the field of view. In the case of crowded fields, it is preferable to use the median image in order to avoid the influence of sources close to the PSF reference stars.

In crowded fields that were observed under bad seeing conditions the wings of the point sources may be so extended that it is difficult to extract a PSF even when using the median of many stars. In that case one can try to model the whole PSF or to extract the kernel of the PSF from the observed stellar field and to model just its wings. A Moffat function is usually well suited for this purpose.

The *StarFinder* software package (Diolaiti *et al.* 2000) [1] is an ideal tool for extracting PSFs in crowded stellar fields. It estimates an initial PSF from the median of several bright, preferably isolated stars chosen by the user. The PSF can then be improved by repeatedly identifying stars in the image and taking them into account when re-extracting the PSF. *StarFinder* can also be used for obtaining fluxes and positions of stars via PSF fitting.

A remaining limitation on the accuracy of the PSF is always the isoplanatic angle. In spite of some ongoing efforts in the community on including the spatial variation of the PSF across an image in deconvolution procedures, no satisfying solution to the problem has been found yet. This means that a certain PSF estimate can only be used for deconvolution of areas corresponding in size to the isoplanatic patch.

[1] *StarFinder* is an IDL code for the analysis of crowded stellar fields, specifically developed for AO imaging. It can be obtained at the *StarFinder* page at the University of Bologna: *http://www.bo.astro.it/~giangi/StarFinder/*.

1.5.4 *Iterative Blind Deconvolution*

Iterative blind deconvolution (IBD) is an algorithm that reconstructs the object distribution in an astronomical image *and* the PSF at the same time (Jefferies & Christou 1993). In order to achieve this aim, the algorithm takes advantage of *a priori* information on the astronomical image. Such constraints on the object distribution, $f(x, y)$ and on the PSF, $h(x, y)$, are for example:

(1) $f(x, y)$ and $h(x, y)$ must both be positive (Positivity Constraint).
(2) They must convolve to yield the measured signal.
(3) They must occupy finite nonzero regions in the image domain.
(4) They must not be the trivial solution of the convolution of the observed signal with a delta function.

Other constraints can also be applied. A very useful one is, for example, that multiple images of the same object are constrained to have a common object. Hence, effective use can be made of the multiple images usually delivered by speckle imaging or adaptive optics observing techniques.

In IBD an error metric function is introduced to measure the violations of the above constraints.

$$\varepsilon = E_{im} + E_{conv} + E_{bl} + E_{FM} \tag{1.54}$$

E_{im} is the image domain error. It can be described by the sum of the squares of the negative pixels in the estimate of the object and of the PSF. E_{conv} is the convolution error. It measures the quadratic deviations of the convolution of the object and PSF estimates from the measured image. E_{bl} is the band-limit error and measures the power that is contained in the PSF estimate at frequencies higher than an appropriate cutoff frequency (because the PSF cannot be known to arbitrarily high frequencies). The Fourier modulus error, E_{FM}, measures whether the estimated object contains information at frequencies higher than the diffraction limited cutoff.

Convergence is reached via the minimization of the error metric ε. For this purpose, the algorithm uses the conjugate gradient method as described in Press *et al.* (1989). In order to apply the conjugate gradient method, the derivative of the error metric with respect to each variable, i.e. each pixel in the object and PSF estimates, has to be calculated (see Jefferies & Christou 1993).

Every deconvolution algorithm needs some kind of a priori information about the image. The advantage of the IBD algorithm is that it uses very general constraints. The frequent problem of finding an accurate PSF estimate for deconvolution is avoided in IBD. In principle, one can even start with a white noise PSF, although this would be very time consuming. While we have found IBD to work less well on single images, we have found that it can make effective use of multiple images of the same object, which poses tight constraints on the object distribution. In speckle and AO imaging observations it is common to obtain several tens to hundreds of short exposure images of the same object.

An implementation of the IBD algorithm is the publicly available IDAC program code, developed at Steward Observatory by Matt Chesalka and Keith Hege (based on the earlier Fortran Blind Deconvolution code - IDA - developed by Stuart Jefferies and Julian Christou). We would like to refer the interested reader to the IDAC home page *http://babcock.ucsd.edu/cfao_ucsd/idac/idac_package/idac_index.html.*

As for the disadvantages of IBD, using this method effectively requires considerable experience of the user. Also, artefacts in the deconvolved images cannot be completely avoided. Above all, one has to mention the strong demand of computing power. We have found that deconvolution with current average PCs or workstations is not practical for arrays larger than about 256×256, when using more than about 100 input images.

1.5.5 *Comparison of Deconvolution Algorithms*

In the upper left panel of Fig. 1.17 we show a simple shift-and-add (SSA) image of about one hundred 0.5 s-exposure speckle imaging frames of the GC IRS 16 cluster observed with SHARP at NTT in 2001. A linear deconvolution of this image can be seen in the upper right panel. The lower left and right panels show a LR and an IBD deconvolution, respectively, of the SSA image. It can immediately be seen, how deconvolution facilitates the detection of sources that are located in the broad PSF wings of bright stars.

The maps resulting from the different deconvolution algorithms compare very well among each other and with the SSA image. The advantage of deconvolution becomes clear because the deconvolved maps are much richer in detail. In the linearly deconvolved map the "ringing" effect can be seen around the brighter sources. Some artefacts can also be seen around the brighter sources in the LR and IBD deconvolved images, but they are less grave than in the case of linear deconvolution. A typical problem is the blank areas around bright sources, where faint sources cannot be recon-

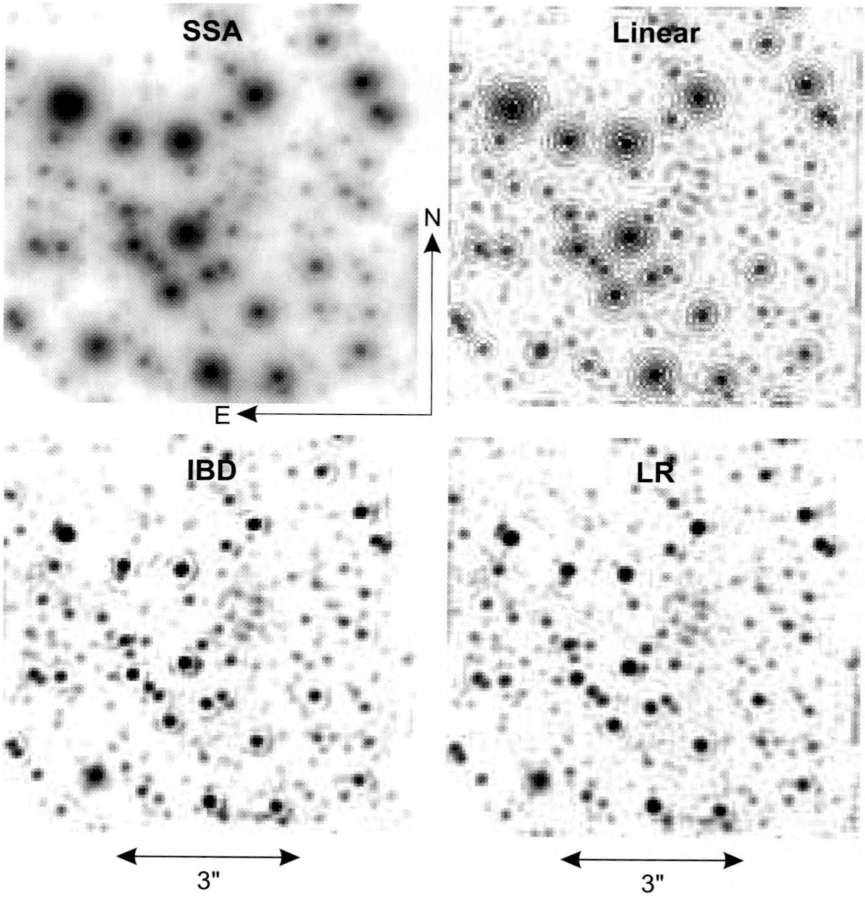

Fig. 1.17 The effect of image deconvolution demonstrated for the central 6.4"×6.4" including the IRS16 complex at the Galactic Center. Upper left panel: SSA image of the IRS 16 cluster as seen by SHARP at NTT In 2001. Upper right panel: linear deconvolution. Lower left panel: deconvolution with IBD (IDAC). Lower right: LR deconvolution. The LR and IBD deconvolved maps were smoothed with a Gaussian of FWHM 100 milli-arcseconds.

structed reliably. The rings around the brighter sources in the IBD map are probably caused by restricting the initial estimate PSF to a limited size with a circular aperture, i.e. a sharp cut off at high frequencies.

For a comparison of the results of deconvolution methods and an assessment of their reliability applied to various data sets, we show in Fig. 1.18

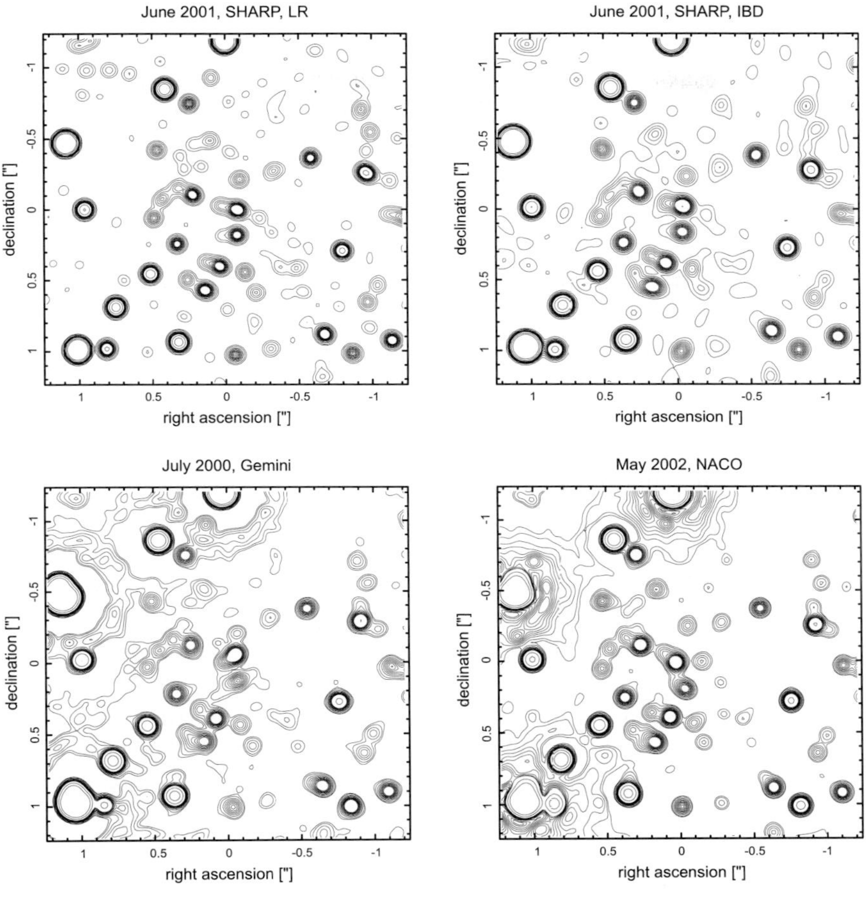

Fig. 1.18 Comparison of the final image processing results for different epochs and data from different telescopes. A 2.5"×2.5" area centered on the position of Sgr A* is shown. The FWHM of the restoring beam was approximately 100 mas in all the images. Upper left: SHARP/NTT 2001, deconvolved with IBD; upper right: SHARP/NTT 2001, deconvolved with LR; lower left: Gemini 2000 (LR); lower right: NACO 2002 (LR). The bright sources at the left and upper edges of the field were saturated in the latter two data sets and therefore show ample deconvolution artifacts in form of rings. Contours were drawn at 10, 20, ...90, 100, 200, 400% of the peak flux of S2, the source closest to the center in all images. Some small differences between the maps arise due to the fact that they cover different epochs and that the stars near the center have fast proper motions.

Gemini 2000, SHARP 2001 (IBD and LR), and NACO 2002 maps, all of them restored to a FWHM of ~100 mas. It can be seen that the maps compare very favorably with each other, apart from artefacts introduced by source saturation in the Gemini and NACO images. Since the maps cover different epochs, some differences due to fast stellar proper motions near the center of the images can also be seen.

1.6 Future IR Interferometry

High-resolution imaging at the diffraction limit of 4 m to 8 m-class telescopes using speckle interferometric techniques as well as adaptive optics has provided some of the strongest evidence for the presence of a massive black hole at the center of the Milky Way. In addition, the investigation of the stellar populations within the central stellar cluster has resulted in a number of unsolved questions mostly related to the presence of apparently young and luminous stars and the related possibility of star formation in that region. In order to work towards a solution of these questions and to explore and utilize the laboratory to probe physics in the direct vicinity of the supermassive black hole an even higher angular resolution is desirable. Given that both high angular resolution and sensitivity will be required, this can only be achieved by interferometry with large apertures or by even larger single telescope apertures than are available today. The large apertures are needed in order to study in detail the rapidly variable infrared emission from the source at the position of Sgr A* (see section 2.13.4).

The diameters of large single dish apertures will be in the range of 30 to 100 m. The most advanced studies of extremely large telescopes (ELT) are those for the OverWhelmingly Large telescope (OWL; Gilmozzi *et al.* 2001), the California Extremely Large Telescope (CELT; Nelson 2000b, 2003), and the Euro50 Extremely Large Telescope (Andersen *et al.* 2003). The interferometric Large Binocular Telescope (LBT, see below; e.g. Hill 1997; Herbst *et al.* 2000; Straubmeier *et al.* 2003) can be regarded as one of the first ELTs that is already under construction.

While single extremely large telescopes (ELTs) are currently being designed, a number of interferometers working in the infrared have already started operation or are close to that point. Here we outline some of their technical specifications and how they can be used to improve our knowledge about the Galactic Center. Before continuing, we would like to point out to the reader an important difference between interferometry at radio to sub-millimeter wavelengths and at infrared (and optical) wavelengths: in the radio domain, coherent detection of the signals with heterodyne receivers is possible. The correlation is then done after mixing the signal to a lower intermediate frequency. In the infrared domain, however, interferometry is done with Michelson interferometers. This means that the light beams from different apertures must be combined *before* they reach the detector. The accuracy and high transmission needed for the respective beam delay lines makes interferometry at infrared (and optical) wavelengths so challenging.

Future IR interferometric observations (e.g. Eckart *et al.* 2002b, 2003a; Paresce *et al.* 2003) will allow us to resolve several individual sources in the central stellar cluster that already appear to be partially resolved with large single aperture telescopes. Here especially the interferometric investigation of individual sources in the MIR domain will be (and is already) possible (see also results in Fig. 1.20 and section 2.12.1). This will cover mostly the dust embedded objects like IRS 1, 3, 13, and 21. Studies on the binarity of stellar systems may be important to analyze their formation and the overall dynamics. The bright blue He-star IRS 16SW has already been identified as a possible binary star (Ott *et al.* 1999). Tracing high velocity stars (Ghez *et al.* 2003a; Schödel *et al.* 2003) — as well as finding them in the extremely crowded cusp region around the central massive black hole at the position of Sgr A* (Genzel *et al.* 2003b) — is definitely a goal for interferometry. For the Galactic Center it is especially the future dual-feed capability of the fully mature VLTI system via the PRIMA instrument, the Keck interferometer, as well as the large FOV interferometric imager at the LBT, LINC/NIRVANA (see references below) that will allow us to obtain valuable results of unprecedented quality. IR interferometry will also reach full maturity at a time when other interferometric instruments at different wavelengths will be fully operational, e.g. the Atacama large mm-wavelenght array ALMA. The means it will have passed most of the upgrades to shorter and longer wavelengths and 4 auxiliary telescopes (AT) of 1.8 m diameter will have been installed.

1.6.1 *VLTI*

The VLTI will allow sensitive interferometric observations with a very large collecting area and a suite of high-throughput beam combiner instruments. In the NIR domain the VLTI will allow for baselines of the order of up to 200 m and an angular resolution of the order of 1 milli-arcsecond (mas) (Fig.1.19). A combined observation using the four 8.2 m diameter unit-telescopes (UT) as well as four 1.8 m diameter auxiliary telescopes (AT) will provide a total collecting area of approximately 220 m^2 (e.g. Eckart *et al.* 1997; Glindemann *et al.* 2003). In total the site can host up to 8 telescopes with 8 delay lines. In addition to VLTI several high angular resolution, high sensitivity interferometric facilities will go online. Currently AMBER (e.g. Richichi *et al.* 2000) allows for three beam interferometry in the NIR (1-2.5 μm) with expected limiting magnitudes of K=14 and K=20 with a bright reference source. Interferometric spectroscopic observations with R=10,000 will be possible. In the MIR MIDI (e.g. Leinert & Graser 1998) is operational with an angular resolution of 20 mas on a 100 m baseline covering the the N-band (8-13 μm, expandable to Q-band 17-26 μm). Here limiting magnitudes are N=4 and 5.8 (with reference source). For both instruments the field of view (FOV) is of the order of ±1".

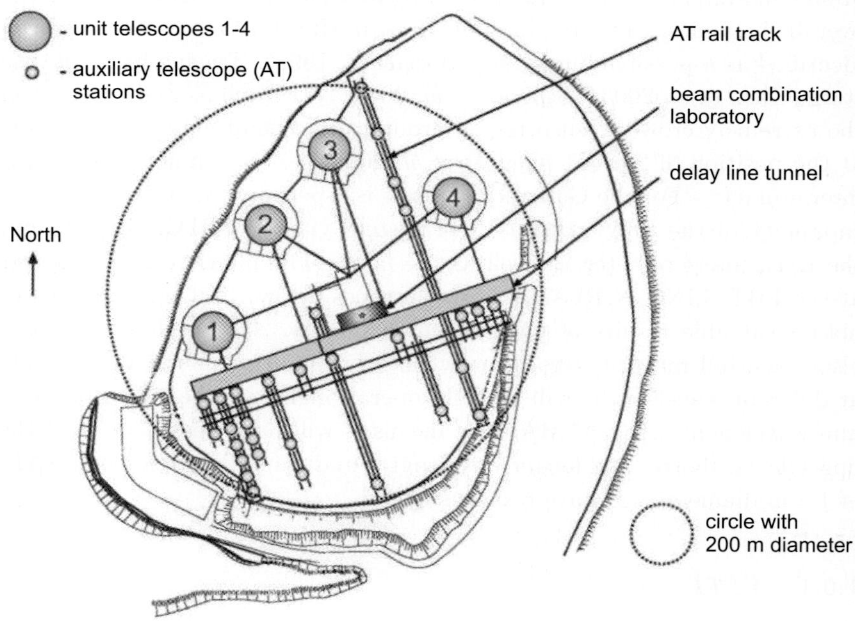

Fig. 1.19 Schematic view of the ESO Paranal site showing the locations of the 8.4 m diameter UTs and the possible locations of the ATs along with the delay line tunnel.

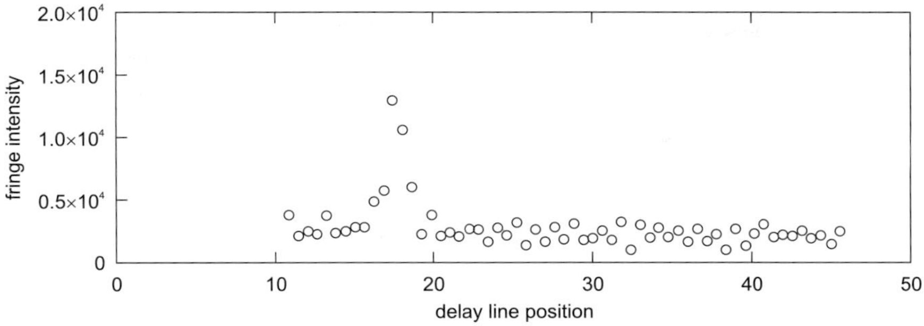

Fig. 1.20 Fringe amplitude as a function of delay channel obtained with MIDI and the VLTI on IRS 3 in the central stellar cluster (Eckart, Pott et al., in preparation).

The effective FOV will be significantly increased with PRIMA within the next few years (e.g. Quirrenbach *et al.* 1998; Glindemann & Lévêque 2000).The objective of PRIMA is to enable simultaneous interferometric observations of two objects — each with a maximum size of 2 arcsec — that are separated by up to 1 arcmin, without requiring a large continuous field of view. PRIMA will enhance the VLTI performance in several ways. The limiting magnitude will be shifted to K=15-20, allowing imaging of faint objects with high angular resolution (<10 milli arcsec) and high precision (<10 μas) astrometry.

On March 17, 2001, the VLT interferometer saw for the first time interferometric fringes on sky with its two 40 cm diameter test siderostats on a 16 m baseline. Seven months later, on October 29, 2001, fringes at a wavelength of 2.2 μm were found with two of the four 8.2 m Unit Telescopes (UTs), named Antu and Melipal, spanning a baseline of 102 m (Schoeller *et al.* 2003).

Recently — on 8 July 2004 — Eckart, Pott et al. have carried out first mid-infrared interferometric observations of a number of bright 10 μm sources within the central stellar cluster using MIDI and the 47 m UT2/UT3 baseline (see section 2.12.1). In Fig.1.20 we show the fringe amplitude that was obtained on the source IRS 3 as a function of the delay channels.

1.6.2 *Keck Interferometer*

The KI will operate over a maximum baseline of ~85 m with a resolution of the order of 6 mas in the NIR (e.g. Colavita 2001). Its collecting area will be dominated by the two 10 m class telescopes and will amount to about

160 m^2. Auxiliary telescopes are foreseen as well. Similar to the VLTI a two channel capability will allow measurements of a bright reference source for fringe tracking over angles of $\approx$2 arcminutes.

First fringes of the system using the 40 cm test siderostats were achieved in February 2001, followed by first fringes using the two 10 m Keck telescopes with their adaptive optics systems on the night of March 12, 2001 (Colavita 2001).

1.6.3 *LBT LINC/NIRVANA*

The LBT consists of two 8.4 m diameter mirrors in one single mount over a fixed baseline of 14.4 m. First light with a single mirror is anticipated to be in 2005. Interferometric first light (with two mirrors) will probably be around 2006. The 110 m^2 collecting area samples virtually all baselines ranging from 0 to 22.8 m and the resolving power at 1.25 μm wavelength will be $\sim$9 mas. The NIR beam combination will be carried out by LINC (e.g., Herbst *et al.* 2000; Straubmeier *et al.* 2003). The LBT as a Fizeau interferometer will have an exceptionally large imaging FOV of the order of at least 1 arcminute. Using MCAO techniques (multi conjugate adaptive optics) this will be extended to at least 2 arcminutes.

The LBT will be ideally suited for a number of GC investigations. The entire central cluster can be imaged at a high angular and high time resolution in snapshot mode. Deep studies of the distribution of stellar populations, the variability of individual sources, including Sgr A*, and a search for stars orbiting the central black hole at small distances (see Rubilar & Eckart 2001) will be possible with the LBT interferometer in a very efficient way.

Chapter 2

Observational Results

The compact emission from the astronomical source Sagittarius A* (Sgr A*) was first discovered at radio wavelengths in 1974. It was almost immediately associated with a potential massive black hole at the center of the Milky Way. The evidence for that grew stronger as the knowledge about the mass distribution in the central parsec improved via studies of gas and stellar dynamics. Observations in the infrared domain allow one to penetrate the ~25–30 magnitudes of extinction present at visible wavelengths. Measurements with the most recent observatories and cameras show that the distribution of mass is consistent with a compact supermassive object of $3 - 4 \times 10^6$ $M_\odot$. In chapter 3 we will discuss the astrophysical reasoning that leads to the conclusion that this mass concentration cannot be stable in any physical configuration other than a massive black hole. As expected by theory this most unusual object was found to be located in a small central region with a cusp-like over-density of high velocity stars that are moving at speeds up to several 1000 km s^{-1}. We now know that the central object is associated with an X-ray and infrared source that is highly variable in time showing numerous flares of emission on a daily basis. In this chapter we outline the observational results that have lead to our current knowledge about the properties of the stellar cluster, the stars, the interstellar medium, and the physics of the black hole at the center of the Milky Way.

2.1 The Discovery of Sagittarius A*

The first detection of a compact source at the center of our Galaxy was achieved in the radio domain by Balick & Brown (1974). Details of the history of that discovery are given in Goss & McGee (1996) and Goss *et al.*

(2003). An inspiring theoretical prediction that led to that discovery was made by Lynden-Bell (1969) and Lynden-Bell & Rees (1971). They pointed out a possible similarity between distant quasars (i.e. the active nuclei of galaxies at cosmological distances) and the activity at the center of the Milky Way. The authors realized that radio interferometry would be a well suited tool to study the proposed similarities and to search for a possible black hole at the center of the Galaxy. The discovery of Sgr A* itself dates back to February 1974. Bruce Balick and Robert L. Brown used a combination of the three-element Green Bank interferometer with a 45 foot antenna located at a distance of 35 km in Huntersville West Virginia. At wavelengths of 11 and 3.7 cm the obtained angular resolutions were 0.7" and 0.3", respectively. The sensitivity of, and the baseline between the combined telescopes were more than adequate to resolve the ~25 Jy of extended emission from the Sgr A West complex, which led to the discovery of a compact source with a flux of 0.6 Jy and 0.8 Jy at 11 cm and 3.7 cm, respectively. Brown (1982) first used the name Sgr A* (spoken "Sagittarius A star") in order to distinguish the compact source from the more extended emission of the Sgr A West complex.

The first successful detection of Sgr A* with Very Long Baseline Interferometry (VLBI) was made by Lo *et al.* (1975) in May 1975 using a 242 km baseline between the Owens Valley Radio Observatory (OVRO 40 m antenna) and the NASA Goldstone 64 m antenna at a wavelength of 3.7 cm. The angular resolution of this constellation was of the order of 50 mas and the derived upper limit on the source size assuming a single component was 20 mas. The measured flux density of the compact source was 0.6 Jy at 3.7 cm. Combining the results of various measurements the authors suggested that the emission of Sgr A* may be variable in time. Lo *et al.* (1975) also found that the compact radio emission was "nearly centered" on the stellar cluster observed by Becklin & Neugebauer (1975) at 2.2 μm with an angular resolution of 2.5 arcseconds.

After this detection it was still unclear whether the compact radio source was really associated with the nucleus of our Galaxy. The possibility of a background source, associated with an extragalactic object, was successfully excluded after the detection of the secular motion of Sgr A* by Backer & Sramek (1982). Secular motion manifests itself as an apparent movement of Sgr A* in the Galactic Plane with respect to extragalactic (i.e. geometrically much more distant) sources and is a direct consequence of the motion of our sun around the center of the Milky Way. Current measurements of the proper motion of Sgr A* were obtained with two different experiments. Reid *et al.* (1999) used the VLBA over a period of 2 years and Backer & Sramek (1999) the NRAO Very Large Array over a period of 16 years. Both groups found very consistently that the apparent motion of Sgr A* lies almost entirely in the plane of the Galaxy. At the position of the sun, which lies in the Galactic Plane at about 8 kpc from the dynamical center, the observed proper motion of Sgr A* relative to extragalactic sources of 6 ± 0.4 mas yr^{-1} translates into a circular rotation speed of 220 ± 20 km s^{-1}. Therefore the apparent proper motion of Sgr A* was found to be consistent with Sgr A* being at rest at the center of the Milky Way. The obtained data show that any residual proper motion of Sgr A*, with respect to extragalactic sources, is less than about 20 km s^{-1}. This slow velocity strongly supports the assumption that the source Sgr A* is directly located at the dynamical center of the Milky Way, since an eventual foreground source (i.e. a non-thermal radio source located between the solar system and the Galactic Center and circling the center of the Milky Way at a similar speed as the sun) would have a different projected velocity. In a next step, assuming dynamical equilibrium between Sgr A* and the fast moving stars in the cluster in which it is embedded, the very small residual proper motion of Sgr A* leads to a conservative lower limit of the mass of Sgr A* of more than 1000 $M_{\odot}$. More information on the size, spectrum, and polarization of Sgr A* in the radio domain is given in chapter 2.13.

2.2 Large-Scale Structures at the Galactic Center

The large-scale structures in the GC can be studied most conveniently in the radio and X-ray wavelength regimes, where the high extinction does not interfere with the observations. LaRosa *et al.* (2000) presented a wide-field VLA image of the Galactic Center region at a wavelength of $\lambda =$ 90 cm (Fig. 2.1). Centered on Sgr A, the image covers an area of 4×5 deg^2 with an angular resolution of 43 arcseconds (see also Pedlar *et al.* 1989; Anantharamaiah *et al.* 1991). At a wavelength of $\lambda = 90$ cm the observations are sensitive to both thermal and non-thermal emission and result in an image of unprecedented completeness of the large-scale radio structures in this region.

The most notable large scale radio structure within the Galactic Center region is the Sgr A complex, which has a diameter of approximately 40 pc. It consists of the compact synchrotron source Sgr A* (e.g. Balick & Brown 1974; Beckert *et al.* 1996) and the thermal spiral feature Sgr A West (Ekers *et al.* 1983; Lo & Claussen 1983). Sgr A East, a non-thermal shell source, encloses the Galactic Center in projection (Ekers *et al.* 1983). It is located mostly 1 to 2 arcminutes east of Sgr A* and may be the remnant of an explosion involving about 40 times the energy of a single supernova explosion (Mezger *et al.* 1989; Khokhlov & Melia 1996). Newer results, however, indicate that a single supernova might have been sufficient for the creation of Sgr A East (see Sakano *et al.* 2004).

About 15 to 20 arcminutes north of Sgr A (50 pc in projection) we find the Galactic Center radio arc, which consists of a large number of narrow filaments (Yusef-Zadeh *et al.* 1984). These filaments are most likely non-thermal synchrotron sources and could be magnetic flux tubes filled with relativistic electrons. In the immediate vicinity of the radio arc one finds a number of prominent H II regions that appear to interact with the filaments (e.g. Serabyn & Morris 1994). Several other isolated individual filaments can be found within the central half degree of the Galactic Center (see section 2.5), which presumably trace the large-scale magnetic field in this area (Morris 1994). The most prominent of these structures is the Sgr C filament, which is located about half a degree (75 pc in projection) south of Sgr A. Both the Galactic Center radio arc and the Sgr C filament are apparently part of an even larger scale ($\sim 200\ pc$) structure called the "Omega lobe" or GC lobe (Sofue & Handa 1984; Sofue 1985). These larger scale structures may be remnants of past or ongoing activity at the Galactic Center.

Recent X-ray satellite missions have revealed the large scale distribution of hot gas and high energy sources. The diffuse X-ray emission within the central parsec traces dense hot gas inside the circum nuclear gas and dust disk. On a scale of $\sim$ 10 pc the diffuse X-ray emission is mostly concentrated in mainly thermally emitting bipolar X-ray lobes, which are oriented perpendicular to the Galactic Plane and appear to be centered on the position of Sgr A*. Latest Chandra and XMM observations have revealed a highly variable X-ray counterpart of the Galactic Center black hole (see section 2.13.3) and delivered strong support for the supernova remnant hypothesis of Sgr A East (Maeda *et al.* 2002). The X-ray measurements have also resolved about 2400 X-ray point sources, which follow an overall distribution similar to the large-scale distribution of stars in infrared surveys, i.e. their spatial density falls off with radius as r^{-2} (Muno *et al.* 2003). The diffuse X-ray line emission extracted from these observations is described by Park *et al.* (2004).

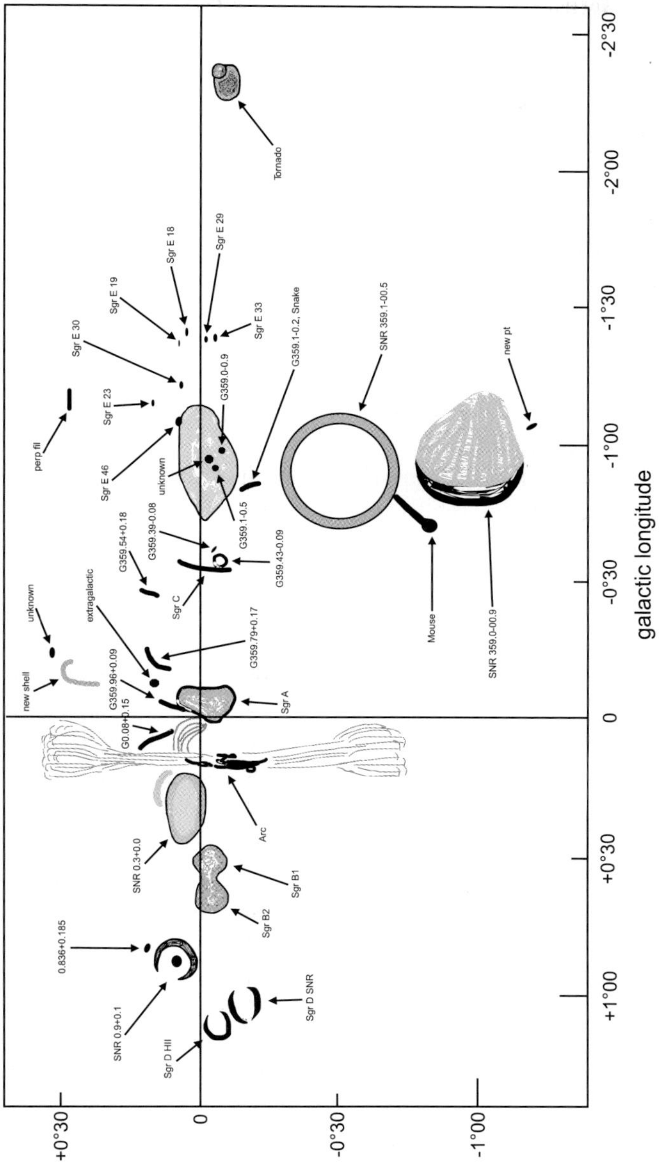

Fig. 2.1 Schematic diagram of the prominent and extended sources seen in the VLA image of the GC at a wavelength of 90 cm as shown in LaRosa *et al.* (2000).

2.3 The Circum Nuclear Disk

Major building blocks of the central few parsecs of the Milky Way are the Circum Nuclear Disk (CND), a central cavity that is surrounded by the CND, and a massive stellar cluster inside this cavity, which holds the black hole at its center (see Fig. 0.1). The Circum Nuclear Disk was originally discovered by Becklin *et al.* (1982). It rotates around Sgr A* and exhibits a sharp inner edge that encloses the central cavity with a diameter of about 3 pc (see Rieke *et al.* 1989; Chan *et al.* 1997; Scoville *et al.* 2003). High angular resolution spectroscopic observations have confirmed the presence of this dense and clumpy molecular ring consisting of warm dust (Zylka *et al.* 1995) and neutral gas (Guesten *et al.* 1987; Jackson *et al.* 1993; Wright *et al.* 2001; Herrnstein & Ho 2002) surrounding a short-lived central cavity of much lower gas density.

While discovered via its far-IR continuum emission, spectral line observations in the millimeter to far-infrared region showed that the gas in the ring is warm, dense, and turbulent. The total mass of the ring out to 4 pc radius is about 10^4 $M_\odot$, and inside of the ring lies the "mini-spiral" (see section 2.4), a group of gas and dust streamers. The ring/mini-spiral morphology is also well illustrated by the high resolution KAO images at 31 and 38 μm in which all of these features are clearly visible (Latvakoski *et al.* 1999).

Vollmer & Duschl (2002) addressed the question of the dynamics of the interstellar medium in the inner 50 pc of the Galactic Center (see also Vollmer *et al.* 2004) and investigated the three dimensional cloud-cloud collision rate in the CND. They found that the whole disk-like structure of the CND can form a stable configuration for several million years and that the current mass accretion rate within the CND is of the order of 10^{-3} $M_\odot$/yr. The authors furthermore speculate that the Sgr A West cloud complex may be the result of a prograde encounter of a $\sim 10^4$ $M_\odot$ cloud with the previously existing CND.

At the outer tips of the projected ellipse of the rotating CND two bright lobes are visible at most tracer wavelengths whose presence is mostly interpreted as being due to limb brightening of the torus of ~ 0.5 pc radius (e.g. Gatley *et al.* 1986; Burton & Allen 1992; Yusef-Zadeh *et al.* 2001). Gatley *et al.* (1984, 1986), Burton & Allen (1992), and Yusef-Zadeh *et al.* (2001) report $H_2(1-0)$ $S(1)$ molecular hydrogen line emission from the central few parsecs, which is mostly associated with the CND, where it is concentrated on the northern and southern lobes. In these areas the line intensity is 4 times higher than the mean intensity towards the CND (2 to 3×10^{-15} erg $s^{-1}cm^{-2}arcsec^{-2}$).

Stacey *et al.* (2002) investigated the molecular gas excitation and gas dynamics within the ring by observing the CO(7-6) rotational transition at 371.651 μm using the South Pole Imaging Fabry Perot Interferometer (SPIFI) mounted on the 15 m JCMT on Mauna Kea (Bradford *et al.* 2002), and by comparing the results to measurements of the dust continuum and other emission lines from that region (see also Stacey *et al.* 2004). The far-infrared emission arises from dust heated by far-ultraviolet starlight from the central cluster. It is associated mainly with the photo-dissociated surfaces of the molecular ring. The far-IR emission therefore is a good tracer for photodissociation regions (PDRs). The good correspondence between far-IR and CO(7-6) maps, particularly at the southwestern portion of the CND, suggests that much of the warm CO resides in the PDRs.

When superposed on a five-arcsecond wide HCN(1-0) map (Guesten *et al.* 1987), it becomes clear that the CO(7-6) emission is concentrated inside the HCN(1-0) emission, which is consistent with a PDR scenario, where HCN exists further away from the ionization source since it is easier to photo-ionize than CO.

2.4 The Mini-Spiral

The so called mini-spiral (Fig. 2.2) consists of ionized gas and connects the Circum Nuclear Disk (CND) to the center of the stellar cluster. Vollmer & Duschl (2000) have re-examined the gas dynamics traced by the [NeII] line emission ($\lambda = 12.81$ μm; Lacy *et al.* 1991) and the $H92\alpha$ radio emission data (Roberts & Goss 1993) and discuss the kinematic structure of the inner $\sim 3 \times 4$ pc^2 of the Galaxy. In a three-dimensional kinematic model of gas streams the bulk of the material is organized in three different planes. A dominant portion of the gas and dust — mostly the northern arm and eastern bar — is located in a main plane which is connected to the inner edge of the CND in the Galactic Center (Vollmer & Duschl 2000). That disk has an inclination of $\sim 25°$ to the line of sight and a position angle of 28° on the sky (N to E). Using NIR imaging spectroscopic data on Brγ emission, Paumard *et al.* (2003) investigated the structure of the mini-spiral in detail and identified nine different structures. They performed Keplerian orbit fitting and derived a detailed interpretation of the dynamics of the mini-spiral that is largely consistent with the model of Vollmer & Duschl

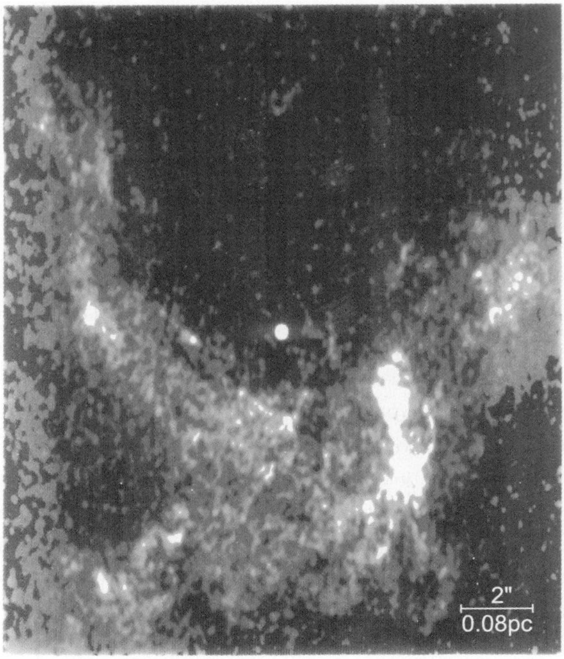

Fig. 2.2 VLA image of the mini-spiral at a wavelength of 1.3 cm by Roberts & Goss (1993). See also color Fig. A.1.

(2000).

A spatial comparison of a velocity integrated CO(7-6) emission map with a far-IR map shows clear correspondence between the tracers over much of the CND (Stacey *et al.* 2004). The strong clumpy features and the bright emission from the northern arm of the mini-spiral are consistent with a PDR interpretation, in which the source of heating is the central stellar cluster. The CO emission from the northern arm of the mini-spiral is offset to the east with respect to the far-IR image of the arm. The CO(7-6) emission from that region strongly suggests that warm molecular gas is flowing into the cavity. Surprisingly the east-west bar is not detectable in CO, and there are also strong depressions in the emission on both the eastern arm and the position of the IRS 16 cluster. The total gas mass in the northern arm as derived from the CO(7-6) line is of the order 5 to 50 $M_\odot$ or about 10% of the mass traced in the [OI] 63 μm line and far-IR continuum (Jackson *et al.* 1993; Latvakoski *et al.* 1999).

2.5 Radio Filaments

In the inner region of the Galaxy a number of unusual non-thermal filaments (hereafter NTFs) have been observed with the Very Large Array (VLA). These peculiar NTFs share the following characteristics (see reviews by Morris 1996; Morris & Serabyn 1996):

- They are unique to the Galactic Center and are found only within a projected distance of 150 pc from Sgr A*.
- Their lengths are tens of parsecs, yet their widths are less than 0.5 pc.
- The orientation of the NTFs is essentially perpendicular to the Galactic Plane to within 20° .
- They exhibit strong linear polarization (up to 50%) and spectral indices which are negative, flat, or in some cases even slowly rising (Anantharamaiah *et al.* 1991; Morris & Yusef-Zadeh 1985). As suggested first by Yusef-Zadeh *et al.* (1984) this is an indication of the synchrotron nature of their emission.
- Their rotation measures are often greater than 1000 rad m^{-2} and their intrinsic magnetic fields are aligned along the long axis of the filament (Tsuboi *et al.* 1986, 1995; Yusef-Zadeh & Morris 1987; Reich 1994).
- Observations have also shown that the NTFs appear to be associated with the ionized edges of molecular clouds (e.g. see Morris & Yusef-Zadeh 1989; Serabyn & Morris 1994).

Recent high-resolution and polarimetric studies of a number of NTFs, e.g. G359.1-0.2, the"Snake", by Gray *et al.* (1995), and G359.54+0.18 by Yusef-Zadeh *et al.* (1997), have revealed fine sub-filamentation along the lengths of these NTFs, or even a bifurcated NTF associated with the Sgr C HII complex (Liszt & Spiker 1995).

The Galactic Center NTFs raise a number of fundamental questions which are not yet solved. So, the source of the relativistic particles that illuminate the NTFs and the mechanism for their acceleration is currently unclear. Furthermore, the apparent relation of the NTFs to both the ionized and the molecular gas is not understood. It is possible that this association is related to — or even necessary for — the generation of NTFs. The NTFs also appear to trace the overall magnetic field configuration at the Galactic Center. However, it is unclear how such narrow structures can be confined.

A broad variety of models were proposed (see Morris 1996) that aim at solving these questions. Electrons may be accelerated by the $v \times B$ electric field, which is induced at the surfaces of molecular clouds that are moving at high velocities through a uniform magnetic field. Benford (1988) and Morris & Yusef-Zadeh (1989) point out that these fields may be sufficiently strong to accelerate electrons to relativistic energies. Reconnection of magnetic field lines was proposed as a way of generating NTFs as well (Heyvaerts *et al.* 1988), or may be responsible for the required acceleration of electrons (Lesch & Reich 1992). NTFs may also be the result of the interaction of fast, ionized stellar winds with the ambient magnetic field (Rosner & Bodo 1996). These winds may arise from the atmospheres of stars with high mass-loss rates.

More recent work on the NTFs concentrated on determining the nature of the interactions between the NTFs and the associated ionized and molecular gas. High-resolution observations of high density tracers like the CS(2-1) line (e.g. see Serabyn & Guesten 1991; Serabyn & Morris 1994) suggest that the source of the energetic electrons and their acceleration region in the Radio Arc lie at the intersection of the NTFs with the Sickle HII region and the underlying molecular cloud.

These observations show that the distribution of the molecular gas is clumpy at locations where both the NTFs and ionized gas are present, and where changes in both brightness and continuity of the Radio Arc NTFs occur (Yusef-Zadeh & Morris 1987). Serabyn & Morris (1994) propose that reconnection of the strong external field with the internal magnetic field lines of molecular clouds occurs at these positions, and that this process may accelerate the electrons to relativistic energies, followed by their diffusion along the external field lines. As a result the NTFs would be illuminated by synchrotron radiation and act as tracers of the external field lines.

Current studies of the distribution of the radio continuum and the emission of recombination and molecular lines along NTF structures indicate that the model of Serabyn & Morris (1994) is very successful in explaining the properties of many of the peculiar filaments (see also Liszt & Spiker 1995; Uchida *et al.* 1996; Staguhn *et al.* 1998).

2.6 Near-Infrared Images of the Central Stellar Cluster

2.6.1 *A Brief History of NIR Imaging*

First attempts to detect near-infrared emission from an astronomical source at the position of the Galactic Center started as early as 1945 (Stebbins & Whitford 1947; Moroz 1961). Due to a combination of low sensitivity and

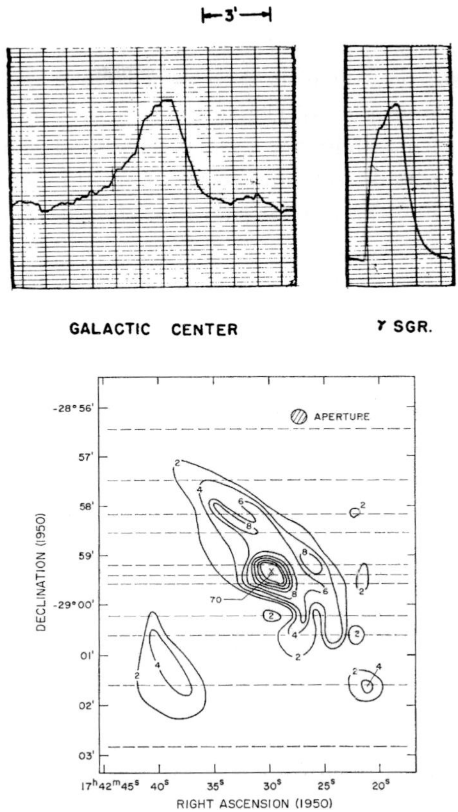

Fig. 2.3 Top: 2.2 μm strip-chart recording in right-ascension of the unresolved star γSgr and the Galactic Center, showing for the first time the presence of the central stellar cluster (Becklin & Neugebauer 1968). The data was recorded using a 0.08' diameter circular aperture. Bottom: first contour map resulting from those strip-charts. The angular resolution of the map is 0.25'. The contour lines are labeled in units of 1.6×10^{-19} W m^{-2} Hz^{-1} $sterad^{-1}$. The unlabeled contours in the center have values of 10, 12, 14, and 16, respectively. The dashed lines represent the positions of the individual scans on the sky.

coarse spatial resolution these initial efforts were not successful.

The first detection of structured emission was achieved by Becklin & Neugebauer (1968) at a wavelength of 2.2 μm in scans with 0.25 and 0.08 arcminute diameter apertures, corresponding to linear resolutions of 0.62 and 0.2 pc (see Fig. 2.3). These scans for the first time revealed the compact nuclear stellar cluster.

In the following years a number of authors obtained single detector maps with higher spatial resolution, which resolved the central cluster into individual bright complexes. With the semi-conductor technology progressing, the introduction of multiplexed, near-infrared array detectors allowed more efficient mapping with even higher angular resolution. The first maps of the Galactic Center using this technique were obtained by Forrest *et al.* (1986), and, for the first time, the central IRS 16 complex could be resolved into many individual sources. Lunar occultation measurements (Adams *et al.* 1988; Simon *et al.* 1990; Simons *et al.* 1990) demonstrated that the brightest sources of the IRS 16 complex (IRS 16NE, 16C, 16SW and 16NW) were very compact with projected diameters of less than 100 AU, and were most likely individual or multiple stars, but not large clusters. For the orientation of the reader, the names and positions of the most prominent bright stars in the GC cluster are indicated in Fig. A.2 in Appendix A.

The first diffraction limited maps (0.15" angular resolution, corresponding to 0.006 pc at the distance of the GC) of the central 20" × 20" at 2.2 μm and 1.6 μm were obtained by speckle imaging (see Eckart *et al.* 1992, and follow-up papers). In these maps the central cluster — including several compact stellar complexes like IRS 1, IRS 13 and IRS 16SW — was resolved into about 600 individual stars (Eckart et al. 1995). They revealed, for the first time, a complex of near-infrared sources very close to the radio position of Sgr A* (see e.g. Eckart *et al.* 1995; Genzel *et al.* 1997).

These results were confirmed by repeated speckle observations at the New Technology Telescope (NTT) of the European Southern Observatory (ESO) in Chile (e.g. Eckart & Genzel 1996, 1997), by seeing-limited imaging under excellent conditions (Herbst *et al.* 1993), as well as by first tip-tilt measurements at 1.6 μm with an angular resolution of 0.3" (Close *et al.* 1992). The latter two observations, however, were not able to resolve the very central stellar cluster.

More recently, the central parsec and the compact stellar cluster around the radio position of Sgr A* were repeatedly observed by speckle and adaptive optics imaging with the 10 m diameter Keck telescope (e.g. Klein *et al.* 1996; Ghez *et al.* 1998, 2000; Tanner *et al.* 2002; Gezari *et al.* 2002), the ESO Very Large Telescope (VLT) (e.g. Eisenhauer *et al.* 2003b; Genzel *et al.* 2003a,b; Schödel *et al.* 2002, 2003; Eckart *et al.* 2004b), the CFHT (Davidge *et al.* 1997; Paumard *et al.* 2001), the ESO 3.6 m telescope at La Silla (Clénet *et al.* 2001), and the Gemini North telescope. These observations confirmed the original detection of the Sgr A* cluster and revealed an increasing number of details hand in hand with the progress made in observation and data reduction techniques. Repeated observations at the NTT, VLT, and Keck telescopes allowed measurements of the velocities of stars in the central parsec. Adaptive optics observations provided the first clear identification of the compact radio source Sgr A* with a variable near-infrared source (Genzel *et al.* 2003a; Ghez *et al.* 2004).

2.6.2 *Diffraction Limited Images*

With the development of improved observation and analysis techniques NIR images continue to give new insight into the distribution and nature of stars within the central star cluster of our Galaxy. They allow studying the variability, possible extension, and proper motions of stellar objects, as well as the investigation of the central arcsecond and the phenomena that are linked to the near-infrared counterpart of the compact radio source Sgr A*.

Although providing lower angular resolution and not quite the same sensitivity, the speckle imaging mosaic images obtained by the SHARP camera at the ESO NTT (Eckart & Genzel 1997) compare well with more recent NIR adaptive optics images (e.g. see Schödel *et al.* 2003). The faintest reliably identifiable sources correspond to K-band magnitudes of 15^m to 16^m, which implies a dynamic range of the images of more than 8 magnitudes with respect to the brightest source, IRS 7 ($K = 6.8^m$).

Already the initial 1991 SHARP speckle image reconstructions (Eckart *et al.* 1992) revealed a source close to the position of the compact radio source Sgr A*, which was resolved into a cluster of individual stars (Eckart *et al.* 1995) in the following observing runs. Within a projected distance from Sgr A* of less than 2" these images show about 20 stars of K-band magnitudes between 13^m and 16^m and a magnitude dispersion of 3 stars with $m_K < 14$, 9 stars with $m_K < 15$ and 20 stars with $m_K < 16$.

In general, the Speckle observations at the NIR-band of $\lambda \leq 2.2\ \mu$m resolved the emission from the central parsec fully into individual stars and groups of stars. The central IRS 16 complex splits up into a multitude of sources, and the extended ridge in IRS 16SW-E found by other authors (Simon *et al.* 1990; Depoy & Sharp 1991b) was resolved into at least 5 bright stellar sources. IRS 13 was found to be a ≈ 0.5" radius cluster consisting of at least 5 to 10 members (see also Maillard *et al.* 2003). In particular the SHARP speckle imaging through narrow band line filters (Eckart *et al.* 1995) clearly showed that IRS 16NE, IRS 16C, IRS 16SW and IRS 16NW were truly HeI emission line stars and not part of an extended distribution. As proposed by Krabbe *et al.* (1991, 1995) the IRS 16 complex

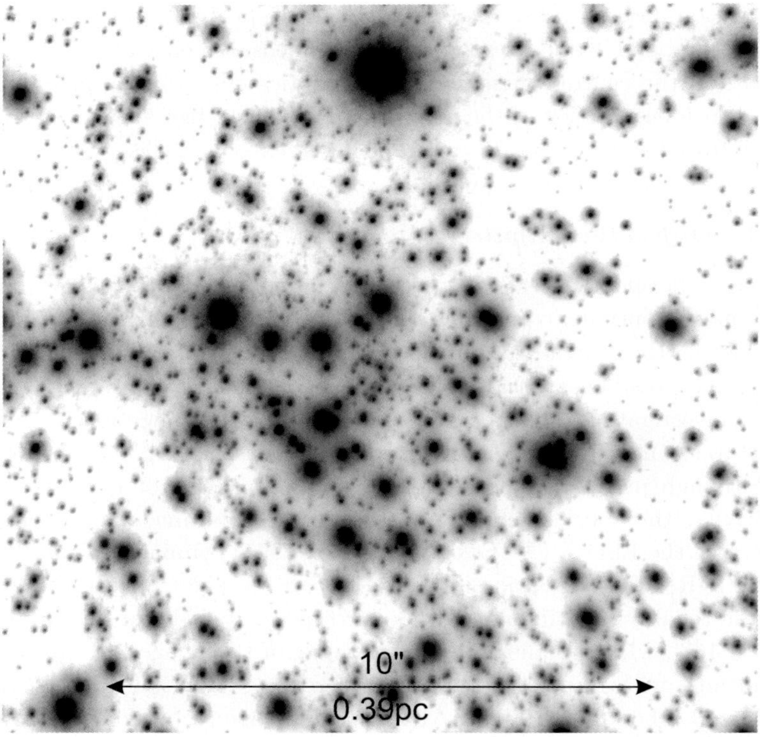

Fig. 2.4 NAOS/CONICA H-band AO image of the GC stellar cluster, observed on August 30, 2002 during science verification of the instrument at the ESO VLT. The total integration time of the image is 1500 s. Exposures at various dithered positions were combined to a final mosaic. The grey scale is logarithmic. Sgr A* is located near the center of the image. The AO was locked on IRS 7, the bright supergiant near the top of the image. North is up and east is to the left.

thus was identified as the central concentration of the hot star cluster. Lunar occultation observations showed that the brightest of these objects were likely single or multiple stars but not large clusters, as their sizes were found to be less than 100 AU (Adams *et al.* 1988; Simon *et al.* 1990; Simons *et al.* 1990).

The flux density calibration of the SHARP images was derived from published flux densities of bright members of the central cluster like IRS 7, IRS 16NE, IRS 16NW, and IRS 16C (Rieke & Rieke 1988; Depoy & Sharp 1991b; Blum *et al.* 1996; Ott *et al.* 1999; Clénet *et al.* 2001).

Within the uncertainties of the flux density calibration, less than a few

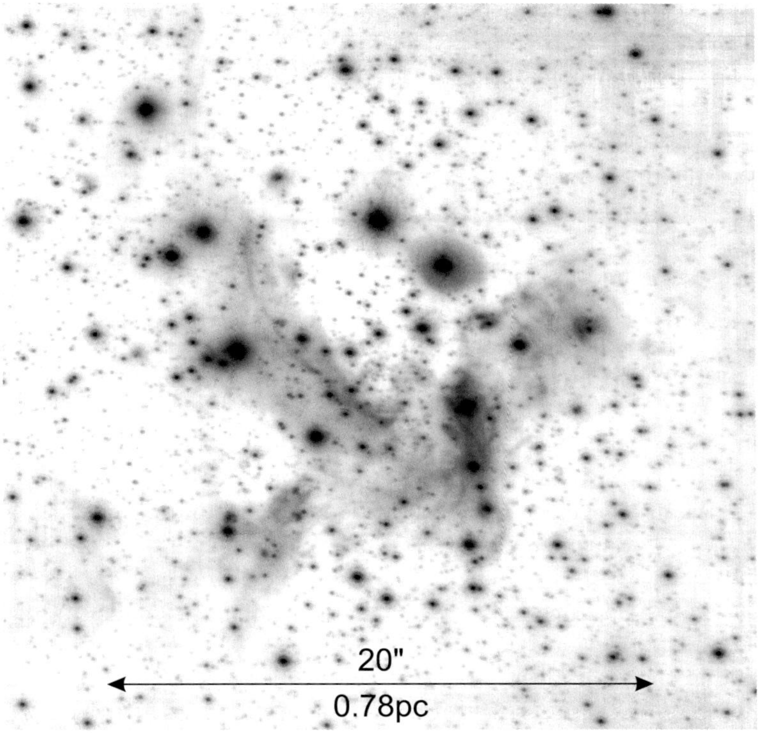

Fig. 2.5 NAOS/CONICA L-band AO image of the GC stellar cluster, observed on August 30, 2002 during science verification of the instrument at the ESO VLT. The total integration time of the image is 2250 s. Exposures at various dithered positions were combined to a final mosaic. The grey scale is logarithmic. Sgr A* is located near the center of the image. The AO was locked on a visual guiding star located about 30" NE of the center. North is up and east is to the left.

percent of the more than 1000 stars with $6.8^m < m_K < 16^m$, which could be identified in the SHARP high resolution K-band images, emit the full flux density derived on the basis of earlier seeing limited images. Two thirds of the flux is contributed by only the 29 brightest sources ($K < 10.5^m$). A comparison of the speckle data and the most recent deep adaptive optics images with seeing limited long exposure images shows that most of the apparent diffuse emission seen in images with an angular resolution of $\geq 1''$ is due to a combination of smearing of faint sources and the broad wings of the point spread functions of bright sources.

Due to its location in the southern hemisphere, where the GC passes close through zenith, and its unique capabilities (i.e. its near-infrared wavefront sensor, see section 1.4.3.1) NAOS/CONICA at the ESO VLT is currently the best instrumentation for deep high-resolution observations of the GC at NIR wavelengths. Figures 2.4 and 2.5 (see also Schödel *et al.* 2002; Genzel *et al.* 2003b; Eckart *et al.* 2004b) show diffraction limited maps of the central 0.5–1.0 pc of the Milky Way observed with NAOS/CONICA at the VLT at wavelengths of 1.6 μm (H-band) and 3.8 μm (L-band). The faintest sources in the presented images have an H-band magnitude of 19–20^m and an L-band magnitude of 15–16^m.

2.7 The Radio and Infrared Positional Reference Frames

The unambiguous identification of the radio source Sgr A* with an individual object in near-infrared images of the central arcsecond had been a major objective since the first observations of this kind. However, Sgr A* was found to be surprisingly faint at infrared wavelengths. Its faintness in combination with the extreme density of stellar sources in its immediate surroundings made the task therefore extraordinary difficult. A successful identification requires the precise knowledge of its radio position with respect to other stars in the field of view and — even more important — in a common radio-infrared positional reference frame. Until the early 90s the astrometric position of Sgr A* in the infrared images was only known to within about ±0.3", which did not allow one to differentiate between the approximately 10 sources in that area known at that time. However, a combination of speckle/AO NIR high resolution images with interferometric radio images allowed one to improve the positional accuracy by more than one order of magnitude (Menten *et al.* 1997; Reid *et al.* 2003a).

The key to this experiment is the fact that many red, cool giant and supergiant stars exhibit strong hydroxyl (OH at 1.663 GHz), water (H_2O at 22.235 GHz) and/or silicon monoxide (SiO at 43.122 GHz) maser emission. Very Long Baseline Interferometric (VLBI) observations (Diamond *et al.* 1994; Miyoshi *et al.* 1994; Greenhill *et al.* 1995a) indicate that SiO masers arise from the innermost regions ($\approx 4 - 8$AU radius) of the circumstellar envelopes of these stars. Similarly, the H_2O maser radiation originates in the stellar atmospheres of many OH/IR stars (infrared bright, short-lived stars in a high mass-loss phase, see e.g. Winnberg *et al.* 1985; Lindqvist *et al.* 1992) and M-type supergiants (e.g. Martinez *et al.* 1988; Levine 1995). Thus, for a typical Mira variable located at the GC, the SiO maser emitting

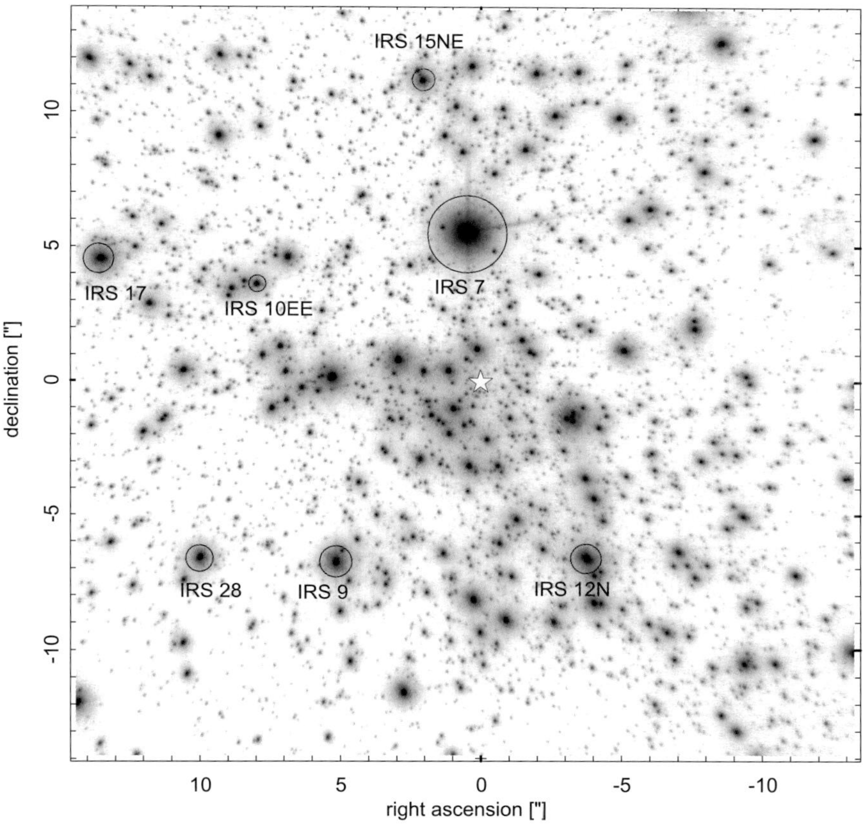

Fig. 2.6 A K-band NACO/VLT adaptive optics image like the one presented here was used by (Reid *et al.* 2003a) to establish the astrometric positions of stars relative to Sgr A* in the infrared frame. The 7 maser sources in the field-of-view are marked by circles. The position of Sgr A* is marked by an asterisk. Offsets from Sgr A* in arcseconds are marked on the x- and y-axes, positive to the north and to the east.

region of its atmosphere suspends an angle of ≤ 1 mas (Reid *et al.* 2003a). This fortunate circumstance allows measuring the position of maser stars at the Galactic Center with great accuracy.

The first experiment of this kind was performed by Menten *et al.* (1997). They determined the position of Sgr A* on SHARP images with an accuracy of 30 mas and found no NIR source corresponding to Sgr A*. Measurements of stellar accelerations in the following years (Ghez *et al.* 2000; Eckart *et al.* 2002a) indicated that Sgr A* was positioned 40-50 mas east of the location given by by Menten *et al.* (1997). The analysis of Menten *et al.* (1997) was mainly limited by the fact that they could use just two maser stars because the were limited by the smaller field-of-view of SHARP ($\sim 13''$), which allowed them to derive only a single pixel scale and the image rotation for the NIR image.

NACO at the VLT offers a field-of-view of $27''$ when operating in a diffraction limited mode at 2.2 μm. A much larger region can be mapped by composing a mosaic of several images with different initial pointings. When Reid *et al.* (2003a) repeated the alignment experiment (see also Schödel 2004), they used such an image and identified seven SiO maser stars in it (circled in Fig. 2.6), which can be easily found, because they are bright AGB stars. Therefore the radio positions of these stars, which can be measured with the VLA to an accuracy of ~ 1 mas with respect to Sgr A*, and on the NIR image with an accuracy of a few mas, can be used to align the infrared and radio positional reference frames relative to each other.The higher resolution and larger field-of-view of NACO and the longer baseline of the radio data enabled Reid *et al.* (2003a) to solve for higher order terms needed in the alignment process. Therefore they could determine the position of Sgr A* in the NIR frame with an uncertainty of less than 10 mas. The new alignment was consistent with the position of the dark mass inferred from the stellar acceleration measurements of Ghez *et al.* (2000) and Eckart *et al.* (2002a) and also with the position of the focus of the orbit of the star S2 (Schödel *et al.* 2002; Ghez *et al.* 2003b; Schödel *et al.* 2003). However, the faintness of Sgr A*, confusion in the dense cluster in which it is embedded, and the proximity of the bright star S2 between 1999 and 2002 inhibited the detection of a NIR Counterpart of Sgr A* until 2003, when Genzel *et al.* (2003a) reported flaring and quiescent emission at wavelengths of 1.6, 2.2, and 3.8 μm (H, K, and L-bands) at the position of Sgr A*. Variable emission from Sgr A* at 3.8 μm was also found in 2003 independently with the Keck telescope by Ghez *et al.* (2004) (see section 2.13.4).

2.8 Number Density Counts

The overall structure of the central stellar cluster can best be derived from diffraction limited NIR maps. Early investigations into the structure of the cluster did not have a very high spatial resolution and sensitivity so that it was not possible to clearly separate individual stars in that region, and the radial distribution of the light emitted by the overall cluster was used to determine its shape. The center of light, however, lies approximately on the bright IRS 16 complex, located 1" to 2" east of the Sgr A* position. This finding was initially contrary to expectations and had even been considered as evidence that Sgr A* could not be a supermassive black hole (Allen & Sanders 1986). In contrast to these early attempts at determining the structure of the stellar cluster, the first high-resolution speckle imaging data enabled Eckart *et al.* (1992) to derive from the median of the relative source positions that the distribution of stars is centered within $\pm 1"$ (3σ) on Sgr A*. Scoville *et al.* (2003) showed that the extinction coefficient follows a smooth distribution across the central 10 to 20 arcseconds with no indication of variations on spatial scales of about 1" to 2". Therefore the offset concentration of the overall cluster NIR light emission is a phenomenon that is liked to the inhomogeneous distribution of various populations of stars. Initial attempts of deriving the stellar distribution did not take into account that the cluster luminosity is completely dominated by a few dozen bright supergiants, which are — in part due to their small number — not evenly distributed across the field. In the central 5 arcseconds Krabbe et al. (1991) found a cluster of very luminous, massive stars with prominent HeI line emission, while outside of that area the emission is dominated by late-type giants with deep CO absorption bands (e.g. Rieke & Rieke 1988; Sellgren *et al.* 1990; Haller *et al.* 1996; Krabbe *et al.* 1995; Genzel *et al.* 1996). If one wants to get a clear image of the structure of the underlying stellar cluster, it is therefore more appropriate to investigate the distribution of stars via their number density rather than their brightness, assuming that the observed stars in the GC cluster can be used as tracers for the underlying distribution of fainter stars.

First coarse stellar number density counts were presented in Eckart *et al.* (1992). The data presented by Eckart *et al.* (1995) suggested an excess of stars above a flat core near Sgr A*. Alexander (1999) analyzed stellar counts from three different data sets and also found indications of a stellar cusp around Sgr A*. However, the existence of the cusp, as inferred from the star counts available at that time, still remained somewhat inconclusive.

Imaging observations with NACO at the VLT in 2002 provided much better suited images for this type of analysis: they were taken at the diffraction limit of an 8 m-class telescope, were two to three magnitudes deeper than any previous images due to the excellent quality of the AO correction, and covered a large field-of-view of approximately $20'' \times 20''$. Genzel *et al.* (2003b) used this data for obtaining completeness corrected num-

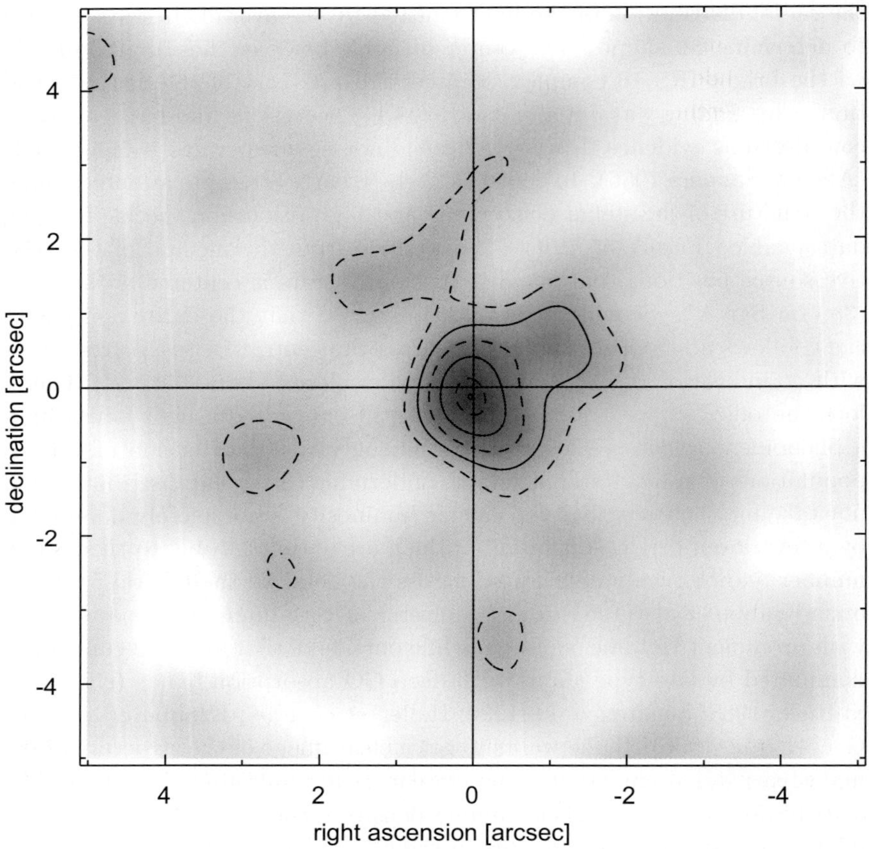

Fig. 2.7 Representation of the two-dimensional surface density of stars in the central 5" × 5" of the GC stellar cluster. The stellar number counts were extracted from NACO/VLT K- and H-band images and corrected for crowding and completeness. The faintest K-band magnitude used was $m_K = 17$. The corrected surface density was smoothed with a Gaussian of 1" FWHM. The mean surface density and its standard deviation were calculated for a field-of-view of 9" × 9". The contours delineate the areas where the surface density exceeds the mean density by 2, 3, 4, 5, 6, and 6.3 σ. The 3, 5, and 6.3 σ contours are marked by straight lines. For more details and a similar figure see Genzel *et al.* (2003b).

ber density counts of the stars in the central parsec. They used H- and K-band images because practically the same stellar population is sampled at both wavelengths, but errors can be minimized by combining images from different band passes. The so-called completeness correction accounts for the difficulty of detecting close companion stars and faint stars in the wings of the instrumental Point Spread Function (PSF) of brighter objects. The magnitude of the correction can be evaluated by first adding artificial stars at different positions in the stellar field and subsequently trying to re-identify them. In that way maps can be produced that reflect the probability of identifying a star with a certain magnitude at a given position. For weak stars, these maps resemble a negative picture of the stellar cluster, i.e. probabilities are low where the luminosity is high. Finally, one can count sources of a given magnitude present within annuli of increasing radii around Sgr A* and correct these counts for completeness (this procedure assumes symmetry of the cluster with respect to Sgr A*, of course).

A significant problem when doing the completeness correction is the increasing uncertainty of large corrections, a fact which may become problematic especially for the innermost annuli, where both the completeness and the number counts are low for faint stars (due to the high source density and the small surface of the annuli). In order to avoid possibly uncertain completeness corrections significantly larger than a factor of 2, Genzel *et al.* (2003b) only used stars with K-band magnitudes brighter than 17^m for their analysis. Based on this principle, they derived completeness corrected K-band luminosity functions (KLFs) from differential source counts for the overall cluster (in a $9''$ radius around Sgr A*) and for the cusp region ($1.5''$ around Sgr A*).

Genzel *et al.* (2003b) combined the completeness corrected stellar number counts from the NACO/VLT images with earlier SHARP/NTT data that were taken at larger (up to approximately 100", see Genzel *et al.* 2000) separations from the center. The results of these latest number density counts (see Fig. 2.7 and 2.8) show that stars in the central stellar cluster with magnitudes brighter than $m_{2.2\mu m} = 17^m$ can be described very well by the number density distribution expected for an isothermal stellar cluster with a core radius of approximately 0.34 pc, similar to what has been found by an earlier analysis of lower quality SHARP/NTT images by Eckart *et al.* (1995). The exciting new result of Genzel *et al.* (2003b) was, however, that they could clearly identify a stellar cusp, i.e. an excess over the flat core of the outer isothermal cluster, that is centered at the position of Sgr A*. The following subsections summarize the results of Genzel *et al.* (2003b) on the structure and luminosity function of the stellar cluster in the central parsec of the Milky Way.

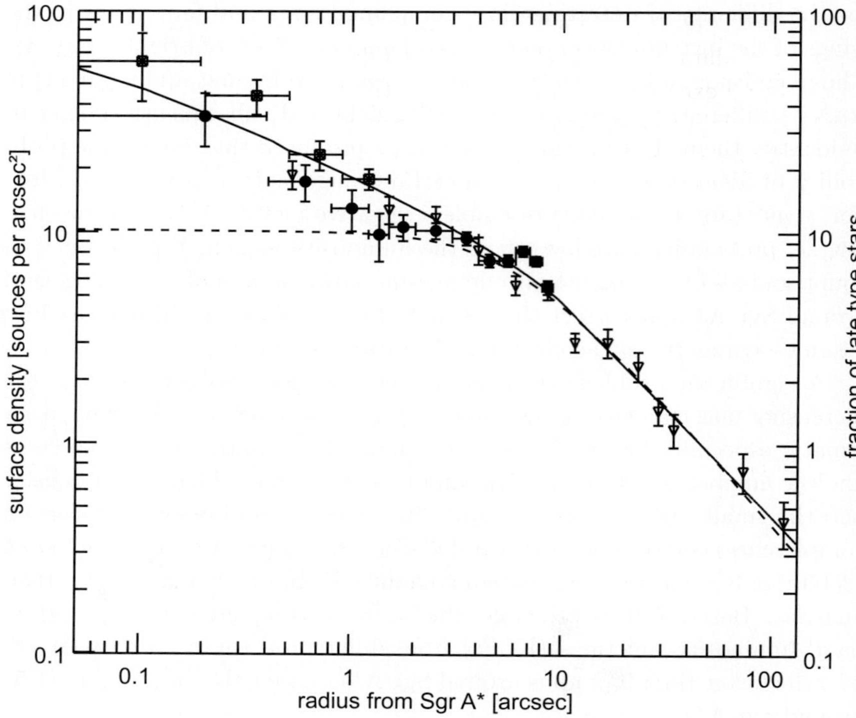

Fig. 2.8 Radial average of the crowding corrected stellar number density in H- and K-band as shown by Genzel *et al.* (2003b). The data shows clear evidence for a cusp component above the distribution expected for an isothermal stellar cluster, marked with a dashed line.

2.8.1 *Evidence for a Central Stellar Cusp*

Figure 2.8 shows the completeness corrected, binned stellar surface number density distribution for stars brighter than $m_K = 17$ (and brighter than H= 20 for the central arcsecond) as a function of projected distance from Sgr A* (analogous to Fig. 7 of Genzel *et al.* 2003b). Towards larger radii the NAOS/CONICA data was expanded with shallower ($K \leq 15$) number counts of SHARP/NTT (see Genzel *et al.* 2000) that cover fields at larger distances from Sgr A*. In the overlap regions these counts were scaled for the best match with the deeper NAOS/CONICA data. The stellar number density counts for projected separations from Sgr A* of more than ~2-3 arcseconds resemble a flattened isothermal sphere of core radius 0.34 pc (see Genzel *et al.* 1996, 2000, and references therein). In contrast to earlier studies, the data obtained with NAOS/CONICA/VLT shows a clear indication for an excess in number density in the central few arcseconds, a fact which was already indicated, but could not be proved conclusively by the earlier SHARP/NTT and Keck data (Eckart *et al.* 1995; Alexander 1999).

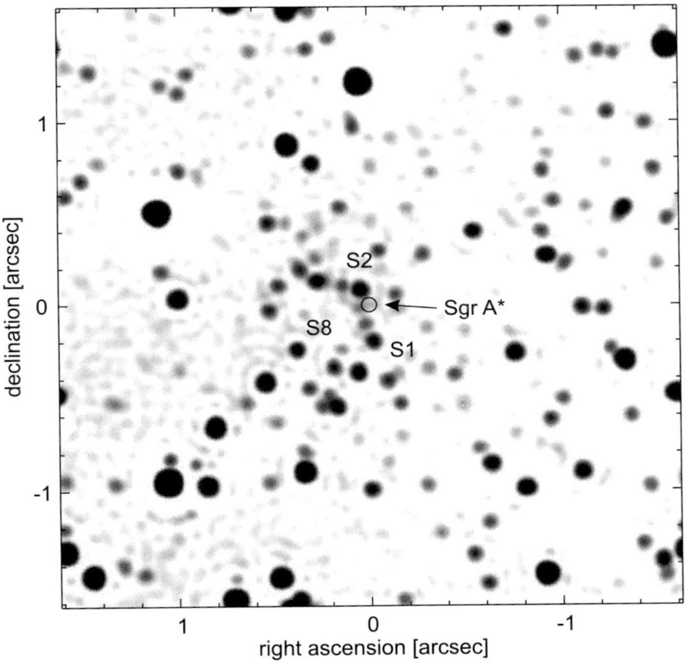

Fig. 2.9 Deep K-band image of the cusp around Sgr A*. The image resulted from a Lucy-Richardson deconvolution and subsequent beam restoration (with a Gaussian of 60 mas FWHM) of a K-band NACO/VLT image from May 2003. North is up and east is to the left.

Hence, the deep AO images confirm the existence of a cusp around Sgr A* (see Fig. 2.7 and 2.9), since – even independent of any incompleteness correction – the surface density of faint stars increases with decreasing separation from the radio source Sgr A*. There is no indication for a flat core as would be expected for a relaxed, isothermal cluster. The peak of the cusp coincides with the position of Sgr A* [$(\Delta\mathrm{RA}, \Delta\mathrm{Dec})$=(+0.09", −0.15")] with an uncertainty of ± 0.2" (see Fig. 2.7 and Genzel *et al.* 2003b).

In order to describe the radial dependence of the stellar mass density in the Galactic Center, Genzel *et al.* (2003b) tied the number counts to the radial mass distribution of Genzel *et al.* (1996). Assuming a dynamical (stellar) mass of 3.2, 8.4, and 27.3×10^6 $M_\odot$ at distances $R = 1.9$, 3.8, and 11 pc from Sgr A* and following Alexander (1999), Genzel *et al.* (2003b) derived the following broken-power law for the stellar mass density in the Galactic Center:

$$\rho_\star(R) = 1.25 \times 10^6 \left(\frac{R}{10''}\right)^{-a} \quad [M_\odot \mathrm{pc}^{-3}], \tag{2.1}$$

with $\alpha = 2.0 \pm 0.1$ at $R \geq 10$", and $\alpha = 1.4 \pm 0.1$ at $R < 10$". Concentrating on only the central 4 arcseconds a generalized Maximum Likelihood (ML) analysis similar to what is presented in Alexander (1999) results in a power-law cusp slope of $\alpha = 1.3 \pm 0.1$ (see Fig. 2.8 and Genzel *et al.* 2003b). A calculation of the stellar density of the cusp based on these relations leads to 3×10^7 $M_\odot$ pc^{-3} at $R = 1$", and 7×10^8 $M_\odot$ pc^{-3} at $R = 0.1$".

An alternative description of the data is a localized cusp on Sgr A* superposed on a larger-scale, isothermal cluster with a distinct break in between. Assuming such a distribution, Mouawad *et al.* (2003b) get similarly large values for the stellar mass densities in the cusp.

The above analysis critically depends on the assumption that the ratio of number counts to total stellar mass (most sources are faint and cannot be directly observed) is constant with radius and environment in the Galactic Center. However, this assumption cannot be entirely valid, since the analysis of the stellar luminosity function (see following section) shows that the stellar population changes to a certain degree in the very dense inner cusp region. Recently, there has been much interest in the possibility that a cluster of dark stellar remnants, e.g. stellar black holes or neutron stars, may have accumulated at the bottom of the gravitational potential well (see Morris 1993; Miralda-Escudé & Gould 2000). A tool for testing this idea is examining deviations of stellar orbits near the supermassive central black hole Sgr A* from the Keplerian case. First studies indicate an upper limit on the mass of a hypothetical dark cluster near Sgr A* of about 4×10^5 $M_\odot$ (see discussion in section 3.7.6 and Mouawad *et al.* 2003a, 2004). This caveat has to be kept in mind when considering the stellar density analysis. However, since the number counts are dominated by faint stars and since the overall slope of the luminosity function in the cusp is similar to the large scale cluster it appears justified to assume that the stellar number counts describe the underlying stellar population reasonably well.

The observed stellar density distribution is consistent with theoretical models for stars surrounding a massive central black hole, which predict the formation of a power-law cusp. The expected radial slopes range between 0.5 and 2. The exact value depends on the formation scenario of the cusp and on the density and stellar-population-dependent importance of inelastic stellar collisions. Relaxed, single mass stellar cusps have a steep slope of $\alpha \sim 7/4$ (Bahcall & Wolf 1976, 1977), while un-relaxed clusters around an adiabatically growing hole have a shallower slope of $\alpha \sim 3/2$ (Young 1980). In multi-mass, lower density cusps ($\rho_\star < 10^7\ M_\odot\ pc^{-3}$) the model calculations by Murphy *et al.* (1991) predict a steep slope ($\alpha \sim 7/4$) too, but higher density cusps ($\rho_\star \sim 10^8\ M_\odot\ pc^{-3}$) have flatter central slopes because of the increasing importance of stellar collisions ($\alpha \geq 1/2$).

The models also predict that a large fraction of the cusp stars near the black hole should be gravitationally bound to it, because

$$\sigma^2 = v_c^2/(1+\alpha) < v_c^2, \tag{2.2}$$

where $v_c^2 = GM_{BH}/R$ is the square of the escape velocity (Alexander 1999). The fraction of bound stars increases with the steepness of the power-law cusp, which is described by the parameter α. Most models, with the exception of the one by Young (1980), predict that the stellar cusp should be dynamically relaxed. Young (1980) assumes that the time scale on which the central black hole grows, is short compared to the stellar relaxation time scale of the overall cusp region, a requirement that is not applicable to the Galactic Center, where the relaxation time of the cluster is short (of the order 10^8 yr, see section 3.6.5) and the mass accretion rate onto the black hole is very low ($< 10^{-7}$ M$_\odot$ yr^{-1}, see section 3.8.3).

2.8.2 *K-band Luminosity Function*

In addition to the number density distribution of stars it is of great interest to know the distribution of stars as a function of luminosity. This measure can be obtained from the completeness corrected differential source counts. The luminosity function allows us to draw conclusions on the stellar population present in the central star cluster of the Milky Way.

The results for circular regions within a radius of 1.5", i.e. the cusp region, and for the overall nuclear cluster within a radius of 9" of Sgr A* are presented in Genzel *et al.* (2003b). Since the NAOS/CONICA counts are incomplete for the very brightest magnitudes because of saturation effects, they have combined the NAOS/CONICA number density counts with the SHARP/NTT counts to a common KLF. Here, we reproduce the results of Genzel *et al.* (2003b) in Fig. 2.10, which shows the KLF for the overall nuclear cluster, Fig. 2.11, a color-magnitude plot of the same region, and Fig. 2.12, the KLF for the central cusp-like region. The error bars take into account the effects of the number density correction. At brighter magnitudes the NAOS/CONICA data is in very good agreement with the earlier Keck and NTT data, but now extends the KLF to K-band magnitudes of $18-19$, about 3 magnitudes deeper than the earlier measurements. The Galactic Center KLF therefore samples all giants and supergiants, as well as the main sequence to spectral types A5/F0 ($\sim 2\mathrm{M}_{\odot}$).

From the observed KLFs Genzel *et al.* (2003b) draw the following conclusions: the shape of the KLF in the central $p < 9"$ (0.35 pc) region (here p is the distance from Sgr A* in the plane of the sky) can be described in a first approximation by a power law ($\mathrm{d}\log\mathrm{N}/\mathrm{dK} = \beta \sim 0.21 \pm 0.02$). In the range $14 \leq \mathrm{K}_s \leq 19$ the overall KLF has a somewhat flatter, but generally similar slope compared to the KLF of the Bulge of the Milky Way several degrees from the center ($\beta \sim 0.3$, e.g. see Tiede *et al.* 1995; Zoccali *et al.* 2003), and the KLF on a scale of several tens of parsecs around the center (Figer *et al.* 2004). A $\beta \approx 0.3$ power-law is well matched by the theoretical KLF of an old stellar population with a Salpeter initial mass function (IMF) and a continuous star formation history (see also Alexander & Sternberg 1999; Figer *et al.* 2004). The flatter slope of the Galactic Center KLF compared to the Bulge is mostly caused by an excess of the counts at $\mathrm{K} \leq 14$, by a factor of 1.4 to 2. This excess at the bright end of the KLF can probably be attributed to young early and late type stars (see also Lebofsky & Rieke 1987; Blum *et al.* 1996; Davidge *et al.* 1997; Blum *et al.* 2003). The KLF of the overall cluster shows a prominent excess hump centered at $K_s \sim 16$, a factor of about 2 above the power law. This hump can also be found in KLFs of the Bulge (Tiede *et al.* 1995; Zoccali *et al.* 2003), and on scales of tens of parsecs around the Galactic Center (Figer *et al.* 2004). It is the region in the KLF, where the so-called horizontal branch (HB)/red clump (RC) stars (the name derives from their position in color-magnitude diagrams) accumulate. These are old and metal rich, core He-burning stars with characteristic masses of 0.5 to 3.5 $M_{\odot}$. The red clump stars can be seen in the color magnitude plot at K-magnitudes around 16 (Fig. 2.11). The HB/RC excess relative to the number of gi-

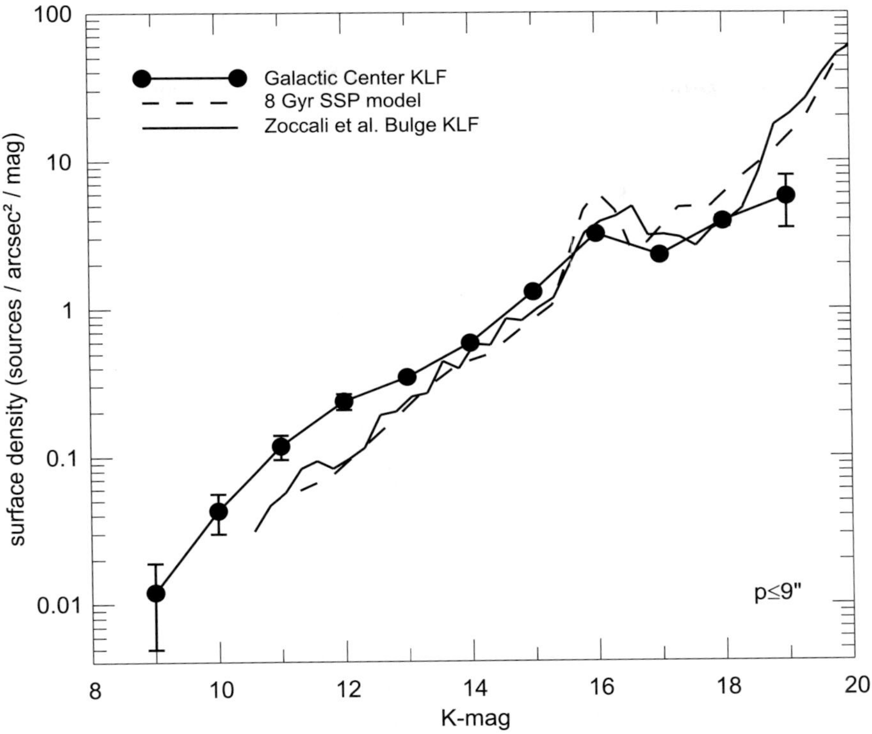

Fig. 2.10 KLF for the central 9″ region as shown by Genzel *et al.* (2003b): K-band luminosity function (KLF) in sources per square arcsecond per magnitude shown by filled circles with 1σ error bars. The graph is a result of the combination of NAOS/CONICA and SHARP/NTT data corrected for incompleteness. The continuous curve is the KLF of the Galactic Bulge on scales of degrees and the dashed curve is a single age (8 Gyr) stellar population model of the Bulge (from Zoccali et al. 2003). Both were scaled vertically to match the center data, and corrected horizontally to the same K-band extinction. The excess at K$\sim$ 16 is due to stars on the horizontal branch/red clump, i.e. old metal rich, low mass stars.

ant stars making up the power law component suggests that the Galactic Bulge on average has close to solar metalicity (Tiede *et al.* 1995). An old ($\sim$10 Gyr), single age (Single Stellar Population - SSP) model with a bulge metalicity distribution deduced from the color-magnitude properties of the Bulge appears to give a fairly good representation of the hump. In summary, the KLF in the central parsec of the Galactic Center thus resembles one of an old cluster with an admixture of bright young stars, a finding

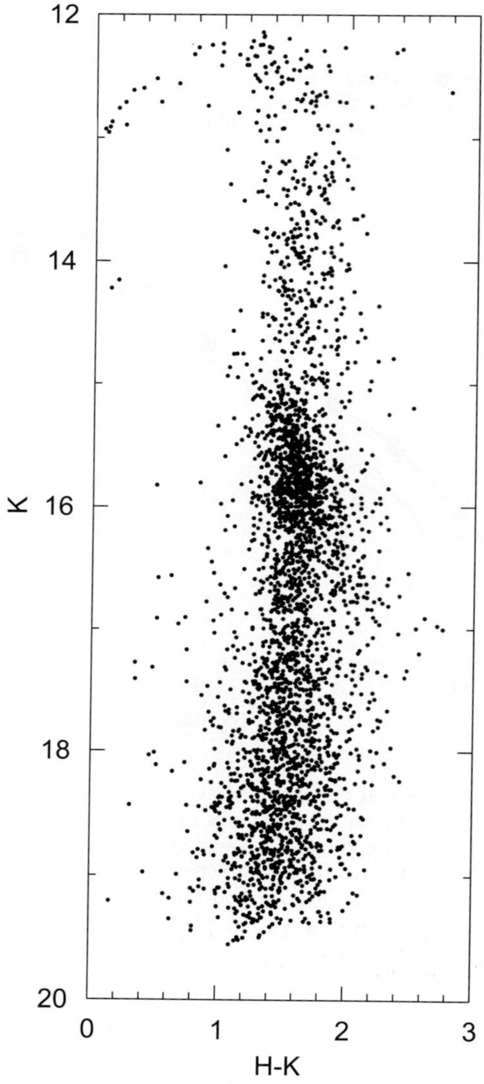

Fig. 2.11 Color-magnitude diagram for the central region of radius 9" in projection (as in the KLF shown in Fig. 2.10), showing prominently the horizontal branch/red clump stars (Genzel *et al.* 2003b). Most of the very blue stars at K< 13 are probably caused by saturation in the K-band image, which results in their K-magnitudes being underestimated.

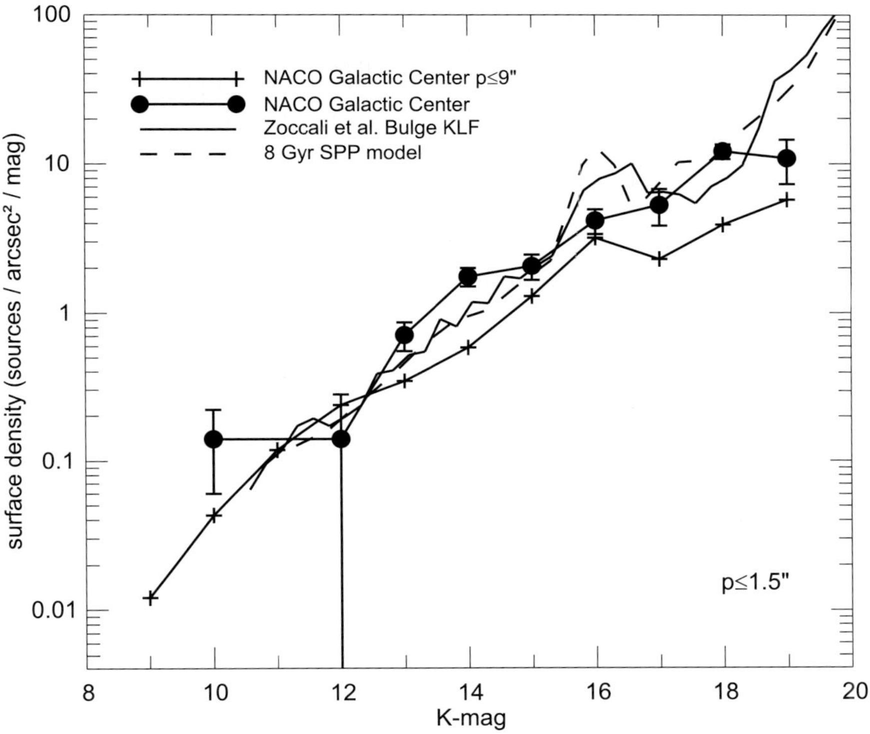

Fig. 2.12 K-band luminosity function (Genzel *et al.* 2003b) — sources per square arcsecond per mag — of the central cusp region (filled circles, projected distance to Sgr A* $\leq 1.5''$,$\pm 1\sigma$ error bars). Other symbols and curves are as in the caption of Fig. 2.10.

that agrees with earlier investigations (e.g. Lebofsky & Rieke 1987; Krabbe *et al.* 1995; Blum *et al.* 1996; Davidge *et al.* 1997; Morris & Serabyn 1996; Figer *et al.* 2004).

The situation is different for the KLF of the cusp (Fig. 2.12), for which Genzel *et al.* (2003b) found a similar slope compared to the overall cluster, but noted the absence of the bump due to horizontal branch/red clump stars. Also, an excess of stars with magnitudes between 13 and 15 can be noted in the cusp. This can be interpreted as an increased presence of early-type stars, a lack of late-type giants, and/or the absence of old, low-mass stars (the horizontal branch stars).

The distinct properties of the stellar population in the cusp may be due to mass segregation effects. Mass segregation should work effectively in the dense cusp environment and lead to an expulsion of low-mass stars. An alternative interpretation is that stars ascending the red giant branch are stripped of their envelopes by close interactions of stars in the cusp (Alexander 1999; Genzel *et al.* 2003b). In this way, the helium cores of these stars would not reach the critical mass for ignition and they would directly evolve into helium white dwarfs.

2.9 Polarization of Filaments and Stars

The polarization of electromagnetic radiation allows us to draw important conclusions on either the radiation mechanisms or the absorption effects due to the light propagation through a medium. Towards the Galactic Center the corresponding investigations can be carried out for the extended thermal and non-thermal radio emission and mid-infrared emission of the gas and dust as well as the radiation originating from the stellar sources within the central cluster.

2.9.1 *Extended Filaments in the Sgr A Complex*

In the mid-infrared the radiation coming from the northern arm and east-west bar of Sgr A West shows an appreciable amount of linear polarization. This observed polarized emission is produced by the thermal emission of magnetically aligned dust grains. Aitken *et al.* (1991, 1998) presented arcsecond-resolution imaging polarimetry of these ionized filaments at a wavelength of 12.5 μm. The observed amount of polarization places a lower limit of $\sim$2 mG on the magnetic field in the northern arm. The polarimetry images of Aitken *et al.* (1991, 1998) and Glasse *et al.* (2003) clearly show that the magnetic field is a property of the diffuse material of the large-scale filaments in the Sgr A west complex (the northern arm and east-west bar) rather than of the embedded sources in that area (Fig. 2.13). Furthermore, the magnetic field strength is apparently independent of the changes in density and temperature in the vicinity of these compact embedded sources. In addition to an overall distribution of polarized flux density there are variations on small scales both of the strength and the direction of the magnetic field. In some areas of the east-west bar the magnetic field direction is rather complex and southwest of IRS 1 there is an abrupt decrease of polarization combined with a change of sign, which may either indicate a rapid change of magnetic field direction close to the dynamical center, or be due to a superposition of nearly orthogonal fields in the IRS 1 region. However, while the field strength shows great variations, the field direction changes smoothly across IRS 1.

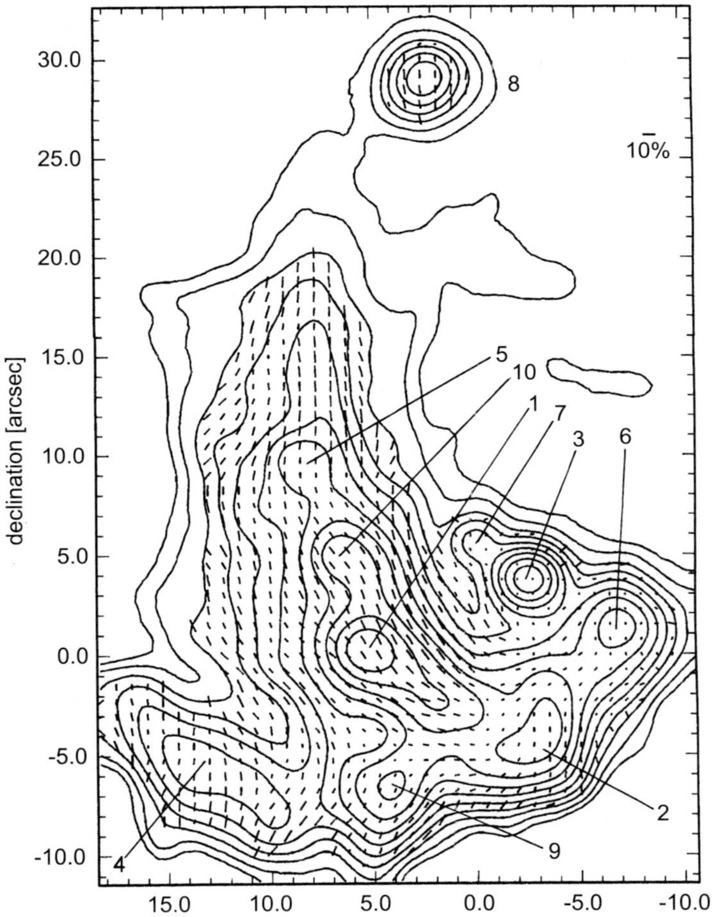

Fig. 2.13 Polarization of the mid-infrared light at a wavelength of 12.4 μm (Aitken *et al.* 1991). Some prominent sources are indicated with their IRS designations. The emission is dominated by dust in the mini-spiral filaments.

Aitken *et al.* (1998) show that the physical conditions in the ionized filaments lead to a very uniform grain orientation along the local magnetic field. The authors also note that, independent of the initial magnetic field configuration in a cloud, tidal stretching will ultimately cause the field lines to be roughly parallel to the direction of the shear. Since the position angle of the polarized emission is at right angles to the projection of the field direction on the plane of the sky, the observed degree of polarization provides some measure of the component of the field along the line of sight, but it is also influenced by factors such as grain shape, the extent of grain alignment, the grain composition, and local field inhomogeneities. The polarized light, however, allows one to at least partially reconstruct the overall field in three dimensions, and the results can be compared with the H92α observations of those structures by Roberts & Goss (1993). Rather than interpreting the northern arm and east-west bar as an orbital path or a spiral arm, Aitken *et al.* (1998) offer a new interpretation from this comparison: their data implies that these filaments represent a tidally stretched structure in free fall about Sgr A* with significant deviations from a single plane. The observed distribution of polarized dust and H92α emission gas may be interpreted as the inner ionized rim of a more extended neutral cloud (see also Paumard *et al.* 2003).

At longer wavelengths a smoothly varying polarized flux density is observed (Werner *et al.* 1988; Morris *et al.* 1992). Werner *et al.* (1988) presented the first detection of linear polarization of the far-infrared (100 μm) emission from the about 3 parsec diameter dust ring surrounding the Galactic Center and reported an observed degree of polarization of the order of 1 to 2 percent at three different positions. Since Werner *et al.* (1988) found that the dust ring is optically thin at 100 microns, the observations sample dust emission throughout the entire cloud, and the data provides first information on the configuration of the magnetic field in the dust ring. In follow-up observations, Hildebrand *et al.* (1990) found that the inferred magnetic field lines lie approximately in the plane of the CND (for a more detailed descriptions, see also Wardle & Konigl 1990; Hildebrand *et al.* 1993).

Morris *et al.* (1992) measured the polarization of the 100 μm continuum emission at 14 positions in the dense and warm molecular cloud associated with the arched filaments (also called the "bridge") of the radio arc near the Galactic Center and obtained degrees of polarization of up to 6.5%. As in the mid-infrared this polarized flux density is most likely produced by thermal emission from elongated interstellar grains oriented along the local magnetic field. In the arched filaments the magnetic field is very uniform on scales of 1-10 pc and is aligned parallel to the long dimension of the thermal radio filaments.

2.9.2 *Polarized Emission from Stars*

Due to anisotropic scattering processes by interstellar or circumstellar dust particles one might expect that most of the sources in the central stellar cluster show some degree of polarization, either because of polarization along the line of sight or because of intrinsic polarization of the individual objects. First low angular resolution studies of the polarized near-infrared flux density in the central stellar cluster were performed by Knacke & Capps (1977) and Lebofsky *et al.* (1982). The first observations at the diffraction limit of a 4 m-class telescope were conducted by Eckart *et al.* (1995) (see also Ageorges 1995). For the latter observations the authors installed a rotating wire grid in front of the SHARP/NTT speckle camera and used the polarization of the sources IRS 1 and IRS 7 for calibration (Lebofsky *et al.* 1982). Within the Galactic Plane there is an average polarization of about 4% at a position angle of $\sim 30°$ due to interstellar dust grains. Ott *et al.* (1999) found a flux-weighted polarization in the entire region of 4.1% $\pm$ 0.6% at P.A. 30° $\pm$ 10°, parallel to the Galactic Plane within the error bars. As there is no reason for local anisotropies to be aligned in any special direction, the overall polarization can be explained best by anisotropic scattering at aligned particles in the spiral arms of the Milky Way along the line of sight from the Earth to the Galactic Center.

In addition to the average polarization there is an unexpected excess polarization of a few sources, which is of the order of 10% to 20% and/or at an angle different from the general alignment of the stellar polarization vectors in the cluster. This is for example the case for the sources IRS 21 and IRS 1W, which are also among those sources that show an excess of emission at a wavelength of 10 μm and are extended. Furthermore, IRS 21 shows an almost featureless very red spectrum that is dominated by thermal dust emission (Fig. 2.14; Gezari *et al.* 1985; Ott *et al.* 1999; Tanner *et al.* 2002).

One plausible explanation that has been offered for the observed excess in polarization is that it is due to Mie-scattering in an extended, asymmetric dust shell around these stars, a phenomenon which is also observed in a number of galactic bipolar outflow sources (e.g. see Bastien & Menard 1990; Ageorges *et al.* 1996a; Ott *et al.* 1999; Ageorges *et al.* 1996b). Due to the distribution of dielectric dust grains of sizes comparable to the wavelength of the incident (unpolarized) radiation polarized light is scattered

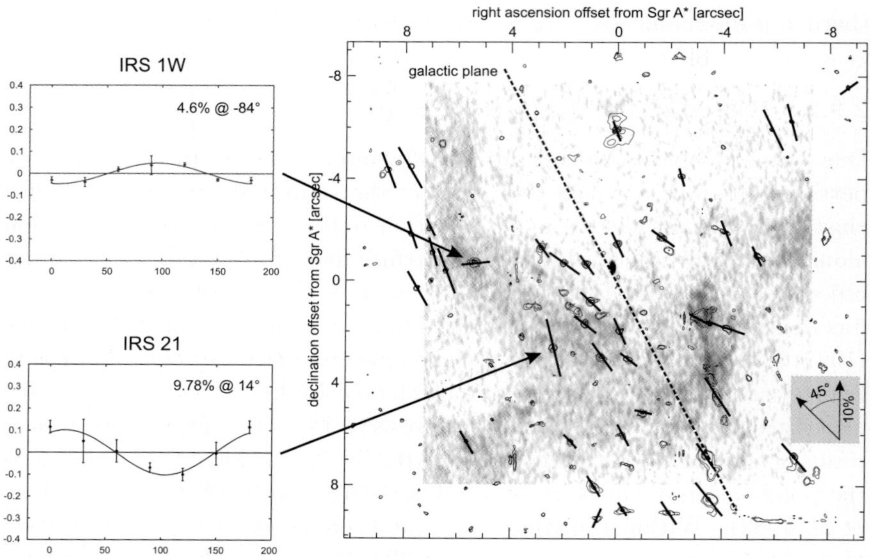

Fig. 2.14 Right: polarization of sources in the central $12.8'' \times 12.8''$ as obtained from SHARP data (Ott *et al.* 1999). The mean flux-weighted polarization in the field is 4.1% at 30°. Typical uncertainties are $\pm 1.9\%$ and 18°. As an overlay the flux at a wavelength of 1.3 cm shows the location of the mini-spiral. Left: the polarization data shown for the two sources IRS 1W and IRS 21 that exhibit extended emission. The curve represents the cosine function that was fitted to the data. A zero phase shift of the cosine function would indicate a north-south direction of the E-vector for the polarized radiation. In the map, north is up and east is to the left. (Offsets are given in arcseconds; south is positive and east is positive.)

perpendicular to the line of sight between the radiation source and the dust grain (with the E-vector also perpendicular to that connecting line). If the dust is distributed evenly in a circularly symmetric pattern around the star no integrated polarization can be observed. On the other hand, if this symmetry is broken in the form of a bipolar outflow or any other kind of non-circular symmetric distribution, a net polarization typically ranging between 10% and 30% of the integral near-infrared flux is observed.

The degree of polarization reported for IRS 21 is at least twice the value of the mean polarization in the whole field. Ott *et al.* (1999) report a polarization of this source of $10\% \pm 1.3\%$ at a position angle of $14° \pm 10°$ in a 0.5" aperture, while Eckart *et al.* (1995) derived in their diffraction limited image a polarization of 17% at P.A. 16° in a 0.25" aperture. The difference in the degree of polarization most likely results from the different aperture sizes. The observed degree of polarization of IRS 1W is 4.6% ± 2.5% and thus comparable to the overall polarization in the Galactic Center region, however, with −84° ± 8°, its P.A. differs significantly from the overall position angle (Eckart *et al.* 1995; Ott *et al.* 1999). In the mid-infrared at a wavelength of 12.4 μm (Aitken *et al.* 1991) the general orientation of the polarization vectors of these two sources agree to within 20° with the orientation derived from the 2.2 μm Speckle results obtained with the SHARP camera. The mid-infrared emission is clearly due to emission from the dust.

The agreement between the mid- and near-infrared polarization vectors therefore implies that the polarized radiation is most likely emission from warm dust (rather than scattered light at 2.2 μm wavelength) with an orientation of the position angle perpendicular to the magnetic field lines, which follow the northern arm of the mini-spiral.

Since several sources in the Galactic Center cluster with a deviation in their polarization properties are located along the so-called mini-spiral (e.g. IRS21 and IRS1), it is very likely that their polarization is influenced by local interstellar dust present within this gaseous feature. Tanner *et al.* (2002) and Tanner *et al.* (2003) suggested that these sources create bow-shocks when they plow through the dense medium of the mini-spiral.

2.10 Stellar Velocities and Orbital Accelerations

An important goal of the near-infrared high angular resolution imaging is the investigation of the dynamics of the central star cluster. This allows determining the mass profile in the Galactic Center, i.e. the profile of mass enclosed in circular apertures centered on the position of Sgr A*. Early attempts to measure the enclosed mass used gaseous tracers of the velocity field, a method which is not necessarily free from magneto-hydrodynamical influences and which uses only line-of-sight velocities, i.e. which does not probe the full velocity field. Stars, on the other hand, can be regarded as ideal frictionless test particles of the gravitational potential and their full three-dimensional velocities can be measured, i.e. motion along the line-of-sight and on the plane of the sky (proper motion).

Determining the velocities of stars in the central cluster in combination with their density distribution allows testing our theoretical understanding of the dynamics of dense clusters around supermassive black holes, such as the formation of a cusp, mass segregation effects, or isotropy of such systems. Knowledge of the velocity field also allows the identification of individual stellar populations with potential common origin, which is important for the analysis of the dynamical evolution and star formation history of the central stellar cluster.

Over the past decade two major programs to determine the proper motions of stars in the Galactic Center were carried out. Since 1992 a team of astrophysicists from the the Max-Planck-Institute of extraterrestrial Physics (MPE; A. Eckart, R. Genzel, T. Ott, R. Schödel, and others, with A. Eckart and R. Schödel now at the University of Cologne) have been carrying out Speckle and Adaptive Optics observations with ESO telescopes at La Silla and Paranal in Chile. Starting a few years later, in 1995, a group of astrophysicists of the University of California Los Angeles (UCLA; A. Ghez, M. Morris, E. Becklin and others) have been using the Keck telescope for the same purpose.

The most recent analysis of stellar dynamics in the central parsec, done by the MPE group (Ott 2004; Genzel *et al.* 2003b), lists about 100 stars with full three-dimensional velocity information and about 1000 stars with measured two-dimensional proper motion velocities.

2.10.1 *Measuring Stellar Proper Motions*

A detailed description of the derivation of proper motions is given in Eckart & Genzel (1997), Ghez *et al.* (1998), and more recently in Ott (2004). In this section we only give a short overview dealing mostly with the SHARP and NACO data sets explaining some of the fundamental steps in the analysis. The addressed algorithms and essentials are similar for the analysis and results of the Keck data (see all references to Ghez et al.).

The data base used consists of a few hundred shift-and-add images (SSA) from Speckle observations with the SHARP/NTT camera (covering one to two observing runs per year between 1992 and 2001) and was extended since the beginning of 2002 with the AO imaging data obtained with the system NAOS/CONICA at the ESO VLT UT4.

In a first step, one has to measure the positions of point sources in the images. For the 13/27 mas (NACO) to 50 mas (SHARP) pixel scale images PSF cross-correlation, Gaussian fitting and aperture centroid measurements on sources in the raw SSA images, the LUCY deconvolved images, and the diffraction limited restored maps all give the same positions to within 1-10 milli-arcseconds, depending on the quality of the images (speckle or AO, good or bad seeing, quality of AO correction, pixel scale) and on whether stars are bright and/or isolated or very faint and/or close multiple sources.

The pixel positions of the stars will of course depend primarily on the instrument used, on the pixel scale, the image rotation, the pointing of the telescope for the individual images and on other parameters, such as e.g. optical image distortions, which might even vary between observing runs. Therefore, as a second step, the stellar positions measured in the hundreds of individual images must be transformed to a common coordinate system. For this purpose, the imaging parameters for each coordinate and image with respect to a reference frame are determined. A correction to second order terms was found to be sufficient. Here, one assumes that the pixel coordinates (x_i, y_i) of the i-th star can be written in terms of the corrected offset coordinates (Δx_i,Δy_i) from the base position as

$$x_i = a_0 + a_1 \Delta x_i + a_2 \Delta y_i + a_3 \Delta x_i^2 + a_4 \Delta x_i \Delta y_i + a_5 \Delta y_i^2 \tag{2.3}$$

and

$$y_i = b_0 + b_1 \Delta x_i + b_2 \Delta y_i + b_3 \Delta x_i^2 + b_4 \Delta x_i \Delta y_i + b_5 \Delta y_i^2. \tag{2.4}$$

The 0th order is the base position (a_0, b_0), the 1st order (proportional to Δx and Δy in each coordinate) relates to a variation in pixel scale as well as a rotation of the camera, and the 2nd order parameters give the image distortions (proportional to Δx^2, Δy^2 and $\Delta x \Delta y$ for each coordinate). The 2×6 instrumental parameters (a_0, b_0, ... a_5, b_5) were determined for each data set by comparison to a reference frame (a NAOS/CONICA image with a sufficiently large field-of-view), which in turn was linked to the positional radio reference frame as described in Eckart & Genzel (1997), Menten *et al.* (1997), Reid *et al.* (2003a), Ott (2004), or Schödel (2004).

The 2×6 imaging parameters are obtained by solving an over-determined nonlinear equation for N stars via orthonormalization of the $12 \times N$ matrix, a procedure which turns out to be very stable in terms of selection of the N stars, as long as the stars used for the transformation are sufficiently bright and isolated and $N \gtrsim 50$. The position of the centroid of the N stars and the camera rotation angle are the main fit parameters. However,the second order parameters must not be neglected if one intends to attain accuracies of the order of a milli-arcsecond for the stellar positions. With the highest quality data (NAOS/CONICA and similarly for the Keck data) a baseline as short as 2 years is sufficient to detect proper motions larger than about one hundred kilometers per second. Assuming a distance of 8 kpc to the Galactic Center a proper motion of 1 mas/yr corresponds to a velocity of approximately 39 km/s.

2.10.2 *Motions in the Outer Cluster*

For stars with separations from the compact radio source Sgr A* between 0.035 pc and 0.35 pc (i.e. between 0.9 and 9.0 arcsec) the intrinsic proper motion velocity dispersion per coordinate (R.A. or Dec.) is about 160 ± 15 km/s at a mean projected distance from Sgr A* of 0.12 pc (Genzel *et al.* 2000; Schödel *et al.* 2003). This value is in excellent agreement with the radial velocity dispersion results. This indicates that — to first order — the stellar velocity field is indeed *close to isotropic* (but see discussion of small deviations from isotropy in sections 3.1 and 3.6.2).

The velocity dispersion determined from proper motion and line-of-sight velocity measurements of stars in the central parsec was found to increase toward Sgr A* in agreement with Keplerian rotation about a point mass. This indicates the presence of a compact mass of about $2 - 3 \times 10^6$ $M_\odot$ (e.g. Eckart & Genzel 1997; Ott 2004). Here compact means: not resolved by the motion of ensembles of stars or even a single stellar orbit.

2.10.3 *Motions in the Inner Cluster*

In the central arcsecond of the small Sgr A* cluster the proper motions are much larger than in the large-scale, isothermal stellar cluster at the center of the Milky Way. Evidence for these motions can already be seen by eye when comparing raw SSA or adaptive optics images from the different observing epochs (e.g. Fig. 2.15, the high-passed SSA images in Fig. 8 and Fig. 9 of Eckart & Genzel (1996) and the maps presented in Schödel *et al.* (2003)). Near Sgr A*, some proper motions exceed 25 mas/yr or approximately 1000 km/s. With the data from 10 years of high-resolution imaging of the central stellar cluster Schödel *et al.* (2003) presented proper motions for more than 40 stars at projected distances of $\leq 1.2''$ from Sgr A*. The data presented by Schödel *et al.* (2003) gives evidence for radial anisotropy of the cluster of stars within an arcsecond around Sgr A* (see also Genzel *et al.* 2000).

In Fig. 2.16 the velocity vectors of the stars in the inner few arcseconds are given as they were determined from NAOS/CONICA images taken in 2002 and 2003. The two most obvious, bright high velocity stars are S1 and S2.

The proper motion velocity of the star S1 was already measured in early

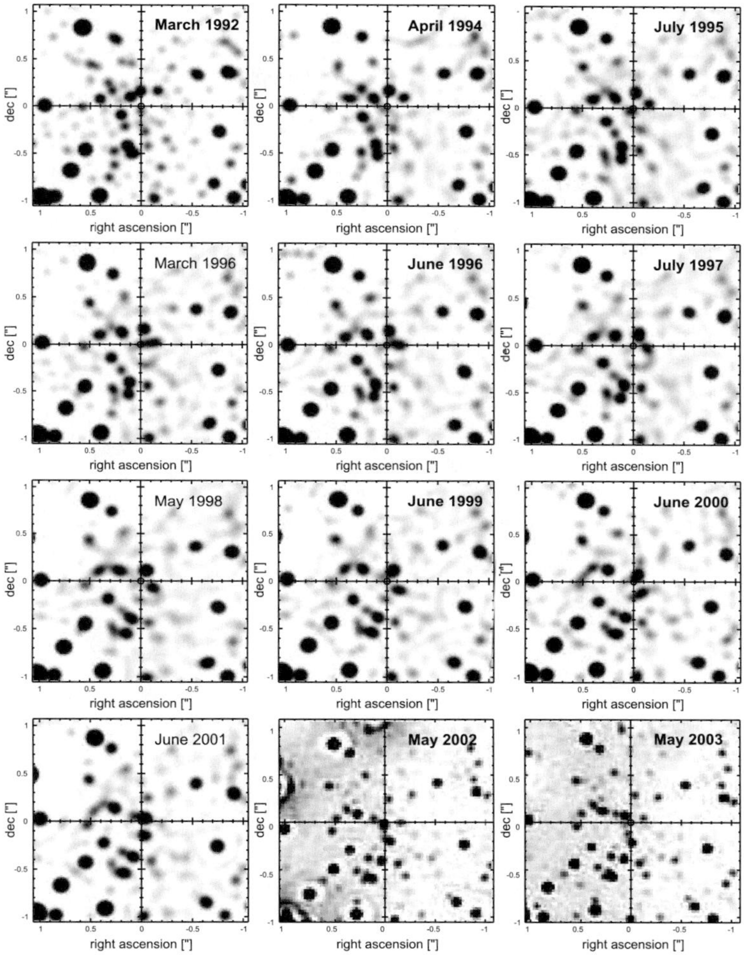

Fig. 2.15 High-resolution maps of the central $2'' \times 2''$ around Sgr A* (see Schödel 2004). Epochs 1992 to 2001 show maps obtained from SHARP/NTT imaging data. All maps resulted after Lucy-Richardson deconvolution of SSA images and subsequent restoration with a Gaussian PSF. FWHM of the PSF on the SHARP maps is 100 mas. Epochs 2002 and 2003 are NACO/VLT images (FWHM of 60 mas). Offsets are given in arcseconds from Sgr A*. North is up and east is to the left.

publications to reach the impressive speed of about 1400 km/s (e.g. Eckart & Genzel 1997; Ghez *et al.* 1998; Eckart *et al.* 2002a), but star S2 even exceeded that value as it passed through the peri-center of its orbit in April

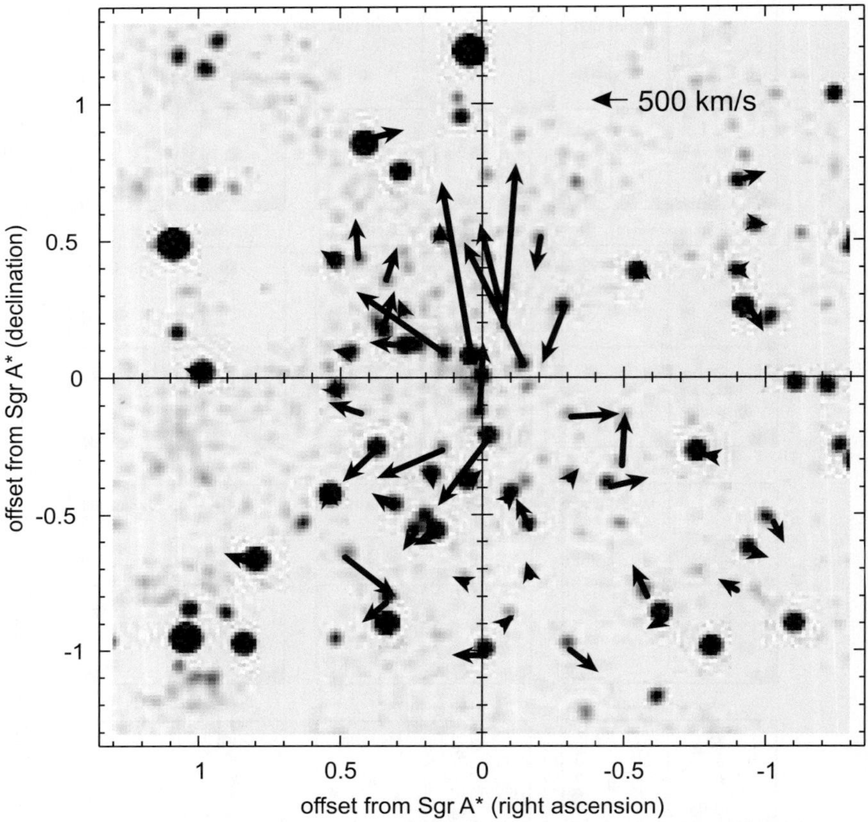

Fig. 2.16 Proper motion of stars near Sgr A*. Stellar velocities within 1.2″ of Sgr A*, derived from the NACO 2002/2003 imaging data, superposed on a NACO 2003.35 LR deconvolved and beam restored K-band image (Schödel 2004). Sgr A* is visible in its flaring state as a point source at the origin.

2002 at a distance of only 17 light hours from Sgr A* with a proper motion velocity of about 6950 km/s (corresponding to a space velocity of about 7300 km/s; Eisenhauer *et al.* 2003b). Such high velocities had already been predicted in some scenarios by Eckart *et al.* (2002a) (see their Fig. 15).

In Tab. 2.1 we summarize the proper motion of some of the central stars as they are indicated in Fig. 2.16. From these velocities one obtains with a Leonard-Merritt mass estimator (see chapter 3.3) an enclosed dark mass of $4.0 \pm 0.8 \times 10^6$ $M_\odot$ within a projected distance of about 0.025 pc (0.6″) from Sgr A* (Schödel 2004). Surprisingly, the stellar velocities within this central cluster show signs of radial anisotropy (see section 3.6.2; Genzel *et al.* 2000; Schödel 2004).

Table 2.1 **NACO/VLT 2002-2003: list of stars near Sgr A*.** *Listed are (from left to right): name, distance from Sgr A*, brightness, offset in R.A. and Dec., velocity and corresponding uncertainty in R.A. and Dec.. The stars are ordered with increasing distance from Sgr A* (second column). Positions are for the epoch 2003.45. Accuracies of all positions are about 1-2 mas. For further details see Schödel (2004).*

Name	R (arcsec)	M_K (mag)	ΔR.A (arcsec)	ΔDEC (arcsec)	$v_{R.A.}$ (km s^{-1})	$dv_{R.A.}$ (km s^{-1})	v_{DEC} (km s^{-1})	dv_{DEC} (km s^{-1})
S2	0.085	13.9	0.038	0.075	410	87	2501	118
	0.123	15.5	0.009	-0.122	-49	76	979	100
S13	0.153	15.6	-0.146	0.046	819	68	1751	64
S14	0.161	15.5	0.135	0.088	1210	73	881	79
	0.200	17.1	-0.075	0.186	-170	132	2330	112
S1	0.221	14.4	-0.024	-0.219	709	60	-966	62
S12	0.286	15.2	-0.060	0.280	267	55	1106	65
	0.299	16.3	0.141	-0.264	899	65	-396	87
S4	0.299	14.1	0.276	0.116	481	61	51	71
	0.344	16.6	-0.313	-0.143	-672	49	40	66
	0.371	16.1	0.280	0.242	64	65	177	96
S10	0.388	13.8	0.052	-0.384	-195	60	194	63
	0.390	15.3	-0.291	0.259	310	52	-815	59
S5	0.392	15.0	0.352	0.172	-161	67	565	115
S9	0.402	14.8	0.181	-0.358	-32	68	-192	67
	0.444	14.9	-0.103	-0.431	-81	57	85	58
	0.447	17.1	0.428	-0.130	451	161	158	93
S8	0.460	14.1	0.377	-0.264	454	64	-450	73
	0.481	16.2	-0.314	-0.364	-106	49	135	60
S6	0.481	15.1	0.473	0.089	218	68	52	77
	0.493	16.4	0.340	0.357	-150	67	467	79
S7	0.523	14.9	0.521	-0.048	-104	65	-122	76
	0.543	15.3	0.147	0.522	9	63	177	65
	0.552	16.5	-0.206	0.512	81	58	-499	60
	0.568	15.3	0.315	-0.472	297	74	206	73
	0.578	15.3	-0.168	-0.553	194	52	444	57
	0.588	17.1	-0.494	-0.320	-41	39	752	53
S11	0.598	14.0	0.165	-0.575	311	78	-153	69
	0.603	15.4	-0.454	-0.397	-521	40	127	50
	0.620	16.4	0.438	0.438	36	75	567	80
	0.620	15.3	0.241	-0.572	154	96	-242	73
W6	0.681	14.1	-0.560	0.387	9	47	160	41
	0.681	14.7	0.526	0.433	175	62	129	77
S0-13	0.698	13.1	0.544	-0.437	32	67	153	76
	0.750	16.8	-0.170	-0.731	61	67	224	74
	0.751	17.5	0.059	-0.748	182	110	80	93
	0.816	16.1	0.484	-0.658	-700	86	-566	81
W9	0.823	13.4	-0.774	-0.281	30	36	6	36
	0.886	16.8	-0.100	-0.880	-55	58	50	63
	0.891	15.4	0.339	-0.824	334	71	-280	84
	0.965	11.8	0.419	0.869	-559	68	138	69
W5	0.978	13.4	-0.945	0.252	-188	60	-345	32
S0-5	0.982	12.3	0.342	-0.921	-160	75	205	64
	0.990	15.7	-0.583	-0.800	203	46	451	43
	0.998	15.3	-0.919	0.388	207	65	0	41
	1.004	12.9	1.004	0.019	191	61	66	87
	1.019	14.5	-0.012	-1.018	465	62	8	54
	1.045	15.8	-0.309	-0.998	-400	52	-324	47
S0-4	1.060	12.2	0.812	-0.681	392	69	155	81
	1.095	13.8	-0.646	-0.884	261	36	-151	39
	1.130	15.3	-0.983	0.557	-95	74	-13	31
	1.157	15.0	-0.961	-0.644	-201	47	-65	35
	1.157	15.5	-1.028	-0.531	-147	51	-304	36
	1.175	15.2	-0.925	0.724	-304	74	98	37
	1.191	17.2	-0.900	-0.779	270	73	163	55

2.10.4 *Accelerations of Stars Near Sgr A**

First measurements of stellar orbital accelerations near Sgr A* were reported by Ghez *et al.* (2000) and later on confirmed by Eckart *et al.* (2002a). The observed projected orbital accelerations represent a change in the proper motion velocities, and in the case of the Galactic Center they are therefore conveniently given in units of milli-arcseconds per square year (1 mas $yr^{-2} = 1.2 \times 10^{-6}$ km s^{-2}). The orbital accelerations of the stars with separations from Sgr A* of only a few milli-arcseconds are of the order of a few mas yr^{-2}, and in fact are comparable to those experienced by the Earth as it orbits the sun. Since the acceleration vectors — even in projection onto the plane of the sky — point at the gravitational center of mass, orbital acceleration measurements can help to localize this dark mass. Ghez *et al.* (2000) and Eckart *et al.* (2002a) both found a discrepancy between the then known nominal radio position of Sgr A* (Menten *et al.* 1997) and the position of the gravity center. This motivated Reid *et al.* (2003a) to repeat the alignment of the radio and infrared frames with improved data, which resolved this inconsistency.

Furthermore, orbital acceleration measurements allow to derive an upper limit on the enclosed mass. For a star at a projected separation R from the center and a projected, observable acceleration a_{obs} the total enclosed mass M can be calculated using the equation

$$M = a_{obs} R^2 G^{-1} cos^{-3}(\theta) \quad , \tag{2.5}$$

with θ being the angle between the radius vector to the star and the plane of the sky containing the central mass. Plotting the lower limits $M\ cos^3(\theta)$ of the enclosed mass as a function of the projected radius is equivalent to the assumption that the stars are in exactly the same plane of the sky as the central dark mass at the position of Sgr A* (see Eckart *et al.* 2002a, Fig. 5), an assumption which is of course not justified in general, and any more realistic approach needs to correct for such geometrical projection effects. The quantity $(cos\theta)^{-1}$ increases monotonically with the distance from the plane of the sky in which Sgr A* is located and its median can be calculated for each given stellar density distribution $n(r)$. A statistical estimate of M can therefore be derived using median values of $(cos\theta)^{-1}$ for any assumed density distribution $n(r)$ (Eckart *et al.* 2002a).

Table 2.2 **Parameters of individual orbits near Sgr A*:** *List of the orbital parameters and their uncertainties for 6 stars near Sgr A* (see Schödel 2004). All values are given for a GC distance of 7.94 kpc (Eisenhauer et al. 2003b). S2: The first line gives the values and uncertainties from Schödel et al. (2002), the second line from Eisenhauer et al. (2003b) and Schödel (2004), respectively. The third line gives the results of Ghez et al. (2003b). Other stars: The first lines list the results of Keplerian fitting with the formal fit errors and the second lines list the values and errors derived from variations of the central mass (S1, S12, S13) or of the focus position (S14,S12) that was fixed during the fitting procedure (see Schödel 2004).*

	M_0 (10^6 $M_\odot$)	P (yr)	T_{peri} (yr)
S2	3.6±0.3	15.56±0.35	2002.331±0.012
	4.0±0.3	15.02±0.72	2002.34±0.01
	3.7±1.5	15.2±1.0	2002.30±0.05
S14	4.4±11.0	14.6±14.1	2000.8±0.6
	4.5 ± 0.2	16.0 ± 2.7	2000.9 ± 0.1
S12	3.2±1.5	46.2 ± 8.7	1995.5 ± 0.06
	3.2 ± 0.05	45.7 ± 1.2	1995.4 ± 0.02
S1	3.6 ± 2.0	87.8 ± 18.8	1997.9 ± 2.6
	$3.6^{4.2}_{3.2}$	$87.8^{70.1}_{114.8}$	$1997.9^{1997.7}_{1998.0}$
S13	3.6 ± 1.3	44.3 ± 6.5	2005.8 ± 0.4
	$3.6^{3.9}_{3.4}$	$44.3^{39.7}_{838.8}$	$2005.8^{2006.3}_{2002.1}$
S8	3.4 ± ∞	90.5 ± ∞	1987.6 ± ∞
	$3.4^{2.9}_{3.1}$	$90.5^{116.3}_{97.6}$	$1987.6^{1982.7}_{1987.8}$

	e	i (deg)	a (mpc)	r_{peri} (mpc)
S2	0.881±0.007	-48.1±1.3	4.63±0.10	0.55±0.03
	0.88±0.01	-46.4±1.7	4.68±0.17	0.58±0.04
	0.87±0.03	±46.0 ± 4.0	4.62 ± 0.58	0.60 ± 0.17
S14	0.87±0.23	±83.1±3.4	4.8 ± 2.3	0.6 ± 1.1
	0.89 ± 0.04	±83.0 ± 0.3	5.0 ± 0.5	0.2 ± 0.4
S12	0.79 ± 0.03	±56.6 ± 2.7	9.2 ± 0.8	1.9 ± 0.3
	0.790 ± 0.005	±56.6 ± 0.2	9.1 ± 0.1	1.9 ± 0.05
S1	0.33 ± 0.08	±56.0 ± 0.0	14.7 ± 1.7	9.9 ± 1.7
	$0.33^{0.22}_{0.44}$	$56.0^{58.0}_{54.0}$	$14.7^{13.2}_{16.8}$	$9.9^{10.3}_{9.4}$
S13	0.41 ± 0.04	±20.0 ± 0.0	9.3 ± 0.6	5.5 ± 0.5
	$0.41^{0.39}_{0.90}$	$20.0^{7.0}_{42.0}$	$9.3^{8.5}_{70.6}$	$5.5^{5.2}_{7.1}$
S8	0.96 ± ∞	±52.0 ± 0.0	14.7 ± ∞	0.6 ± ∞
	$0.96^{0.61}_{0.97}$	$52.0^{77.0}_{46.0}$	$14.7^{16.4}_{14.9}$	$0.6^{6.4}_{0.4}$

2.10.5 *Stellar Orbits*

The first measurements of orbital accelerations for the stars S1 and S2 were consistent with bound orbits around a ~3 million solar mass central object and increased the inferred minimum mass density in the central region of our Galaxy by an order of magnitude relative to previous results that were based on proper motion velocities alone. Assuming that the stars S1 and S2 were on bound orbits, Ghez *et al.* (2000) derived periods of 35–1200 (S1) and 15–500 years (S2) and eccentricities of 0 to 0.9 (S1) and 0.5 to 0.9 (S2), respectively. From the combined SHARP and Keck data Eckart *et al.* (2002a) could show that star S2 was indeed on a bound orbit around Sgr A* and could give first estimates of its orbital elements. That investigation indicated that the stars S1, S2, and S8 were on fairly inclined and eccentric orbits. However, the range of possible orbital elements constrained by the works of Ghez *et al.* (2000) and Eckart *et al.* (2002a) still allowed for a wide range of orbital solutions.

Observations of the peri-center passage of the star S2 enabled Schödel *et al.* (2002) and Ghez *et al.* (2003b) to calculate a unique orbital solution for this star. Schödel *et al.* (2002) found that it circles a central dark mass of $3.7 \pm 1.5 \times 10^6$ $M_\odot$ with a period of about 15 years, an eccentricity of 0.87, and a peri-center distance of about 17 light hours. Ghez *et al.* (2003b) found very similar values and additionally constrained the focus of the orbit with an accuracy of $\lesssim 2$ mas. Subsequently, Eisenhauer *et al.* (2003b) combined the observed time-dependent positions with measurements of the line-of-sight velocity of S2 in order to determine the geometrical distance to S2, i.e. the Galactic Center, and obtained a value of $R_0 = 7.9 \pm 0.4$ kpc, in agreement with earlier, indirect measurements of this fundamental scale (Reid 1993). Ghez *et al.* (2003a) and Schödel *et al.* (2003) concluded that the center of acceleration coincides with the radio position of Sgr A* to within less than about 2 mas, corresponding to 0.08 mpc or 2.2 light hours at a distance of 8 kpc.

Figure 2.17 shows the measured 1992-2003 positions of S2 relative to Sgr A* (e.g. Schödel *et al.* 2003; Eisenhauer *et al.* 2003b). The black ellipse is the Keplerian orbit determined by Eisenhauer *et al.* (2003b). Its parameters along with their 1σ uncertainties are listed in Tab. 2.2. The orbit of S2 probes the concentration of the central dark mass at a distance of 114 AU or only ~1600 times the Schwarzschild radius of a ~3.6 million solar mass black hole. The derived mass coincides to within 15% with the best previous mass estimates, obtained from statistical projected mass estimators, Jeans mass modeling of the ensemble of proper motions stars near Sgr A* (Genzel *et al.* 2000), or mass estimates from the accelerations of the inner most high velocity stars (Eckart *et al.* 2002a).

Ghez *et al.* (2003a) and Schödel *et al.* (2003) found accelerated motion for 6 stars, with 4 stars having passed the peri-center of their orbits during the observed time span. The orbits of these stars are illustrated in Fig. 2.18,

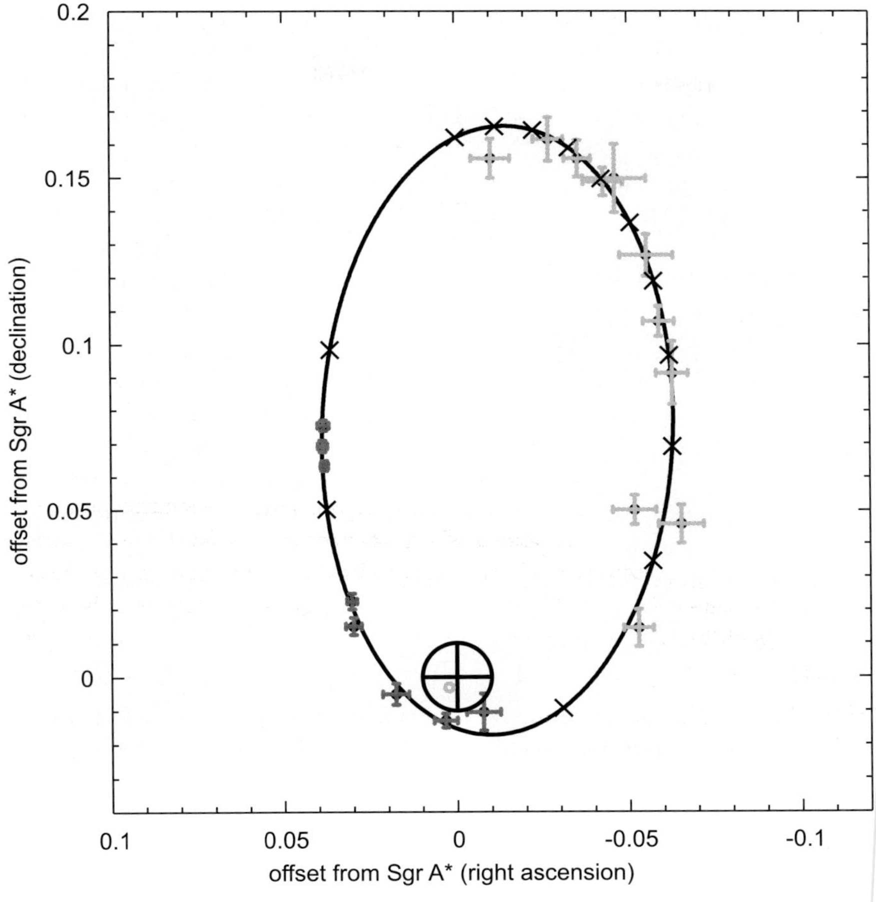

Fig. 2.17 Orbit of the star S2. Light grey: positions measured with SHARP/NTT between 1992 and 2001. Dark grey: positions measured with NACO/VLT between 2002 and 2003. The black crosses give the position of S2 at January 1 of 1992-2004. The black cross and the 10 mas error circle at (0,0) indicate the nominal radio position of Sgr A*. The small circle inside the radio position uncertainty indicates the uncertainty of the position of the focus of the orbit (see also Schödel *et al.* 2003; Eisenhauer *et al.* 2003b; Schödel 2004). See also color Fig. A.3.

with their parameters listed in Tab. 2.2 (see Schödel 2004). Particularly interesting is the case of the star S0-16 (Ghez *et al.* 2003a; Schödel 2004) that passed within just 60 AU of Sgr A*, i.e. even closer than S2 (alias S0-2). However, its orbit is less well determined than that of S2. With

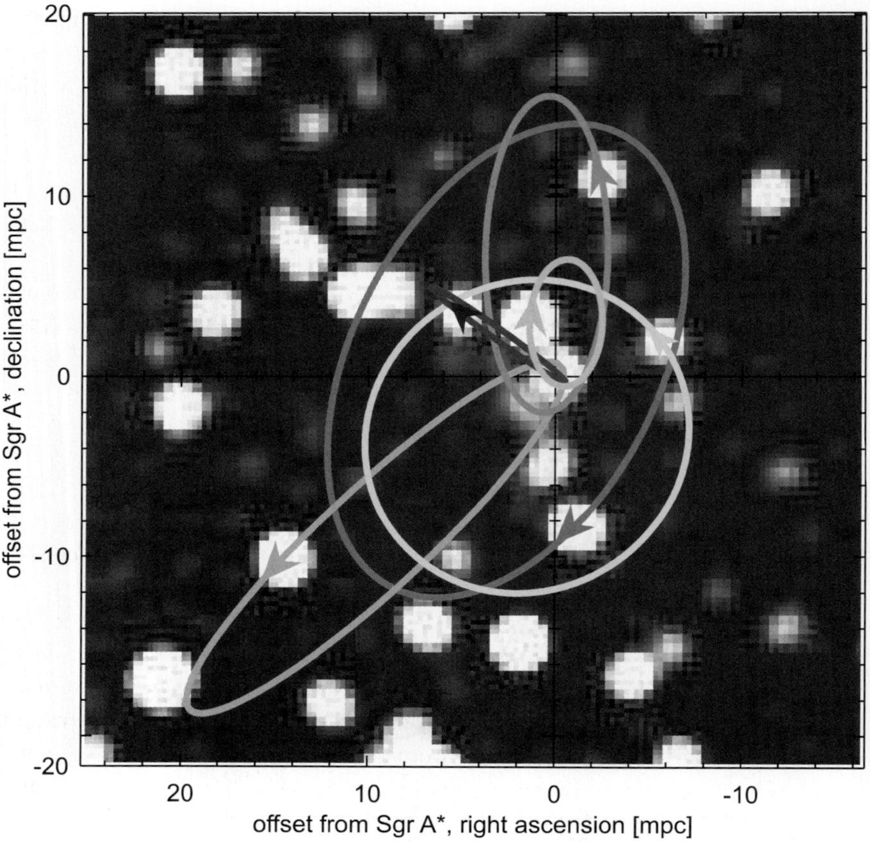

Fig. 2.18 Orbits of stars around Sgr A*. Lucy-Richardson deconvolved and beam-restored high-resolution (FWHM~ 60 milli-arcseconds) near-infrared (2.2 μm) image of the nuclear stellar cluster in the immediate environment of Sgr A*. The image was obtained with CONICA/NAOS at the ESO VLT in June 2003. On this image Sgr A* can be seen in its flaring state (see chapter 2.13.4) as a point source at the origin of the coordinate system. The Keplerian orbits of six stars are over-plotted onto the image (see also Schödel 2004). Arrows indicate the locations of the respective stars and their directions of motion. North is up and east is to the left. See also color Fig. A.4.

observing and data analysis routines improving rapidly, it is expected that soon accelerated motion will be detected for even more stars near Sgr A*.

2.11 Spectroscopy

Similar to imaging observations, stellar spectroscopy can be done most efficiently at near- and mid-infrared wavelengths because of the high extinction towards the center of our Galaxy. Near-infrared flux calibrated spectra and spectroscopically measured line-of-sight velocities were obtained by many authors (e.g. Rieke & Rieke 1988; McGinn *et al.* 1989; Sellgren *et al.* 1990; Lindqvist *et al.* 1992; Haller *et al.* 1996), however, a major portion of the available radial velocity measurements was obtained with the 3D imaging spectrometer built at the MPE (e.g. Krabbe *et al.* 1995; Eckart *et al.* 1995; Eckart & Genzel 1996; Genzel *et al.* 1996, 2000) . In a crowded field like the Galactic Center integral field spectroscopy is clearly an advantage over slit spectroscopy, since the spectra of all sources in a given field of view can be obtained simultaneously and do not have to be collected via multiple slit settings over varying seeing conditions. A new MPE-built imaging spectrometer, SPIFFI (Eisenhauer *et al.* 2003a), that will allow diffraction limited observations at the VLT, is currently being commissioned (see section 1.4.2).

To verify that the spectra of individual isolated stars are not or are only on a very low level contaminated by neighboring objects, the early SHARP/NTT speckle image reconstructions or the more recent adaptive optics images of NAOS/CONICA can be used. In a second step, the radial velocities of the isolated stars are determined via cross-correlation with template star spectra to an accuracy of a few 10 km/s (Krabbe *et al.* 1995; Genzel *et al.* 1996).

Another technique, speckle spectroscopy, applied with SHARP at the NTT allowed retrieving spectroscopic information with very low spectral, but high spatial resolution on stars in the entire field-of-view of the speckle camera (for a detailed description of the technique see section 1.3.5 and Genzel *et al.* 1997). It could not be used for obtaining velocity measurements, but was very useful in determining the nature of stars in the crowded field of the GC. For the high velocity stars in the central arcsecond initial speckle spectroscopy measurements showed that most of these objects are most likely not associated with late type stars, but are more likely early O9 - B0.5 stars with masses of 15 to 20 $M_\odot$. Recently, adaptive optics spectroscopy with 10 m class telescopes has delivered first high resolution spectra, which allow a reliable identification of stellar types and radial velocities (Gezari *et al.* 2002; Ghez *et al.* 2003b; Eisenhauer *et al.* 2003b).

The spectroscopically observed radial velocity dispersion and the velocity dispersion derived from the stellar proper motion data compare very well with each other (e.g. Eckart & Genzel 1997; Genzel *et al.* 2000, and references therein). Both velocity dispersions show a clear Keplerian falloff as a function of projected distance from Sgr A*, an indication that there must be a large compact mass located at the dynamical center, since otherwise the innermost measured velocity dispersion data points would fall significantly below the Keplerian curve. Furthermore, differences in the radial and proper motion characteristics as a function of stellar type indicate a small degree of anisotropy coupled with the evolution of the overall central stellar cluster, e.g. the coherent rotation pattern, counter to galactic rotation, of a group of early-type stars within a few arcseconds of Sgr A* (Genzel *et al.* 1996, 2000, 2003b, see section 3.1 and Fig. 3.1).

2.11.1 *Stellar Populations*

Starting in the late 1980s several international working groups measured NIR spectra of the stars in the Galactic Center stellar cluster (e.g. Rieke & Rieke 1988; Sellgren *et al.* 1990; Krabbe *et al.* 1995; Haller *et al.* 1996). Most of them were found to be red giants and super giants (IRS 7). Allen *et al.* (1990) and Forrest *et al.* (1986) reported broad lines toward the IRS 16 complex that were due to Ofpe/WN9 stars, and a luminous, apparently young object 8" SW of the central cluster — the so called AF star. Shortly thereafter, Krabbe *et al.* (1991, 1995) found a cluster of HeI stars concentrated in the IRS 16 and the small IRS 13 complexes, which are both located within about 4 arcsec of Sgr A*. Detailed atmospheric model calculations of the brightest central hot stars were performed for the first time by Najarro *et al.* (1994, 1997). They found that the HeI stars have effective temperatures of typically a few times 10^4 K and luminosities of the order $2-20\times10^5$ $L_\odot$, and are sources of strong, supersonic winds. Table 2.3 lists the properties of three representative HeI stars (Najarro *et al.* 1997).

An extensive list is provided by Krabbe *et al.* (1995), who list 12 stars of type WN9/Ofpe, of which several objects are pure He I stars (IRS 13E, IRS 15SW, IRS 15NE, IRS 7W, IRS 7E, IRS 16NW, IRS 34W, and IRS 33E) and 5 are WC9 stars (IRS 29N, IRS 6, MPE-1.0-3.5, MPE+1.6-6.8, and MPE+2.7-6.9). The WC9 stars show a broad emission line feature at 2.110 μm as a combination of HeI, CIII, and NIII. For the source IRS 13E a weak 2.189 μm line of He II emission was observed.

Very hot and luminous stars were also found in other regions near the Galactic Center. Nagata *et al.* (1990) and Okuda *et al.* (1990) found the

Table 2.3 Basic Properties of representative hot GC Stars (Najarro *et al.* 1997).

	AF-star	IRS16NE	IRS13E1
$R_*(R_\odot)$	40	85	60
$L_*(10^5\ L_\odot)$	2.0	22.0	22.6
$T_{eff}(10^4\mathrm{K})$	1.93	2.41	2.89
He/H	1.7	1.0	>500
$\dot{\mathrm{M}}(10^{-5})\mathrm{L}_\odot$ /yr	8.7	9.5	79.1
v_∞(km/s)	700	550	1000

Quintuplet cluster, followed by the discovery of Wolf-Rayet and luminous blue variable (LBV) stars in that region by Figer *et al.* (1995), and the discovery of a second cluster of hot stars near the central parsec – the Arches cluster – by Cotera *et al.* (1992) and Nagata *et al.* (1995) (see also Serabyn *et al.* 1998). Figer *et al.* (1998) performed a detailed investigation of the Pistol star, one of the most luminous stars known so far. Figer *et al.* (1999b) found more than a dozen massive stars in early stages of their evolution in the Quintuplet cluster, such as luminous blue variables (LBVs), Ofpe/WN9, and OB supergiants. Figer *et al.* (1999a) presented HST NICMOS observations of both the Quintuplet and the Arches clusters, detecting main sequence stars with masses well below 10 $\mathrm{M}_\odot$ and deriving the slope of the IMF of these two young clusters (Arches ~2±1 Myr, Quintuplet ~4±1 Myr). They found that – compared to other young clusters in the galaxy – the slope of the IMF in these two massive clusters is skewed significantly towards higher masses. The detection of young, luminous stars outside the central parsec with similar properties as the HeI stars in the central cluster strongly suggests that the blue supergiants/HeI stars are not exotic objects such as, e.g. accreting black holes with optically thick atmospheres (Morris 1993), but indeed young massive stars on their way off the main sequence, as proposed by Allen *et al.* (1990), Krabbe *et al.* (1995), and Najarro *et al.* (1997).

Near- and mid-infrared spectroscopy allows for a quantitative analysis of the power budget of the Galactic Center region, and leads to the result that the hot stars essentially determine the radiation field within the central stellar cluster. The Short Wavelength Spectrometer (de Graauw *et al.* 1996b, SWS) on board of the Infrared Space Observatory (ISO) allowed one to observe a wide range of ionic fine structure lines (Lutz *et al.* 1996). Three independent diagnostic line ratios ([SIV] 10.51 μm/[SIII] 18.71 μm, [ArIII] 8.991 μm/[ArII] 6.985 μm, and [NeIII] 15.55 μm/[NeII] 12.81 μm) yielded an average effective temperature of the ionizing stars of 35000 $K \pm 2000$ K, a value which implies only a small contribution from hotter stars like the WN9 to WN5 stars detected via direct near-infrared imaging (Krabbe *et al.* 1995; Blum *et al.* 1995; Genzel *et al.* 1996, 2000).

Using the current theories of stellar evolutionary tracks and hot stellar model atmospheres it is difficult to reproduce the measured line ratios. However, calculations were performed (Lutz *et al.* 1996, and Lutz private communication) for a stellar cluster with initial parameters as suggested by the observations: e.g. for solar metalicity, a star formation time scale of 1 Myr, an upper mass cut-off of 100 $M_\odot$, an ionization parameter of $log\ U = -1$, and a density of 3000 cm^{-3} as derived in Lutz *et al.* (1996). Assuming an age of the most recent starburst of 7 Myr, as suggested by the stellar census, delivers a [NeIII]/[NeII] line ratio of 1 to 2, which is well above the observed extinction corrected value of 0.05, a discrepancy which could arise due to problems with the model atmospheres, as well as with the evolutionary tracks. Already the 7 most luminous and hottest stars (Najarro *et al.* 1997) with $T < 10^{4.5}$ K and $L > 10^{5.75}$ $L_\odot$ provide half of the ionizing luminosity of the central stellar cluster, which is in contrast to current stellar evolutionary tracks, which predict less than 1% of the ionizing luminosity for the same temperature and luminosity range. This result remains unchanged when using tracks with twice the solar metalicity and higher mass loss. As a result, the Galactic Center observations suggest to revise the current evolutionary tracks in the direction of a more pronounced B supergiant phase.

2.11.2 *The Upper Mass Cut-Off*

A critical parameter in the description of the evolution of stellar clusters is the upper cut-off in their mass function. Binary stars are ideal for the determination of stellar masses, and eclipsing binaries can be studied even in more distant stellar systems, since they can be identified quite easily via their light curves derived from variability studies. For the Galactic Center such a study was performed doing aperture photometry on deconvolved images using three different deconvolution methods, and the results are presented in detail in Ott *et al.* (1999). With the time coverage provided by the multi-epoch Galactic Center observations, the authors were able to show that IRS 16SW is an eclipsing binary with a period of 9.72 days.

In the top section of Fig. 2.19, we show the raw light curve as deduced from the SHARP/NTT speckle data. Since the longest observing run in

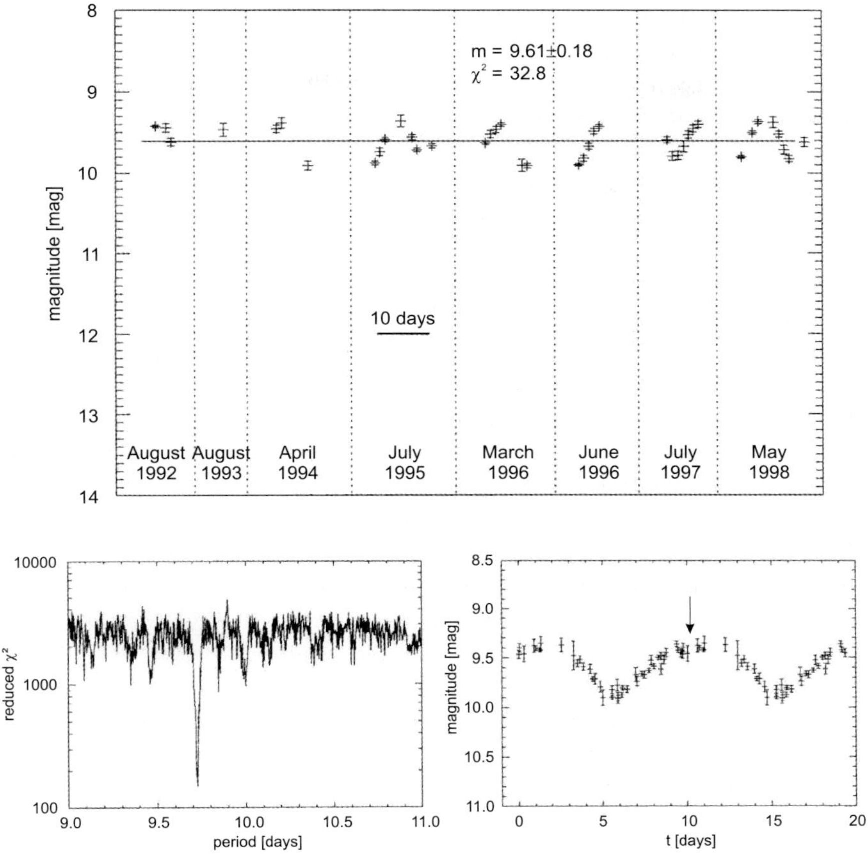

Fig. 2.19 Light curve of IRS 16SW and determination of its period as shown by Ott *et al.* (1999). Top: raw light curve. Bottom left: result of a reduced χ^2 test used to determine the period of 9.72 days. Bottom right: data folded back into a two-period interval. The arrow indicates the possible secondary minimum.

May 1998 was just slightly longer than the period of IRS 16SW, Ott *et al.* (1999) performed a reduced χ^2-test of 1000 periods in the range between 9 and 11 days in order to get a plot of a single period. In the reduced χ^2-test, the absolute minimum corresponds to the period that fits best to the observed data.

The symmetrical shape of the light curve and a hint for a second local minimum suggest that this star is indeed an eclipsing binary. A more detailed discussion on the nature of IRS 16SW is given in Ott *et al.* (1999).

They give an estimation of the mass of IRS 16SW, based on the only assumption that it is a semi-detached β Lyrae system, as it is suggested by the shape of its light curve. Kepler's third law relates the period P, the separation r and the sum of the masses $M_1 + M_2$ of a 2-body system independent of its inclination and ellipticity: $P^2 = 4\pi^2 r^3 G^{-1}(M_1 + M_2)^{-1}$, with G denoting the gravitational constant. Solving for the total mass, we can give a relation between the separation of the components and the mass M of the system, which yields $M = (M_1 + M_2) = \frac{4\pi^2 \cdot r^3}{G \cdot P^2}$. Furthermore, we can give a lower limit to the mass of IRS 16SW when considering the minimum separation, which can be derived in the following way.

The radius of IRS 16SW was estimated by model calculations to be of the order of $\sim 90\ R_\odot$ (Najarro *et al.* 1997). Assuming that the total luminosity originates predominantly from the main component, this would be the minimum separation possible for an eclipsing binary. Inserting this value into the equation above then gives a lower limit to the total mass of $M = 104\ M_\odot$.

Ott *et al.* (1999) calculate some useful limits on the involved quantities by assuming the extreme case that the luminosity is distributed evenly between the two components of the binary star. In that case the stars would have radii of $\sim 65\ \mathrm{R}_\odot$, with a resulting separation of at least $130\ \mathrm{R}_\odot$, and a total mass of the system of $300\ \mathrm{M}_\odot$. The upper limit to the orbital velocity in the system would be 677 km/sec in this scenario. In order to spectroscopically resolve the resulting Doppler-shift in an emission line, a minimum spectral resolution of $\lambda\ /\ \Delta\lambda$ of ~ 1000 is required.

2.11.3 *The Central Arcsecond*

The high velocity stars in the central cusp around the supermassive black hole are of special interest, since the motion of these inner stars allows one to measure the central mass and mass density with high accuracy. Since a significant fraction of the stars in the cusp appear to be early type stars (Genzel *et al.* 1997; Gezari *et al.* 2002; Genzel *et al.* 2003b; Ghez *et al.* 2003b, see also section 2.8.1), it is currently quite unclear how these objects can be present there, how the stellar population in the cusp evolves, and how it is linked to the population at larger radii. Observationally it is very difficult to obtain high quality high resolution spectra of these sources mainly because of two reasons: at first, they are all located within 1" or 2" from the center, so that high angular resolution techniques are needed to separate them, and as a second problem the gas of the mini-spiral emits contaminating Hydrogen and Helium recombination lines.

In the following sections we outline the observational efforts and results that have lead to the first spectra of the brightest stars in the central cluster.

2.11.4 *Results of Speckle Spectroscopy*

The small Sgr A* cluster contains about half a dozen stars brighter than $K = 14.0^m$ and about a full dozen of sources brighter than $K = 14.5^m$ in an area of about 1" to 2" diameter. This cusp is located only about 2" east of the bright $K \leq 9 - 10^m$ members of the IRS 16 complex, which are about 100 times brighter than the stars close to Sgr A*. Despite of that complication early low resolution spectra were obtained with the SHARP/NTT camera and resulted in a valuable estimate of the spectral type of the high velocity sources, information which is essential to determine the masses of these stars and derive a lower limit to the compact dark mass at the dynamical center. The spectroscopic speckle imaging technique, which was used to obtain dispersed stellar spectra at the diffraction limit of the ESO New Technology Telescope on La Silla, involved the use of the speckle camera SHARP, a low-dispersion objective prism, broad, and narrow band filters and allowed to obtain low-resolution spectra of stars brighter than $K \approx 14.5$ over the entire field-of-view of the imaging system. The technique is described in detail in section 1.3.5 and in Genzel *et al.* (1997). In the following sections we will summarize the process of how the spectra where extracted from the objective prism images.

2.11.4.1 *Extraction and Calibration of Spectra*

With the exact relative positions of the stars known from high-resolution imaging, the individual spectra were extracted from the spectrally dispersed images. The relative spectral calibration was done by dividing the extracted spectrum of any star with the raw spectrum of IRS 29N, and multiplying the result with the calibrated (and convolved with the spectral resolution of the speckle spectroscopy data) spectrum of IRS 29N, a well suited source, which shows an almost featureless spectrum at a resolution of $\lambda / \Delta\lambda \sim 800-1000$ (Krabbe *et al.* 1995; Genzel *et al.* 1996). The absolute wavelength calibration was obtained by using the sharp and well defined intensity cutoffs of the K-band filter edges. Due to the almost threefold oversampling of the diffraction limit with the SHARP/NTT camera the accuracy of the wavelength calibration was about three times better than the $\lambda / \Delta\lambda = 35$

spectral resolution of the data. Genzel *et al.* (1997) demonstrated with the spectra of the featureless source IRS 3, of a typical late-type star, IRS 13W, and of two typical HeI emission line stars, IRS 16SW and IRS 16NW, as obtained by speckle spectroscopy, that the overall spectral shape, as well as the prominent CO band head absorption feature, can clearly be discerned in the low resolution spectra. The CO band head feature was also used to verify the wavelength calibration. The narrower He I emission line, however, was too much diluted for reliable detection at the spectral resolution of the speckle spectroscopy data. The detailed procedure of how the spectra of the stars S1, S2, S8 and S11 within the small Sgr A* cluster were extracted is given in Genzel *et al.* (1997).

2.11.4.2 *Spectra in the Central Sgr A* Cluster*

Since the GC stellar cluster is very crowded, most of the spectra are contaminated by other nearby objects. The central Sgr A* cluster, however, is located west of the IRS 16 complex in a region devoid of bright stars, with the separation to any bright star being larger than 1" and the contaminating effects of neighboring sources therefore being comparatively small. About half of the Sgr A* cluster sources were distributed in north–south direction at the time of the speckle spectroscopy experiment, a coincidence which could be exploited to extract low resolution spectra of individual stars in this region.

The resulting spectra of the sources S1, S2, S8 and S11 as obtained with the extraction techniques described in Genzel *et al.* (1997) are fairly flat and increase slowly in flux toward longer wavelength. Since CO bandhead absorption is absent, S1, S2, S8 and S11 cannot be late type giant stars. While appearing somewhat redder than the most prominent He I emission line stars in the IRS 16 complex, their spectral characteristics are consistent with early type stars located in the central cluster. There is also no indication for very strong line emission (or absorption) at the wavelengths of the He I and/or $Br\gamma$ emission lines, which is not surprising, since the narrow He I/Brγ emission lines cannot be observed even in the classical He I emission line stars at this low spectral resolution.

The obtained results are fully consistent with broad band JHK colors and narrow band filter data and were confirmed later by seeing limited spectroscopy using the VLT (see Fig. 2.20 and Eckart *et al.* 1999a) and Keck telescopes (Figer *et al.* 2000). Genzel *et al.* (1997) thus concluded that the $m_K \sim 14.5$ sources in the central Sgr A* cluster are most likely

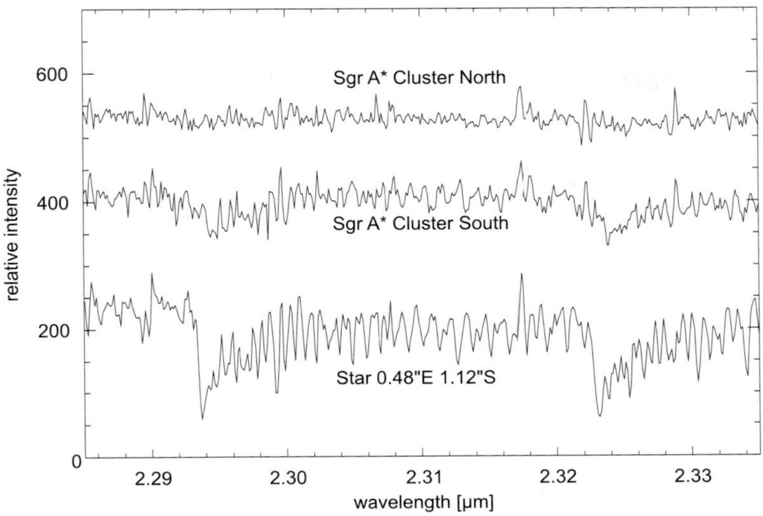

Fig. 2.20 Spectra of the northern and southern part of the Sgr A* cluster as well as the late type star 1.12" south of the center as shown by Eckart *et al.* (1999a). No CO(2-0) 2.293 μm and CO(3-1) 2.323 μm bandhead absorption is measured towards the northern part containing the fast moving sources S1 and S2. The weak absorption features towards the south could be due to contaminating flux density contributions from sources other than S9, S10 and S11. The spectrum of the southern part has been shifted down by 150 units; the spectrum of the late type star was shifted down by 300 units.

moderately luminous ($L \sim$ 5,000 to 10,000 $L_\odot$) early type stars. If these sources are on the main sequence they have to be O9 - B0.5 stars with masses of 15 to 20 $M_\odot$ and ages of less than 20 Myrs.

2.11.5 *Adaptive Optics Spectroscopy of the Central Cusp*

Gezari *et al.* (2002) presented K-band λ / $\Delta\lambda$ $\sim$ 2600 spectroscopy of five stars ($K = 14 - 16^m$) within 0.5" of Sgr A*, applying, for the first time, adaptive optics in combination with spectroscopy on this target. The high spatial resolution of $\sim$ 0.09" and good Strehl ratios of $\sim$ 0.2 achieved with the AO system on the 10 m Keck telescope allowed them, for the first time, to measure individual moderate-resolution spectra of these stars. Two stars (S0-17 and S0-18) were identified as late-type stars by the detection of CO band-head absorption, with the respective absolute K-band magnitudes and CO band-head absorption strengths being consistent with early K giants. Three stars S1 (alias S0-1), S2 (alias S0-2), and S0-16 (alias S14) with

$r_{\rm proj} < 0.0075\ pc$ (~ 0.2") from Sgr A* lack CO band-head absorption, confirming the earlier result of speckle spectroscopy with SHARP/NTT and the seeing limited observations of Eckart *et al.* (1999a) and Figer *et al.* (2000) (Fig. 2.20).

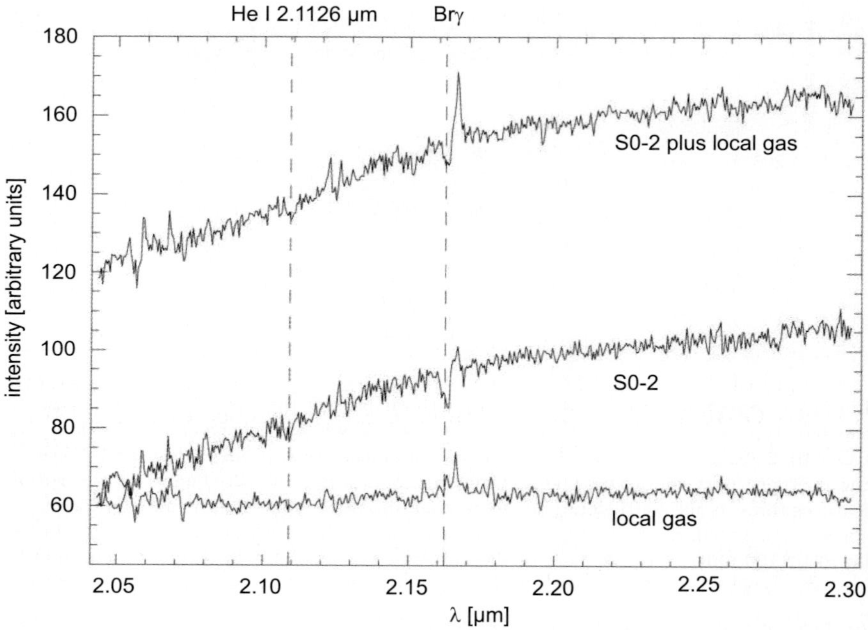

Fig. 2.21 This spectrum was obtained in June 2000 and is shown in Ghez *et al.* (2003b). It is the first spectrum of S0-2 to show detectable photospheric absorption lines (Brγ and He I) and identifies this star as a $\sim$09 main sequence star. The final spectrum (middle) is the raw spectrum (top; with only an instrumental background removed) minus a local sky (bottom). The horizontal dimension was re-binned by a factor of 2 for display purposes only. The vertical lines are drawn at 2.10899 and 2.16240 μm, which correspond to the locations of Brγ and He I for a v_{LSR} of -513 km s^{-1}.

Ghez *et al.* (2003a) reported the first detection of spectral absorption lines in one of the high velocity stars in the vicinity of the Galaxy's central supermassive black hole. Their observations were carried out using the AO system at the 10 m Keck telescope (Wizinowich *et al.* 2000a) and the Near-Infrared Camera 2 (NIRC2, e.g. McLean & Sprayberry 2003b). The AO loop was locked on a $R = 13.2^m$ natural guide star (USNO 0600-28579500; star No.2 in Biretta *et al.* (1982)) located about 30" NNE of S2.

The achieved spectral resolution was R = 4000 (75 km/s) over a spectral range of 0.3 μm, and the accuracy of the wavelength calibration was 9 km/s. Seven 20 minute exposures were obtained with star S2 on the slit of the spectrometer. After calibration of the spectra, a local background near S2 was subtracted in order to remove emission from diffuse gas in the mini-spiral around S2 and possible contamination from the wings of the bright nearby star IRS16C. In the completely calibrated, background subtracted spectrum of S2, $Br\ \gamma$ (2.1661 μm) as well as He I (2.1126 μm) are visible (Fig. 2.21) in absorption with equivalent widths of 0.28 $\pm$ 0.03 nm and 0.17 $\pm$ 0.04 nm. The inferred stellar rotational velocity of 220 $\pm$ 30 km/s is consistent with the value expected for an O8-B0 dwarf. This finding suggests that the star S2 is indeed a massive 15 $M_{\odot}$ young main-sequence star with an age of less than 10 Myr. The radial velocity of S2 in June 2002 was measured to be $v_{LSR} = -510\ \pm\ 40$ km/s (LSR: Local Standard of Rest).

Eisenhauer *et al.* (2003b) obtained spectra of the star S2 in April, May, and June of 2003 with the adaptive optics assisted, near-infrared camera NAOS/CONICA and the near-infrared integral field spectrometer SPIFFI (Tecza *et al.* 1998; Eisenhauer *et al.* 2000) at the ESO VLT. For the SPIFFI observations the spectral resolution was 85 km/s, sampled at 34 km/s, the accuracy of the wavelength calibration was $\pm$ 7 km/s, the effective integration time toward the central part of the mosaic near Sgr A* was about 15 minutes, and the FWHM spatial resolution at 2 μm ranged from 0.25" to 0.3".

For the NAOS/CONICA long slit exposures Eisenhauer *et al.* (2003b) used the infrared wave-front sensor on IRS 7 and a 86 mas slit, resulting in a resolution of 210 km/s sampled at 69 km/s, a spatial pixel scale of 27 mas, and an accuracy of the wavelength calibration of $\pm$ 10 km/s. Here 30 minutes of on-source integration were accumulated. For the 2003 observations the contaminating extended nebular $Br\gamma$ emission from the Sgr A West mini-spiral was far off the stellar absorption line in star S2 (Fig. 2.22). In April the $Br\gamma$ absorption line of S2 was at a velocity with respect to the Local Standard of Rest of $v_{LSR} = -1558\ \pm\ 20$ km/s, and the NAOS/CONICA observations of S2 gave LSR velocities of the $Br\gamma$ absorption line of $-1512\ \pm$ 35 km/s for May and $-1428\ \pm$ 45 km/s for June 2003.

The radial velocity measurements resolved the ambiguity in the inclination of the S2 orbit. In combination with the proper motions of S2 they provided a measure of the black hole mass of $(3.6 \pm 0.6) \times 10^6\ M_{\odot}$ and, for

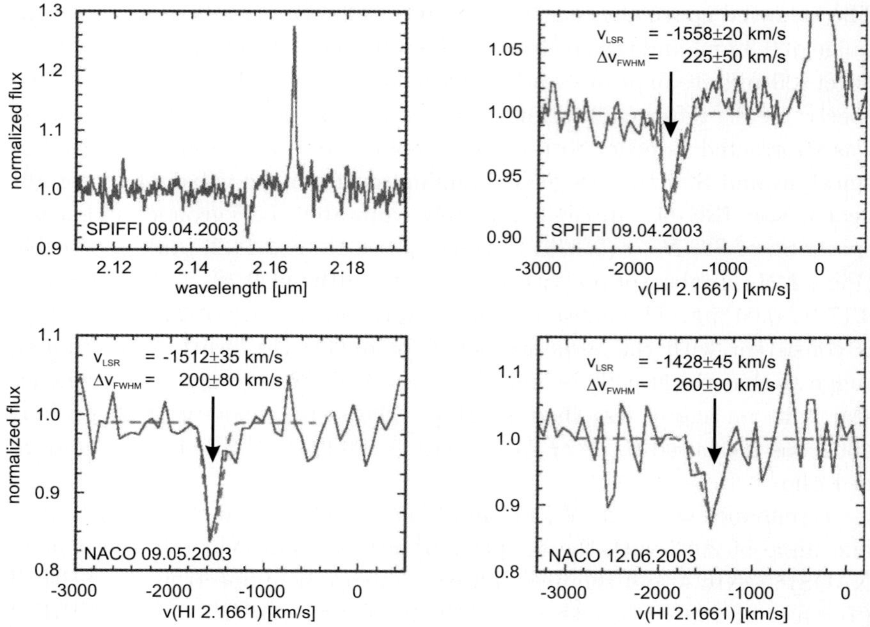

Fig. 2.22 HI Brγ absorption spectra of S2 as shown by Eisenhauer *et al.* (2003b) (for details see there). The data was obtained on 2003 April 8/9 with the SPIFFI integral-field spectrometer (upper two panels), on May 8/9 (lower left panel), and on June 11/12 (lower right panel) with NACO at the ESO VLT.

the first time, a direct determination of the distance to the Galactic Center black hole of $R_0 = 7.9 \pm 0.4$ kpc (Fig. 2.22; see Eisenhauer *et al.* 2003b).

2.12 Star Formation at the Galactic Center?

The luminosity of the entire central parsec of our Galaxy is dominated by a cluster of about two dozen massive stars. The 7 most luminous stars ($L > 10^{5.75}$ L$_\odot$) are moderately hot ($T < 10^{4.5}$ K) blue supergiants and provide half of the ionizing luminosity of that region (Blum *et al.* 1995; Krabbe *et al.* 1995; Najarro *et al.* 1997). Due to their strong 2.06 μm Helium line emission these objects are often referred to as HeI-stars (or briefly He-stars) too.

Such massive, young stars are not exclusively found in the central parsec, but were also discovered in dense stellar clusters in the Galactic Bulge within about 30 pc of the center, for instance the Arches cluster (Cotera *et al.* 1999b; Figer *et al.* 2002, and references therein) and the Quintuplet cluster (e.g. Figer *et al.* 1999b).

In contrast to the Arches and Quintuplet clusters, however, the high number of apparently young, massive, and hot He-stars in the central parsec, where strong tidal forces, magnetic fields and winds pose a serious obstacle for star formation processes, can currently not be explained by normal stellar evolution theories (e.g. Eckart *et al.* 1999b; Thornley *et al.* 2000). In fact, the existence and large number of these stars has been used as an argument against the existence of a supermassive black hole in the GC (Sanders 1992). Their existence and nature are currently one of the greatest enigmas in GC research (see also the discussions on this subject in, e.g. Morris 1993; Morris *et al.* 1999).

There is additional evidence for the presence of early-type stars in the central cluster: the general observation is that in spectroscopic observations there is a drop in the strength of the CO bandhead absorption feature within the central 5-10 arcsec, which is an indication for a reduced percentage of late-type stars and an increased presence of early-type stars in this part of the central cluster. The possible destruction of the envelopes of the brightest late-type giants and supergiants in the central region is discussed, e.g. by Lacy *et al.* (1982), Phinney (1989), Sellgren *et al.* (1990); Haller *et al.* (1996); Genzel *et al.* (1996), and Alexander (1999). An alternative view is that the CO feature becomes diluted towards the center because of the presence of the bright, early-type He-stars (e.g. Sellgren *et al.* 1990; Eckart *et al.* 1995; Krabbe *et al.* 1995; Genzel *et al.* 1996; Haller *et al.* 1996).

In a more recent analysis of the stellar population and evolution in the environment of the central stellar cluster, using stellar number counts and a photometric CO index, Genzel *et al.* (2003b) concluded that the percentage of stars with a CO absorption feature does indeed decrease toward Sgr A*. In combination with the spectroscopic evidence that shows that the stars in the immediate environment of Sgr A*, i.e. in the so-called Sgr A* cluster, might be O/B-type stars, there is hence considerable evidence for the presence of young stars in the central few arcseconds. Their nature is not undisputed, however, because they could also be exotic objects that just appear to be young stars. The issues raised here will be discussed extensively in chapter 3 (see section 3.2).

If the potential O/B-type stars and the He-stars are indeed normal stars and not exotic objects that just appear to be young stars, then they must have formed $< 10^7$ years ago. If star formation is still an ongoing process in the central cluster, then one might expect an even younger population of stars to be present in that region. These very young stars could then still be embedded in the cocoon of dense gas and dust they were formed from.

2.12.1 *Embedded Objects*

In the Galactic Center region there is a population of stars like IRS 21, IRS 1W, IRS 10W, IRS 3, IRS 5, etc., which are very red, dust embedded sources with featureless K-band spectra, and which until recently were the prime candidates for being freshly formed stars. These objects are bright 10 μm sources and are mostly associated with the mini-spiral gas streamers, particularly the northern arm (e.g. Gezari *et al.* 1994; Genzel *et al.* 1996; Tanner *et al.* 2003).

The best studied case so far is IRS 21, which is strongly polarized ($\sim$ 17% at 2 μm; see Eckart *et al.* 1995; Ott *et al.* 1999). Initially, Gezari *et al.* (1985) suggested that IRS 21 was an externally heated, high-density dust clump, but several other classifications were proposed as well, which are able to explain the MIR excess and the featureless NIR spectra, too. Among those theories is an embedded early-type star or a protostar (Krabbe *et al.* 1995; Genzel *et al.* 1996; Clénet *et al.* 2001), while Tanner *et al.* (2002) suggested more recently that IRS 21 is an optically thick dust shell surrounding a mass-losing source, like a dusty and only recently formed WC9 Wolf-Rayet star.

Gezari *et al.* (1985) found IRS 3 to be the most compact and (together with IRS 7) hottest MIR source (T$\sim$400 K) in the central cluster with total integrated flux densities of 32 Jy and 33 Jy at 12.8 μm and 8.3 μm, respectively. With respect to the ISAAC L-band image the photo-center of the M-band image is shifted by $\sim$80mas to the NW. High resolution L- and M-band NAOS/CONICA images show to the SW a sharp interaction zone of the outer part of the dust shell with a wind arising from the direction of the IRS16 cluster. In the late '70s Rieke *et al.* (1978); Becklin *et al.* (1978) argued that IRS 3 is a dust-enshrouded supergiant with a compact circumstellar dust shell. As comparatively young and massive stars are abundant within the central few arcseconds it is, however, also likely that IRS 3 is a massive, luminous and dusty Wolf-Rayet star with strong winds that interact with the GC ISM. This is supported by recent imaging spectroscopy with SPIFFI by Horrobin *et al.* (2004). This would be consistent with the recent VLTI detection of a compact object that contains approximately 30% of the flux density seen in the 8–12 μm MIDI band pass (Eckart, Pott et al., in preparation; see also section 1.6.1).

With the new results from high angular resolution imaging, it has now become very likely that most of the dust embedded sources are in fact associated with luminous stars that closely interact with the gas and dust of the mini-spiral. Cotera *et al.* (1999b) showed that for several of these sources one finds an angular offset between their astrometric position and nearby local maxima in the dust emission and temperature distribution. In the case of IRS 21 the extended dust emission is consistent with a bow shock created by the motion of a hot star through the material of the mini-spiral (Tanner *et al.* 2002). In fact, the bow shock scenario may be applicable to most of the luminous dust embedded sources in the central parsec of the Milky Way. In support of this idea Tanner *et al.* (2003) presented observational evidence that all embedded sources in the northern arm are associated with croissant shaped structures. This picture is consistent with the idea that they are mass-losing young, massive stars, such as those in the IRS 16 complex, that plow through the interstellar gas and dust of the mini-spiral (see Fig. 2.23). Additionally, Rigaut *et al.* (2003) discovered that IRS 8, located in the northern part of the northern arm, is associated with a bow shock structure.

Therefore the interpretation of the bright, extended, dust enshrouded objects in the central parsec as young stellar objects (YSOs) has become very unlikely.

2.12.2 *IR-Excess Objects*

The IRS 13 complex is a small, luminous cluster of stars with a diameter of about 1″ located about 3.5" SSW of Sgr A*, which has recently been resolved into at least 4 bright infrared components (Paumard *et al.* 2001; Maillard *et al.* 2003; Eckart *et al.* 2004b), and which contains one of the brightest hot stars in the entire central parsec (Najarro *et al.* 1997). This complex is associated with compact radio emission, too, that was mapped at 7 mm and 13 mm wavelengths by Zhao & Goss (1998). Eckart *et al.* (2004b) reported the discovery of a small cluster of infrared excess sources located about 0.5" North of IRS 13, and one of several possible explanations for these stars is that they might be young, dusty, embedded stars. This discovery was only possible with the high resolution adaptive optics images obtained with NAOS/CONICA at wavelengths between 1.7 and 3.8 μm.

A thorough comparison of the H-, K-, and L-band images, which are of unprecedented sharpness and sensitivity, reveals a small cluster of pre-

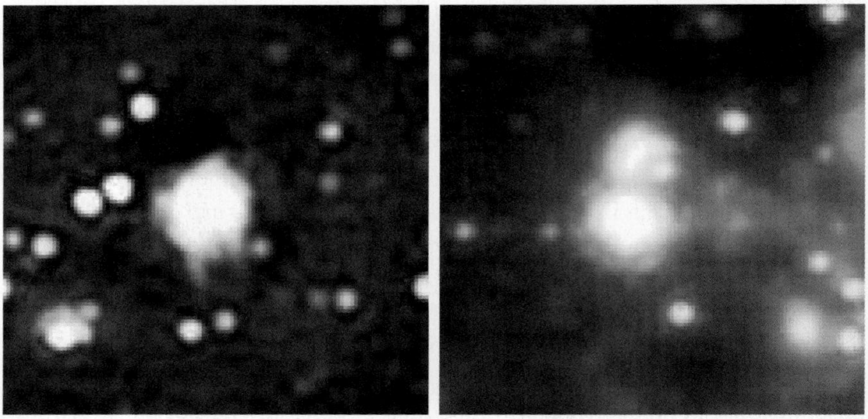

Fig. 2.23 Bow shock structures in the GC stellar cluster. Due to the interaction of windy stars with the interstellar medium in the GC, numerous examples of bow shock sources can be found in the central parsec. This figure shows two exemplary shocks with the convex part pointing towards the NNE and NE, respectively. Left: IRS 1W at 2.2 μm (NACO/VLT May 2003; deconvolved and restored to 60 mas FWHM); the area shown covers 1.6" $\times$ 1.6". Right: bow shock north of IRS 9 (the bright source in the center) at 3.8 μm (NACO/VLT August 2002); the area shown covers 3.2" $\times$ 3.2". See also color Fig. A.5.

viously unknown weak objects, which are labeled α through η in Fig.2.24. Eckart *et al.* (2004b) named this newly found complex of sources IRS 13N — indicating that it is north of IRS 13.

The sources are 10-15 times less luminous than the dusty source IRS 21. The overall IRS 13N cluster shows a diameter of $<$ 0.13 light years, and the individual sources are compact with angular sizes of less than 0.09" (740 AU). The sources are well embedded in the bulk of the ionized gas, which is evident in radio images of the mini-cavity and the bar region, as presented by Yusef-Zadeh *et al.* (1998b) (at λ = 2 cm) and Zhao & Goss (1998) (at λ = 13 mm). Stellar proper motions (Genzel *et al.* 2003b; Maillard *et al.* 2003) also provide evidence that the entire IRS 13 / IRS 13N complex is moving together with the gas of the mini-spiral, and indicate an orbital time scale of the order of $\sim 10^4$ years for the combined motion of the entire complex and the mini-spiral gas.

Fig. 2.25 shows a map of the IRS 13 complex in a 1.5" $\times$ 1$''$ section including the sources in IRS 13N. The entire region is very crowded and the source number densities are of the order of 40 to 80 objects per square arcsecond. With the exception of the sources κ and η the objects of IRS 13N

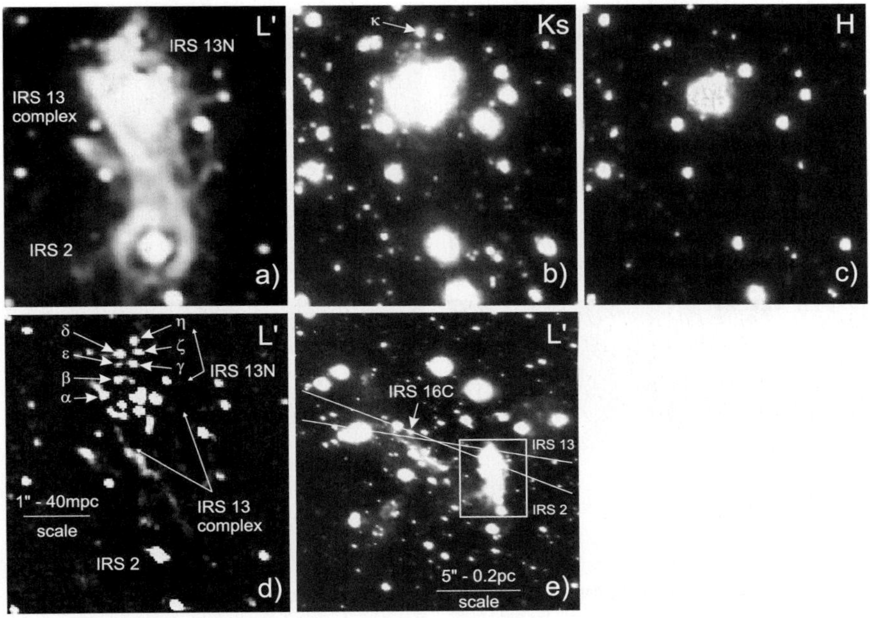

Fig. 2.24 H, K_s, and L' band images of the IRS 13/IRS 2 region as shown by Eckart *et al.* (2004b). The image scales of panels (a)-(d) are given in (d). Panel (d) is a high pass filtered L-image which shows the locations of individual stars, including the newly discovered L-band excess sources α through η. Source κ in panel (b) is located between ζ and δ. Panel (e) is an overview of the region; the box indicates the area shown in panels (a)-(d); slit locations for the spectra shown in Eckart *et al.* (2004a) are also indicated. See also color Fig. A.6.

are all very red (Fig. 2.26) with regard to other sources in the field (see K-L colors given by Clénet *et al.* 2001; Eckart *et al.* 2004b). The objects γ and ϵ are even so red that they could not be identified in the H-band and only lower limits for their H-K colors could be determined so far. The flux density determination of other objects suffers from the high density of stars in that cluster, for instance for source α, the flux density may be affected by a possible overlap with the nearby and bright E1 component of the IRS 13 complex. The blue source κ cannot be identified in the L-band image, since it probably blends with the brighter L-band source ζ, what allows to derive only an upper limit for the K-L color of κ.

A simple, but nevertheless very effective way of comparing source properties and analyzing their continuum spectra is to plot their brightness

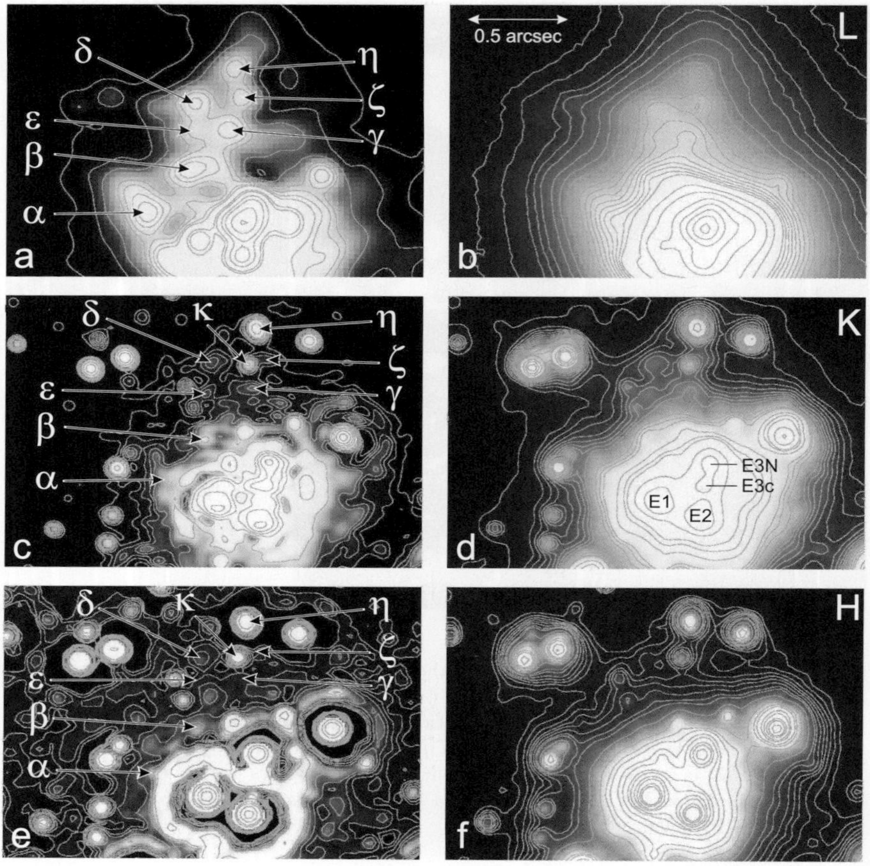

Fig. 2.25 A detailed view of the IRS 13 and IRS 13N complexes with NAOS/CONICA as shown by Eckart *et al.* (2004b). They are shown in the L-band, top images (a) and (b), K-band, center images (c) and (d), H-band, bottom images (e) and (f). The images (b), (d), and (f) are raw shift-and-add images of individual adaptive optics exposures. The images (a), (c), and (e) are deconvolved images restored at the nominal angular resolution of the AO system. The ring-like structures around bright sources are artifacts of the deconvolution process. For clarity in the crowded field we show contours overlaid with a color (grey-scale) version of the same image. The positions as well as magnitudes and colors of individual sources are given in Tab. 2 in Eckart *et al.* (2004a). See also color Fig. A.7.

differences between the various wavelengths in a two-color diagram, which allows one to determine the extinction, possible contribution of hot dust, and (in some cases) to get first information on the spectral type of the stars

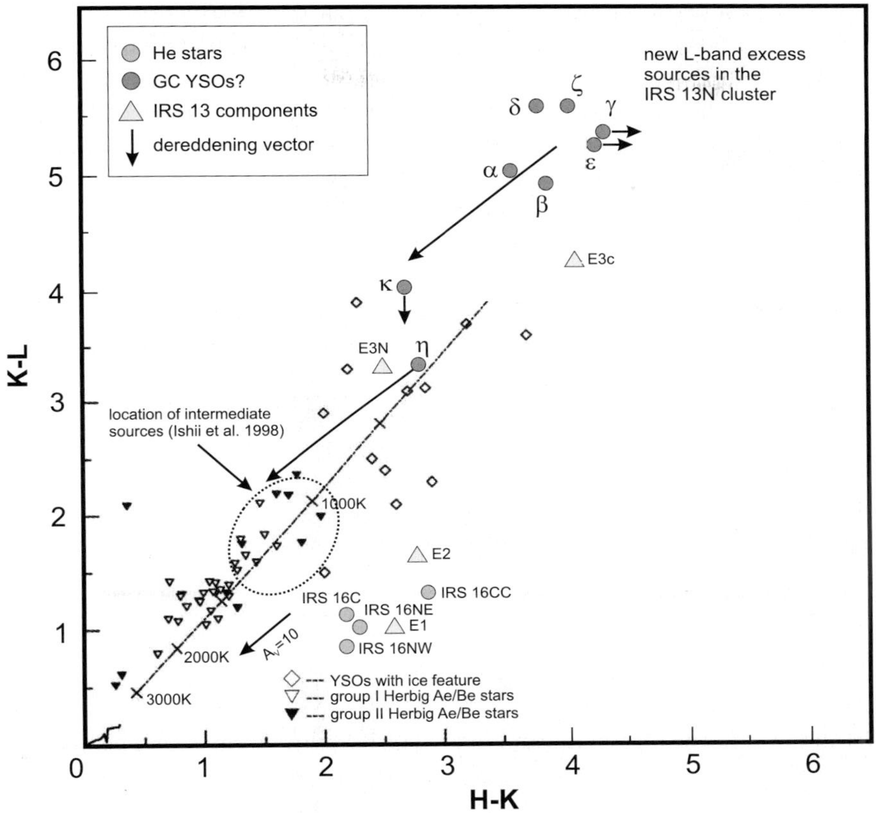

Fig. 2.26 HKL two-color diagram comparing the location of YSOs and Herbig Ae/Be stars to objects in the GC cluster as shown by Eckart *et al.* (2004b). Group I and group II Herbig Ae/Be stars are indicated. Dashed lines show the colors of a single blackbody at different temperatures. A solid line in the lower left corner of the diagram represents the colors of main-sequence stars (Koornneef 1983). Extinction vectors for visual extinctions of A_V=10 as well as A_V=30 are shown. The latter shift the location of the L-band excess sources to the location of YSOs or the intermediate population "N" and "E" sources introduced by Ishii *et al.* (1998). See also color Fig. A.8

(e.g. Glass 1999). In Fig. 2.26 we present an HKL two-color diagram showing the locations of YSOs and Herbig Ae/Be stars, as well as selected objects in the Galactic Center stellar cluster (see Eckart *et al.* 2004b). Sources like IRS 16NE, NW, C and CC are hot and dust free. They are well clustered in the two color diagram and, after taking into account the ≈30 magnitudes of visual extinction towards the Galactic Center, fall close to the location of hot and massive stars near the main sequence. This extinction correction delivers consistent results within the central few arcseconds – irrespective of the detailed position. Combined with the information that on scales of about 1" the extinction variations are small and smooth over that area (Scoville *et al.* 2003), this indicates that this correction can probably be applied to all sources within the central 10" to 20". Any residual extinction of individual sources may then be regarded as a source intrinsic effect and be attributed to individual sources (Moultaka *et al.* 2004).

The colors of the IRS 13N sources α through κ appear significantly redder with respect to the blue He-stars, and their broad band spectra are influenced by the combination of strong extinction and the emission of $T > 500$ K dust. The overall IRS 13N spectrum presented by Eckart *et al.* (2004b) is redder than that of the IRS 13 complex, and suggests that the strong L-band color excess is most likely due to continuum emission of warm $T > 500$ K dust. The IRS 13N sources may be low-luminosity counterparts of bow-shock stars such as IRS 21 (as discussed in section 2.12.1) or – a very intriguing possibility – dust-enshrouded young stellar objects (Eckart *et al.* 2004b).

2.13 Sgr A* Across the Electromagnetic Spectrum

In addition to the total mass and mass density of the compact object associated with Sgr A*, the characteristics of the electromagnetic radiation from that source give us direct information on the physical properties of the immediate surroundings (a few Schwarzschild radii) of the massive black hole at the center of the Milky Way. In the following we summarize some of the observational essentials that are known up to date on the electromagnetic spectrum we receive from that region. Here again high angular resolution is required at almost all frequencies to clearly distinguish between the contribution of the highly variable emission from Sgr A* and the surrounding stars and interstellar medium. The astrophysical implications of these facts will be summarized in chapter 3.

2.13.1 *Size and Spectrum of Sgr A* at Radio Wavelengths*

In Figures 2.27 and 2.28 we show the results of current measurements of the flux density and angular size of Sgr A* in the radio domain. Sgr A* was first identified as an unresolved point source by Balick & Brown (1974) with interferometric observations at 3 and 8 GHz. Early low frequency radio observations of Sgr A* at 408, 960, and 1660 MHz by Davies *et al.* (1976) delivered a non-detection of the source at 408 MHz, and the authors inferred a low-frequency cutoff in the spectrum of Sgr A*, which was considered to be due to free-free absorption in the Sgr A West HII region. Recent 610 MHz measurements by Roy & Pramesh Rao (2004) combined with a detection of Sgr A* at 330 MHz (Nord *et al.* 2004), however, indicate that there is little or no free-free absorption towards Sgr A*.

Davies *et al.* (1976) derived a value of $1.4''$ for the angular size of the source at 1 GHz , much larger than the upper limit of $0.1''$ given by Balick & Brown (1974) at 3 GHz. For comparison, the size of Sgr A* at 330 MHz is between $11''$ and $16''$ with an indication for an ellipticity of ~ 0.7 along a position angle of $35°$ (Nord *et al.* 2004). This decrease in diameter with decreasing observing wavelength can be explained well through the effect of interstellar scattering, a result which was confirmed by Backer (1978). The authors pointed out that all interferometric measurements of Sgr A* are affected by this scattering, which follows a λ^2 law and apparently occurs in the vicinity of the Galactic Center via scattering processes operating on kilometer scale lengths.

VLBI observations of Jauncey *et al.* (1989) showed that the radio image of the non-thermal source Sgr A* is elongated at 8.4 GHz along a position angle of $82 \pm 6°$ with an axial ratio of 0.53 ± 0.10 and a major axis of 17.4 ± 0.5 mas. They confirmed with a wide-band comparison of the angular sizes measured at frequencies from 1 to 22 GHz that the data follows very closely the λ^2 dependence expected from interstellar scattering. However, no similar counterpart to this smaller-scale elongation of the nuclear region was found on arcsecond scales with the VLA, nor are the position angles of Sgr A* aligned with the axis of galactic rotation. Jauncey *et al.* (1989) concluded that the elongated nature of the Sgr A* source structure implies either that the scattering medium is anisotropic, or that some remnant of the intrinsic structure remained visible through the scattering medium.

Marcaide *et al.* (1992) established an upper limit of about 8 mJy at a wavelength of 3.6 cm for any ultra-compact component and determined that the actual size of the source must be less than 15 AU. The first high dynamic range map of the source at 3.6 cm was obtained by Lo *et al.* (1993) based on VLBA measurements and confirmed the result, that at short cm radio wavelengths the (scatter-broadened) source structure of Sgr A* is smooth without detectable fine structure and is well described by an almost perfect

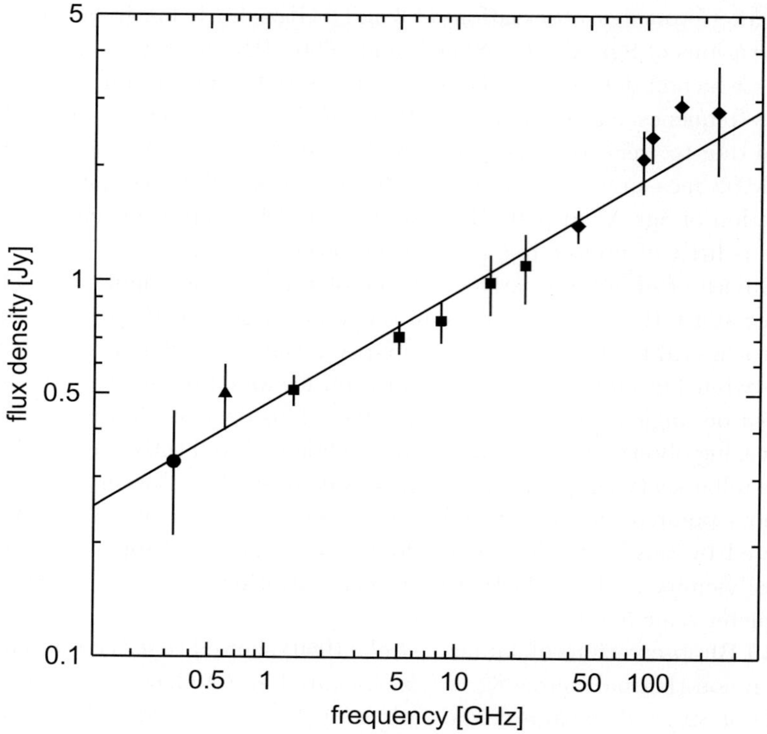

Fig. 2.27 The radio to sub-mm spectrum of Sgr A* ranging from 0.33 to 235 GHz as shown by Nord *et al.* (2004). Bars signify variability limits. The 610 MHz value is from Roy & Pramesh Rao (2003). The 1980-2000 time averaged 1.4 to 23 GHz data are from Zhao *et al.* (2001). The 43-235 GHz data are from Falcke *et al.* (1998) and cover the years 1987-1994. The solid line indicates a spectrum with a spectral index of 0.3.

elliptical Gaussian model (Fig. 2.29).

In the following years the imaging and determination of the source structure of Sgr A* proceeded towards the millimeter domain: Alberdi *et al.* (1993) published a 1.35 cm VLBI map, Krichbaum *et al.* (1993) reported first VLBI detections and source size measurements at 7 mm (43 GHz), 3 mm (86 GHz) maps were obtained by Rogers *et al.* (1994), Krichbaum *et al.* (1998) and Doeleman *et al.* (2001), and even measurements at 1 mm (215 GHz) were performed by Krichbaum *et al.* (1998) (for the radio spectrum of Sgr A* see Fig. 2.27).

The 86 GHz VLBI data of Doeleman *et al.* (2001) could be modeled best

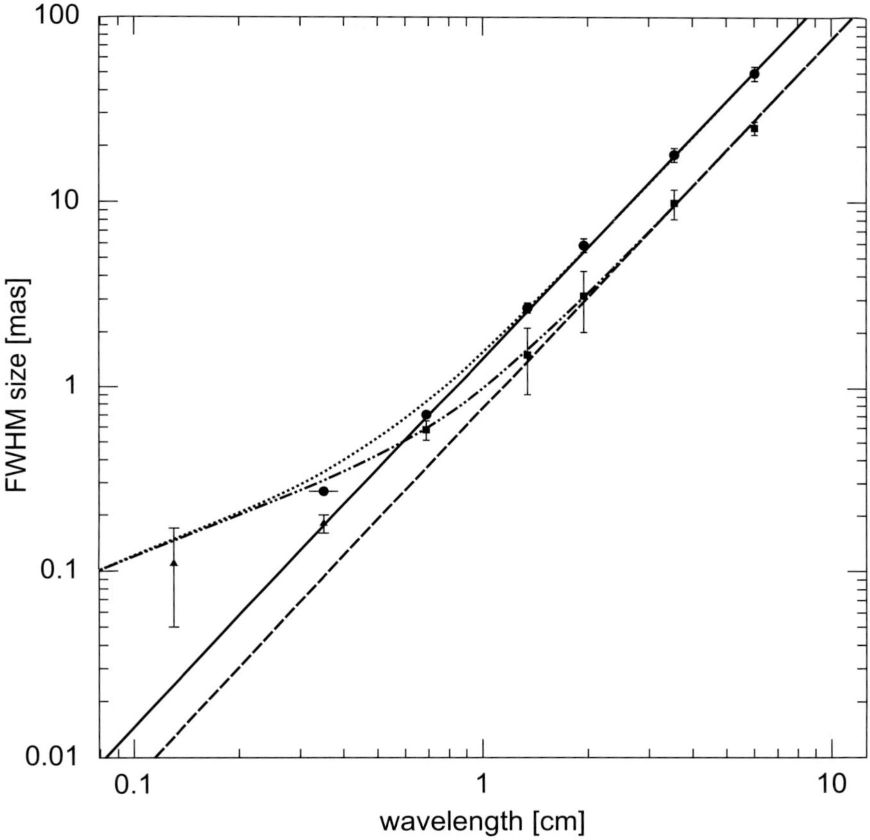

Fig. 2.28 Size measurements of Sgr A* from VLBI at multiple wavelengths as shown by Doeleman *et al.* (2001). Squares are minor-axis sizes and circles are major-axis sizes of best-fit elliptical models at wavelengths less than 3.5 mm from Lo *et al.* (1998). The green and red solid lines are λ^2 scattering relations given by Lo *et al.* (1998). Black triangles show the sizes of circular Gaussian models fit to the 3.5 mm data of Doeleman *et al.* (2001) and to 1.4 mm VLBI data (Krichbaum *et al.* 1998)). Hybrid ADAF model sizes from Özel *et al.* (2000) are added in quadrature to both the minor- and major-axis scattering sizes and are shown as the green, dash-dotted and red, fully dotted lines, respectively. The observed sizes predicted by the ADAF model exceed the 3.5 mm upper limit. The best constraints on the size of Sgr A* were recently presented by Bower *et al.* (2004), who were able, for the first time, to resolve Sgr A* unambiguously (see Fig. 3 in their paper). See also color Fig. A.9.

by a circular Gaussian brightness distribution of FWHM 0.18 $\pm$ 0.02 mas corresponding to a scattered point source. The angular resolution of the

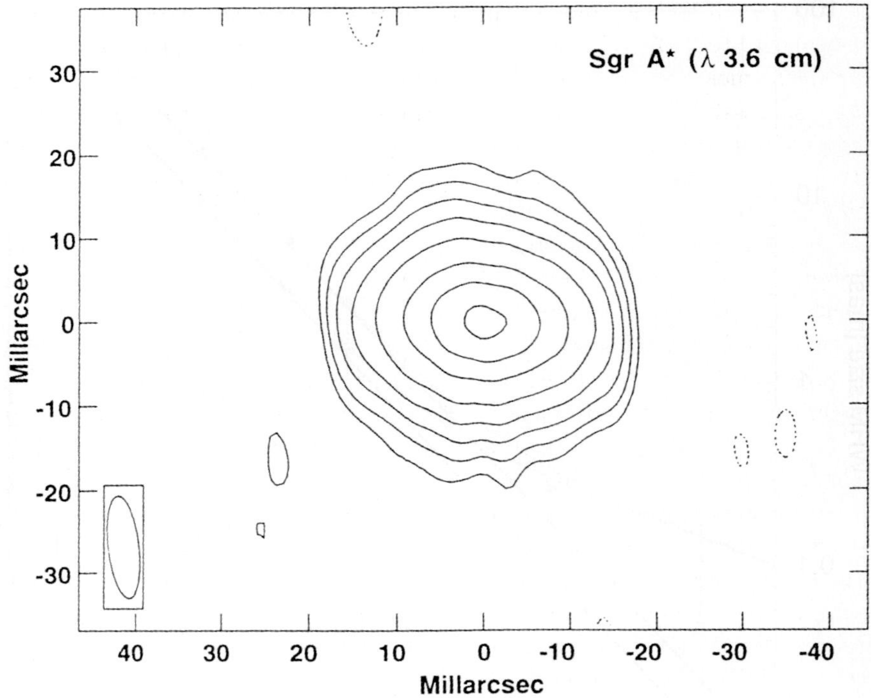

Fig. 2.29 Interferometric image of Sgr A* at a wavelength of 3.6 cm as shown by Lo *et al.* (1993). The beam size is 12.4 mas×6.6 mas. The contour levels are -2, 2, 4, 8, 16, 32, 64, 96, and 128 mJy/beam.

VLBI array is sufficient to constrain the intrinsic extension of the source at an position angle of $-10°$ to be less than 0.25 mas. These results compare well with an extrapolation of intrinsic structure estimates obtained at a wavelength of 7 mm , under the assumption that the intrinsic size of Sgr A* has a greater wavelength dependence than $\lambda^{0.9}$.

In March 1995 Krichbaum *et al.* (1998) detected Sgr A* with the VLBI at 86 GHz and at 215 GHz. At 86 GHz the measured closure phase is almost zero, which is consistent with a point-like or symmetric structure of 190 $\pm$ 30 μas in size. The 215 GHz data showed Sgr A* with a signal-to-noise ratio of 6, allowing to derive a tentative size estimate of 0.11 $\pm$ 0.60 mas, which corresponds to about 1 AU at a distance of 8 kpc and lies well above the scattering size of 0.20 mas at this frequency. Based on these values the derived brightness temperature of the emitting region is in excess of 10^{10} K, and the intrinsic size of Sgr A* appears to be of the order of 10 to 20 Schwarzschild radii of a 3 to 4×10^6 $M_\odot$ black hole.

In the most recent measurement of the size of Sgr A* Bower *et al.* (2004) were successful in determining its intrinsic size by combining 7 mm VLBI observations with the closure amplitude technique. They reported a size of 24 ± 2 Schwarzschild radii (assuming a mass of 4×10^6 $M_\odot$ for Sgr A*) and a $\sim$ 60% increase of the intrinsic size in one of their observing epochs.

To first order the radio spectrum of Sgr A* from centimeter to millimeter wavelengths is described by a power law, $S{\sim}\nu^\alpha$, with $\alpha = 0.2 - 0.35$. With decreasing wavelength this radio emission corresponds to progressively smaller spatial sizes (Melia *et al.* 1992; Melia 1992, 1994; Narayan *et al.* 1995; Falcke *et al.* 1998; Coker & Melia 2000; Liu & Melia 2001). The spectrum turns over in the sub-millimeter range. With an overall spectral index of about -1.3 the spectrum then passes through the near-infrared to the X-ray domain. The information on the local spectral indices in these wavelength regimes is currently uncertain. At shorter radio wavelengths — in the millimeter region, the spectrum shows a "bump" (Zylka *et al.* 1992, 1995). The presence of the spectral bump suggests that a distinct emission component surrounds the central black hole's event horizon. This broad band feature has been confirmed by simultaneous observations of the radio spectrum from the short decimeter (20 cm) to the millimeter (1 mm) region (Falcke *et al.* 1998).

Sgr A* also shows variability at radio frequencies. The cm/mm-band emission varies with an rms amplitude of ~1-20%, with occasional enhancements of a factor of ~2 occurring about 3 times per year (Falcke *et al.* 1998; Zhao *et al.* 2001). Zhao *et al.* (2001) report a 106-day quasi-periodicity at short centimeter wavelengths on the basis of an analysis of 20 years of VLA data. Overall the millimeter/sub-millimeter radiation of Sgr A* was found to be variable on time scales of several days to a few hundred days (e.g. Miyazaki *et al.* 2003; Zhao *et al.* 2003) but not on shorter time scales, with the exception of a single one hour, 30% amplitude spike seen in March 2000 at a wavelength of 2 mm (Miyazaki *et al.* 2003).

2.13.2 *Polarization of Sgr A* at Radio Wavelengths*

The observations show a complex picture of the polarization properties of Sgr A*. Between 1.4 and 112 GHz the emission shows no linear polarization, with upper limits of 0.1%-2% (Bower *et al.* 1999c,a, 2001). At frequencies of 150, 220, 375 and 400 GHz (2000, 1350, 850 and 750 μm) Aitken *et al.*

(2000) reported the first detection of about 10% of linear polarization from Sgr A* with the James Clerk Maxwell Telescope (JCMT) 14 m single dish antenna. However, these observations suffered from the poor angular resolution of the JCMT (22" at 220 GHz). From the lack of polarization at longer wavelengths, the authors concluded that the polarized flux density arises in a compact millimeter/sub-millimeter excess of Sgr A*.

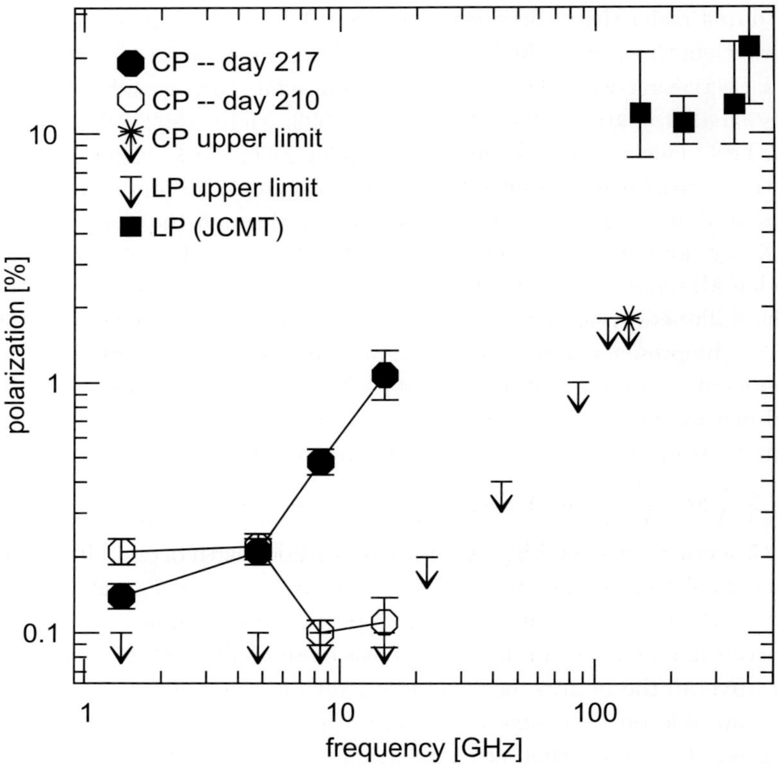

Fig. 2.30 Spectrum of circularly polarized (CP) flux density of Sgr A* from VLA and JCMT measurements as presented by Bower *et al.* (2002).

Subsequently, using the BIMA array with an angular resolution of 3.6"×0.9" Bower *et al.* (2003) measured the linear polarization of Sgr A* to be 7.2%±0.6% at 230 GHz and demonstrated that the polarized flux density is in fact associated with the compact synchrotron radio source and not with a more extended distribution of dust in that area. Combined with the

JCMT polarization measurements the authors derived a rotation measure of $RM = -4.3 \pm 0.1 \times 10^5$ rad m^{-2}. The rotation measure (RM) describes the change in position angle as a function of the observing frequency and is directly related to the electron density along the line of sight. The RM may entirely be caused by an external Faraday screen which is not associated with Sgr A*. However, if it is associated with it, the RM rules out accretion rates greater than $\sim 10^{-7}$ $M_\odot$ yr^{-1}. This is inconsistent with the high accretion rates being necessary in standard advection-dominated accretion flow and Bondi-Hoyle models for Sgr A* (see discussion in section 3.8.3 of this book). This supports the idea that a low accretion rate is an essential fact that may explain the overall faintness of Sgr A*.

Between 1.4 and 43 GHz up to 1% of the emission is circularly polarized, and the spectrum is variable and inverted (i.e. higher flux densities at high frequencies, see Fig. 2.30) with spectral indices of $\alpha \sim 0.5$ up to $\alpha \sim 1.5$ during emission flares (Bower *et al.* 1999b; Sault & Macquart 1999; Bower *et al.* 2002). Models that explain the presence of circular polarization and the absence of linear polarization involve the conversion of linear to circular polarization and a depolarization of the linearly polarized flux across the used bandwidths due to cold electrons (Beckert & Falcke 2002; Ruszkowski & Begelman 2002).

2.13.3 *Sgr A* as a Variable X-Ray Source*

The Galactic Center and Sgr A* have been major targets of all high energy satellites, although this area of the sky is crowded with compact components. Only recently the improved sensitivity and angular resolution of the launched instruments made it possible to identify an X-ray counterpart of Sgr A* and measure persistent X-ray emission from it. In autumn 1999 Chandra observations of the Galactic Center revealed a weak compact source at the position of Sgr A* with an extended X-ray luminosity of $L_{2-10keV} \sim 2 \times 10^{33}$ erg s^{-1} and 0.6 arcsec radius (Baganoff *et al.* 2003).

About one year later Chandra detected for the first time a strong emission flare of Sgr A*, an event which lasted for 2.8 hours and showed an increase in luminosity by a factor of about 50 above the quiescent level, with a peak value of 10^{35} erg s^{-1} (Baganoff *et al.* 2001). In its flaring state Sgr A* appeared to be a point source (Baganoff *et al.* 2001, 2003), and the light curve showed significant variations, including a sudden drop in flux density over a time span of only 10 minutes.

This variation allows constraining the size of the emitting region to less

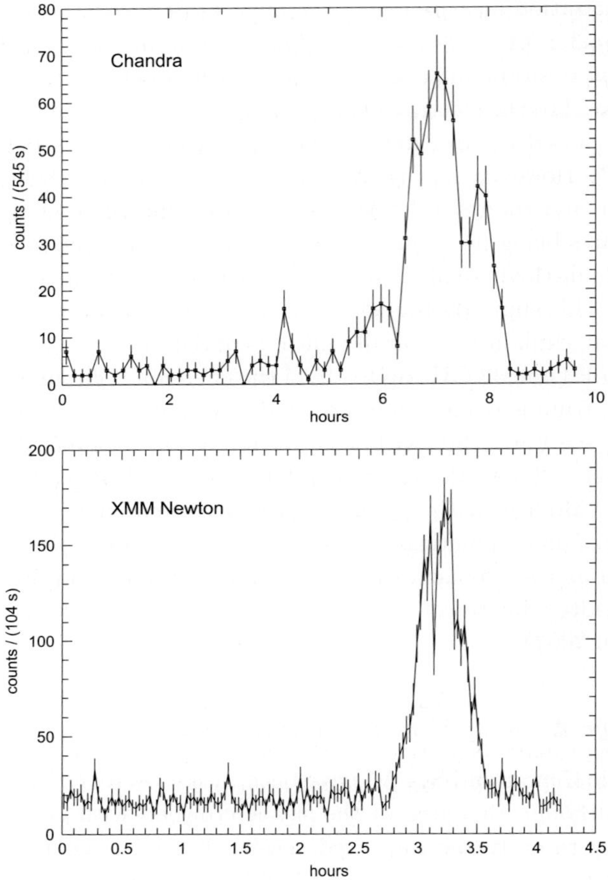

Fig. 2.31 X-ray flares from Sgr A* as shown in Aschenbach *et al.* (2004) (see also Baganoff *et al.* 2001; Porquet *et al.* 2003): ACIS-I light curve of the *Chandra* observation of October 26, 2000 (upper panel) and EPIC light curve of the *XMM-Newton* observation of October 3, 2002 (lower panel).

than about 20 Schwarzschild radii (see e.g. Baganoff *et al.* 2001; Porquet *et al.* 2003) and therefore to the immediate vicinity of the central supermassive black hole. Furthermore, Baganoff *et al.* (2001) showed that the flare spectrum is harder in comparison to the quiescent state, with a slope of the power law photon index of about 2.7 to 1.3 during the flare event.

In 2001 XMM-Newton detected the onset of a similar flare at the end of an observation of the Galactic Center (Goldwurm *et al.* 2003). In this case the flux density of Sgr A* increased by at least a factor of 30 over the quiescent flux density with an estimated peak luminosity of $L_{2-10keV} \sim 6 \times 10^{34}$ $erg\ s^{-1}$. Similar to the Chandra measurements the spectrum hardened during the flare event with a photon index of ~ 1.

A monitoring program of Sgr A* with Chandra in 2002 resulted in a first estimate of the rate of strong flares (with an increase in flux density by more than a factor of 10 above the quiescent state) of 1.2 ± 0.6 per day (Baganoff *et al.* 2003) with indications for even more numerous weaker flares. This shows that the X-ray counterpart of Sgr A* is in fact a highly variable source with powerful emission flare events occurring almost every day.

In October 2002, XMM-Newton recorded the so far most powerful X-ray flare from Sgr A*. This flare lasted only for about 45 minutes but reached an estimated peak luminosity of $L_{2-10keV} \sim 3.6 \times 10^{35}$ $erg\ s^{-1}$, an increase in flux density by a factor of 160 above the quiescent level (Porquet *et al.* 2003). The overall symmetric light curve of this flare showed significant micro-structure on timescales of only a few 100 seconds, and in comparison to previous flares the spectrum appeared to be significantly softer with a photon index of 2.5 ± 0.3.

2.13.4 *Sgr A* as a Variable NIR Source*

One of the primary objectives of the high angular NIR observation programs was the identification of an infrared counterpart of Sgr A*. In 1997 indications for a variable near-infrared source at the position of Sgr A* were reported by Genzel *et al.* (1997). However, due to its faintness and the extreme crowding of stellar sources in the central arcsecond of the stellar cluster, this possible detection could never be confirmed by further NIR speckle observations at the NTT and Keck telescopes (e.g. Ghez *et al.* 1998; Hornstein *et al.* 2002; Schödel *et al.* 2003).

The observation of a possible NIR counterpart at the position of Sgr A* was particularly difficult after about 1999, because stars such as S2 and S14 approached Sgr A* to projected distances of less than 50 to 100 mas (the diffraction limited beam FWHM of the NTT in K-band is 130 mas). Schödel *et al.* (2003) examined ten years of SHARP/NTT data and constrained the peak magnitude of possible variability of Sgr A* on time scales of hours or shorter to less than $K = 13.5^m$.

With the introduction of adaptive optics on both the VLT and the Keck telescopes the situation improved considerably. In 2003 a variable source at the location of Sgr A* was discovered — first at wavelengths of 2.2 and 1.7 μm (Genzel *et al.* 2003a), and then at 3.8 μm (Genzel *et al.* 2003a; Ghez *et al.* 2004). While up to that point NIR observations of the Galactic Center were mainly dedicated to confirm the black hole nature of Sgr A* via stellar dynamics, the properties of the supermassive black hole and its emission and accretion mechanisms can now be studied directly at NIR wavelengths, too.

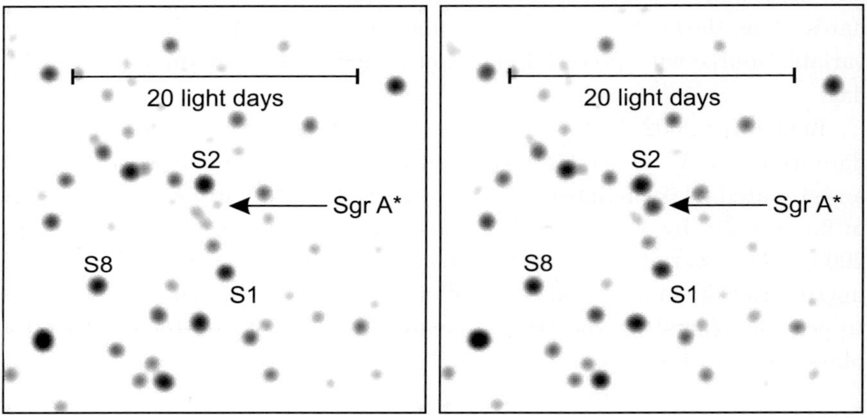

Fig. 2.32 First NIR flare from Sgr A* detected with NACO/VLT in the H-band (1.7 μm) on May 9, 2003. The images of 60 s exposure time each show a region of about 1.3" × 1.3" centered on Sgr A* and resulted from a Lucy-Richardson deconvolution with subsequent beam restoration (40 mas FWHM). The left image was taken at the beginning of the observations and Sgr A* is only visible as a faint point source. The right image was taken near the peak of the flare, about 38 min later, when Sgr A* had brightened by about 2 magnitudes. For more details see Genzel *et al.* (2003a).

During VLT H-band (1.7 μm) observations of the Galactic Center with NACO on May 9th 2003 (Genzel *et al.* 2003a) a variable, flaring source was detected at the position of Sgr A* (Fig. 2.32). The position of the flare was offset only by -1.4 ± 3.0 mas in right ascension and -0.2 ± 3.0 mas in declination from the position of Sgr A* as had been determined from the orbit of the star S2 (Schödel *et al.* 2003; Eisenhauer *et al.* 2003b). The flare in flux density lasted for about 30 minutes with a 5 minute exponential rise and decay time. Since the light crossing time of a 3.6 million solar mass black hole is 35 seconds, this rapid variability corresponds to a spatial scale of less than 10 Schwarzschild radii for such an object.

Following this discovery further observations with NACO/VLT in June

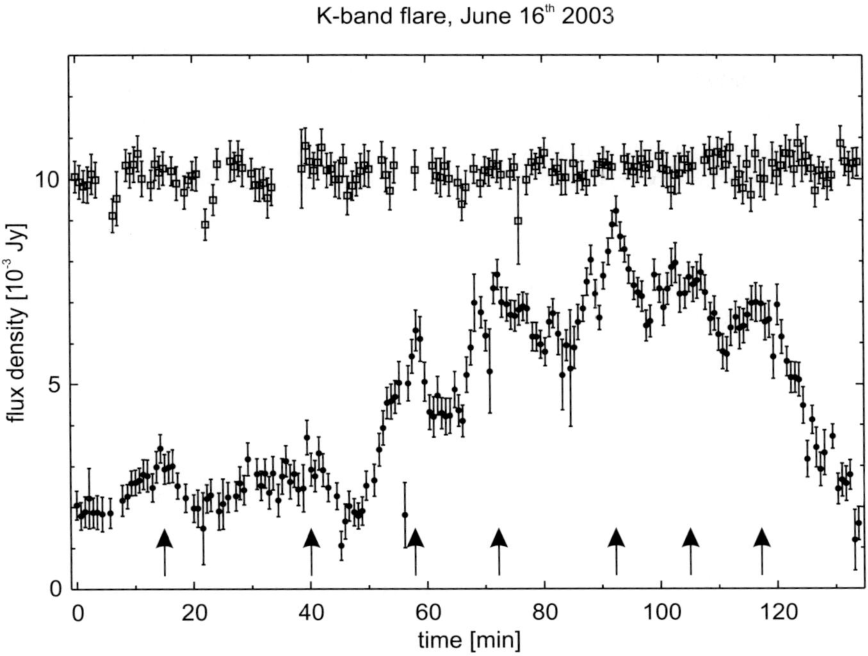

Fig. 2.33 K-band light curve of the NIR flare observed with NACO/VLT on June 16, 2003 (see also Genzel *et al.* 2003a). The highly varying lower curve is the light curve of Sgr A*, while the upper, quasi-constant light curve shows the flux from the comparison source S1, located about 0.2" south from Sgr A*. The fluxes of the sources were measured by PSF fitting with StarFinder (Diolaiti *et al.* 2000) and were calibrated relative to the known magnitudes of four sources located within 0.5" of Sgr A*. The mean position and standard deviation of the flaring source resulting from 179 measurements is 5.1 ± 4.0 mas west and 0.6 ± 5.1 mas north of the nominal position of Sgr A* as inferred from the orbit of S2 (Eisenhauer *et al.* 2003b). Because of confusion with faint stellar sources in the cusp around Sgr A*, its NIR counterpart may easily appear somewhat offset from the nominal position of the black hole.

2003 and an examination of earlier VLT data with respect to variability of Sgr A* resulted in the detection of a quiescent source and several flares at the position of Sgr A* at wavelengths of 3.8, 2.2 and 1.7 μm (Genzel *et al.* 2003a). The flares typically last for half an hour to one hour and are a few times stronger than the detected quasi-quiescent NIR emission of Sgr A*. The best covered flare with the highest quality data from Genzel *et al.* (2003a) is shown in Fig. 2.33. In that figure the arrows in the plot of the K-band flare indicate substructure peaks of the flare. As Genzel *et al.* (2003a) reported, two K-band flares that were observed more than 24 hours apart, and must thus have been unrelated events, showed very similar quasi-periodicity. Both of them showed a significant peak at a frequency corresponding to a time scale of 16.8 ± 2.0 min. A detailed discussion and interpretation of that phenomenon in given in chapter 3.8.

Table 2.4 List of Sgr A* flares observed by Genzel *et al.* (2003a). The offsets in right ascension and declination refer to the astrometric position of Sgr A* as it was inferred from the focus of the orbit of the star S2. The duration of the flares refers to the full width at zero flare flux density, the variability factor is the ratio between the excess emission and the "quiescent" emission.

Band	Date [UT]	ΔRA [mas]	ΔDec [mas]	Duration [min]	Variability
L'	2002.66	0(30)	0(30)	≥ 15	0.7
H	2003.353	-1.4(3.0)	-0.2(3.0)	30	4.7
K_S	2002.455	-2.5(4)	3.4(4)	80	3.1
K_S	2002.457	-6.4(4)	2.5(4)	85	3.2

The NIR flares that were so far observed with NACO/VLT can be characterized by the following constraints:

- **Brightness:** peak brightness in K-band of $K \sim 15.0^m$, comparable to the brightness of the star S1 near Sgr A*.
- **Duration:** duration of $\leq$85 min.
- **Time scales:** short rise and fall time scales of the order 5 min and intrinsic variability during the flares on a similar time scale.
- **Position:** position within a few milli-arcseconds of the dynamical position of the dark mass as inferred from the orbit of S2 (see Tab. 2.4). The position of the quiescent source has a larger uncertainty of the order of 10 mas (mainly because of its faintness and confusion with other stellar sources in the dense cusp), but is still consistent with the dynamical position of the black hole.
- **Intrinsic quasi-periodicity of the flares:** the first two K-band flares, which were completely covered by observations, display a characteristic quasi-periodic substructure, with a period of 16.8 ± 2.0 min. Still under investigation are the indications of quasi-periodicity of the quiescent state of Sgr A*.

It is highly improbable that the observed variability was due to stellar sources because of the extremely short time scales of the flares and because of the astrometric position of the flares, which was within less than 10 mas from Sgr A* at all times: a star close to Sgr A* would have moved by $\sim 20 - 50$ mas during the time interval covered by the four flares reported by Genzel *et al.* (2003a). A star at greater distances from Sgr A* would have an extremely low probability of being located so close in projection to Sgr A*.

Based on the measurements of 2003 the estimated flaring rate is very high: 4 flares were found within a total of 25 hours of observations, which results in about 2 to 6 flares per day when assuming Poisson statistics. This high frequency and the complex temporal substructure of the light curves rule out the possibility that the flares were related to micro-lensing of cusp stars by the black hole (Alexander 1999).

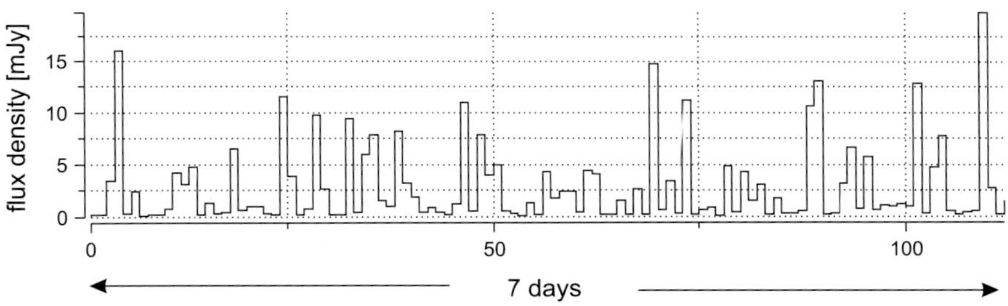

Fig. 2.34 Simulation of the Sgr A* flare activity assuming a power spectrum relation between flare amplitude and the number of flares.

It seems that a possible way to describe the NIR flare activity of Sgr A* is a description via a powerlaw. A first and simple attempt would be to assume that the emission consists of successive flares with a characteristic duration but of varying intensity. In Fig. 2.34 we show a simulation of a seven day period during which the flare amplitudes A follow an expression

$$ln(A) = \xi \; ln(N) + C \quad . \tag{2.6}$$

Here N is the number of flares with a given amplitude, ξ the power spectral index, and C a constant. For a suitable value of ξ and a lower and upper amplitude limit (1.5 and 25 mJy and ξ=−1.0 in the case of Fig. 2.34) it is possible to statistically describe some of the observed NIR flare characteristics. From the currently observed NIR flares these characteristics are a mean flare duration of about 90±20 min, a flare rate of 4±2 day^{-1}, a flare flux of 10±5 mJy and a quasi continuous flux density between flares of 2±1 mJy. In the framework of such a powerlaw description the quasi continuous flux density between flares would correspond to the median flux density of low amplitude flare events.

As for the relation between the NIR flares and variability at other wavelengths, the durations, rise and decay times, and spectral luminosities of X-ray flares are similar (see e.g. Baganoff *et al.* 2001; Porquet *et al.* 2003). The NIR flare rate, however, was almost twice as high as the X-ray flare rate during the Chandra monitoring in 2002 (1.2 ± 0.4 flares per day; Baganoff *et al.* 2003). At the moment, it is not clear whether this points to a physically different nature of the two kinds of flares or whether this was related to an overall change of the variability of Sgr A* at different epochs. Consequently further simultaneous observations are needed to determine the relation between the X-ray and NIR flares.

The variability in the millimeter and sub-millimeter domain – which is likely to be physically linked to the NIR/X-ray activity – is summarized in section 2.13.1.

2.13.5 *How Likely are Strong NIR Flares?*

The most recent AO data from the VLT and the Keck telescope taken in 2004 show that the NIR flare emission found in 2003 is likely to be a persistent phenomenon. Yet NIR flares stronger than the emission from the nearby high velocity stars S1 and S2 have not been detected so far.

Hornstein *et al.* (2002) used the 1995-2001 Keck speckle and adaptive optics data to produce a NIR imaging data set with 3 hours bin length. However, no evidence of a significant excess emission associated with Sgr A* was detected. They found lowest detection limits to the de-reddened emission of Sgr A* at 2.2 μm of about 2.0 mJy. This is close to the quasi-quiescent flux density level of the NIR counterpart of Sgr A* that was found by Genzel *et al.* (2003a). Nevertheless, neither a flare event nor quiescent emission was positively detected, although Hornstein *et al.* (2002) find that statistically (at a $\sim 2\sigma$ level) a flare must have occurred during the total 1995-2002 high angular resolution near-infrared Keck time and derive a probability of 9% for an expected random 3 hour near-infrared flare in excess of 19 mJy (de-reddened).

In May 2002 a simultaneous flux density monitoring program towards Sgr A* was conducted using the Chandra X-ray satellite and the ISAAC instrument of the VLT (Eckart *et al.* 2003b). This campaign lasted for a few days and resulted in an infrared coverage of a flare on May 29th 2002, although the angular resolution of the ISAAC data was only seeing limited. The de-reddened upper flux density limit that could be attributed to a variable source was of the order of 20-100 mJy.

The main objective of that monitoring experiment was the search for strong $\sim$ 10 ksec flares in the X-ray and IR. Following Hornstein *et al.* (2002) the probability $P_{n,m}$ of the successful observation of such a flare, i.e. n detections in m trials, is given by the binomial distribution:

$$P_{n,m} = \frac{m!}{n!(m-n)!}(p^n)(1-p)^{m-n} \quad . \tag{2.7}$$

with p representing the probability of detection in 1 trial (duty cycle) and m being the ratio between observing time and the average flare duration of about 10 ks.

A representative part of the simultaneous ISAAC/Chandra observations resulted in n=3 detected flares over an m=15 period at X-rays, while the IR observations resulted in n=0 flares over a m=50 period. The X-ray observation time was taken from a representative section of the overall X-ray observations for an average X-ray flare rate of one event per day, and other time windows (with additional X-ray flares) would provide similar results. For the ISAAC observations, during which no flare was detected, the observation time was calculated by assuming that the various batches of observations provided a complete coverage of their respective time windows (about 1-4 hours). This is a reasonable assumption since the pauses between the observations within a batch are of the order of minutes and thus short relative to the X-ray flare duration. The total NIR observation time with ISAAC therefore was assumed to be approximately 13 hours.

The X-ray and MIR probability distributions in Fig. 2.35 show a very small overlap. Therefore, if we assume that the X-ray and MIR flares are coupled, we can calculate a common probability for a random simultaneously detected MIR flare in excess of the maximum observed flux of 20 to 100 mJy. That probability has a maximum value of only 0.5%(!). Both the Keck and the ISAAC/Chandra observations had in common that no flare-like emission from Sgr A* on the $> 20 - 100$ mJy level was detected on a timescale of $\sim$ 10 ksec in the IR, although the above discussion shows that the time coverage of the observations was in fact suitable to detect such a variability. However, much shorter flares ($\sim$20 min) of the same strength

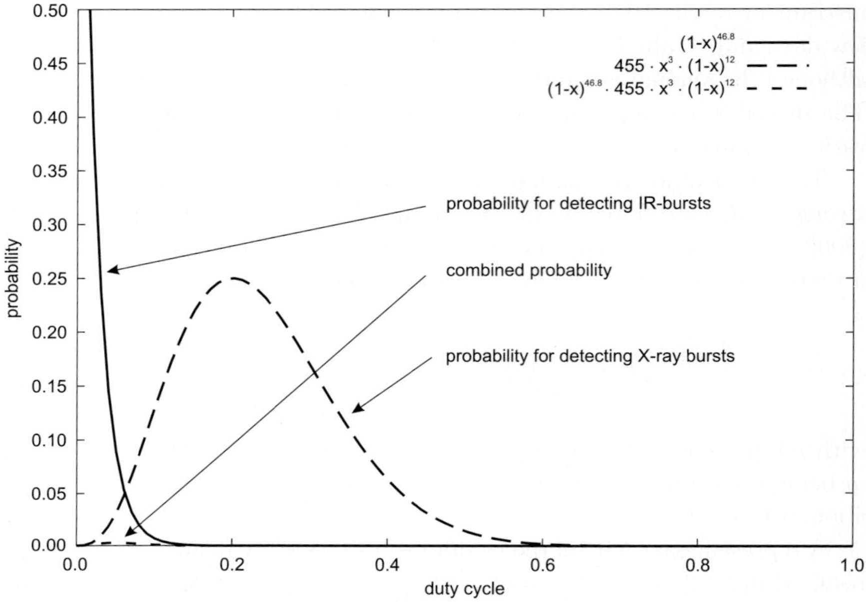

Fig. 2.35 Probability of detecting a ∼10ks flare in the recent ISAAC observations and the Chandra experiment for the corresponding duty cycles (Eckart *et al.* 2003a). The low combined probability suggests that if the flaring activities at near-infrared and X-ray wavelengths are coupled, a flare occurred during the near-infrared experiment and must have been lower than the ISAAC detection limit.

could not be excluded and would remain consistent with the current data.

2.13.6 *First Simultaneous X-Ray/NIR Detection of Sgr A**

Simultaneous multi-wavelength observations are essential for the understanding of the quiescent and flaring state accretion and emission mechanisms of Sgr A*. Eckart *et al.* (2004a) reported on the first simultaneous near-infrared/X-ray detection of Sgr A* (Fig. 2.36) based on their observations in June 2003 with the NACO/VLT adaptive optics instrument and the ACIS-I instrument aboard the Chandra X-ray observatory. A fading flare of Sgr A* with more than 2 times the interim-quiescent flux was detected at the beginning of the NIR observations, and the same flare was also seen in the X-domain with an excess $2-10$ keV luminosity of about 6×10^{33} erg/s. The authors found that the flaring state could conveniently be explained by a synchrotron self-Compton model involving up-scattered sub-millimeter

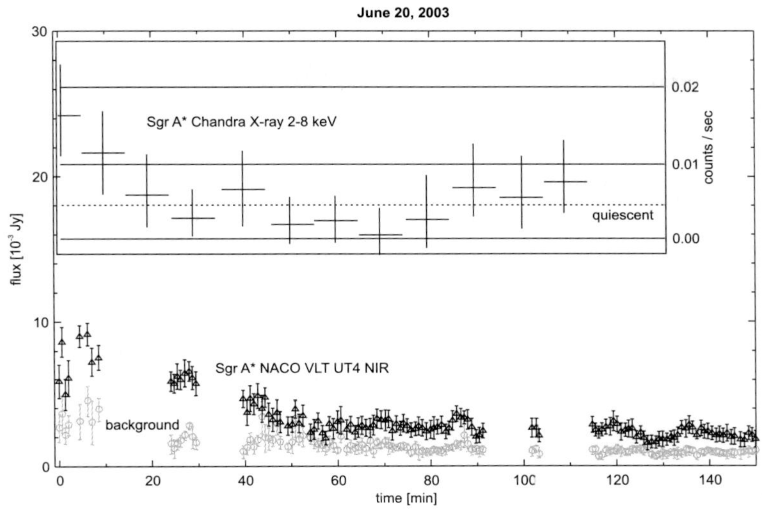

Fig. 2.36 Overlapping parts of the Chandra X-ray and VLT K-band light curves of the first simultaneously detected flare from Sgr A* (Eckart *et al.* 2004a). The X-ray and NIR light curves are plotted with a common time axis. Straight solid lines in the inserted box represent the 0.00, 0.01, and 0.02 counts per second levels. The straight dashed line represents the X-ray IQ-state flux density level. The NIR observations started on 20 June 2003 at 23:51:15 (UT). The NIR data started 0.38 minutes before the midpoint of the highest X-ray data point.

photons from a compact source component allowing for modest bulk relativistic motion. The overall spectral indices $\alpha_{NIR/X-ray}$ ($S_\nu \propto \nu^{-\alpha}$) of the quiescent and flaring states were quite comparable with a value of ~ 1.3. Since the quiescent X-ray emission is extended, the spectral index for the interim-quiescent state is probably only a lower limit for the compact source Sgr A*. The authors estimated a conservative upper limit of the time lag between the end of the NIR and X-ray flare of the order 15 minutes. The observations strongly support the assumption that X-ray and NIR flares are closely related processes. Also, the orders of magnitude higher resolution in the NIR clearly shows that the X-ray flares are associated with Sgr A* and no other object within the $\sim 1''$ beam of the Chandra telescope.

Eckart *et al.* (2004a) also report on quasi-simultaneous observations at a wavelength of 3.4 mm using the Berkeley-Illinois-Maryland Association (BIMA) array. Compared to 8-9 hours before the NIR/X-ray flare they detected a marginally significant increase in the millimeter flux density of Sgr A* during measurements about 7-9 hours afterwards. The increase in the flux density of Sgr A* was 11 ± 3 % compared to the flux density change of the calibration source NRAO 530 by less than $-0.3 \pm 0.4\%$. The slightly higher mm-flux density after the flare may be related to the flare activity observed in the X-ray and NIR domains. There seems to be some evidence for a correlation between flare activity in the mm/sub-mm and the X-ray domain in three cases now (see Eckart et al. 2004). Future observations will reveal whether and how the individual mm-flare events or a general flare activity are related to events in the NIR or X-ray regime. Simultaneous monitoring programs from the radio to the X-ray regime will allow us to investigate the physical processes that are associated with the variable emission from Sgr A* at the position of the massive black hole at the center of the Milky Way.

Chapter 3

Astrophysical Results

The large amount of high quality data on the GC across the electromagnetic spectrum that has been accumulated in the past decades has lead to a detailed picture of the object associated with Sagittarius A* and the stellar cluster at the center of the Milky Way in which it is embedded. The proper motion data on the stars in the GC, in combination with spectroscopically measured radial velocities, have provided a clear picture of the velocity field at the very center of our Galaxy. This data now allows to derive the amount and the compactness of the mass located at the center. A close investigation of the stability of massive and compact clusters of astrophysical bodies leads to the conclusion that such configurations cannot be stable at mass densities as observed in the central parsec and that the dark mass at the center of the Galaxy thus must be present in the form of a massive black hole. The most convincing support for the existence of a supermassive black hole at the center of the Milky Way comes from the orbital motions of several individual stars in its immediate vicinity. Nevertheless, despite the strong dynamical evidence for the black hole nature of Sgr A*, the observed luminosity of this object is too small by about 5 orders of magnitude with respect to Low Luminosity Active Galactic Nuclei (LLGN) and by 8-9 orders of magnitude with respect to Active Galactic Nuclei (AGN). The standard models of accreting black holes were mainly developed from observations of quasars and AGN. However, the theory for weakly accreting black holes, such as Sgr A* is rapidly evolving, relying heavily on simultaneous multi-wavelength observations of the Milky Way black hole. In the radio, near-infrared and X-ray domain Sgr A* can be associated with a source that shows quasi-quiescent emission as well as strong flare activity.

3.1 The Stellar Velocity Field

The diffraction limited NIR maps and spectra of the Galactic Center region obtained over the past decade have lead to a detailed understanding of the stellar dynamics in the Galactic Center. The data resulted in sky and line-of-sight velocities for several 100 stars within the central few arcseconds of the position of the black hole candidate Sgr A* (Eckart & Genzel 1996, 1997; Ghez *et al.* 1998; Genzel *et al.* 2000; Ott *et al.* 2003; Schödel *et al.* 2003).

The overall stellar motions of the stars do not deviate strongly from isotropy. For the bulk of these stars neither circular motion around the center nor radial motion is preferred. Exceptions to this general finding are one, or even two rotating stellar disks and the velocity field of the stellar cusp, a region of increased stellar density around Sgr A*. For more than 50 stars the determination of all three velocity components has been possible so far, and the derived line-of-sight and sky velocities are in good agreement and consistent with a spherical star cluster. Likewise for several 100 stars with known proper motion the sky-projected radial and tangential velocities are also consistent with overall isotropy (Fig. 3.4).

However, most of the bright HeI emission-line stars at separations from 1 to 10 arcsec from Sgr A* are on tangential orbits. This tangential anisotropy of the HeI stars and most of the brighter members of the IRS 16 complex is mainly due to a sky projected clockwise rotation pattern, counter to general galactic rotation. The overall rotation of this young star cluster may be a remnant of the original angular momentum distribution in the interstellar cloud from which these stars were formed.

In contrast to the motion of the bulk of stars in the spherical cluster of about 0.6 pc diameter, the sky-projected velocity components of the stars within 1 arcsec (about 39 mpc) of Sgr A* indicate deviations from isotropy (see Fig. 3.7; Genzel *et al.* 2000; Schödel *et al.* 2003). The orbits of these stars apparently show a preference for more radial or very elliptical orbits. Spectroscopy indicates that they are young, early-type stars. The Sgr A* cluster stars may be those members of the early-type cluster that happen to have small angular momentum, and thus can plunge to the immediate vicinity of Sgr A*. A recently proposed model suggests that they may also have formed at greater distances and then been captured in the vicinity of Sgr A* via 3-body exchange processes and the ejection of stellar mass black holes (Alexander & Livio 2004).

The search for possible anisotropies of the stellar dynamics can be quan-

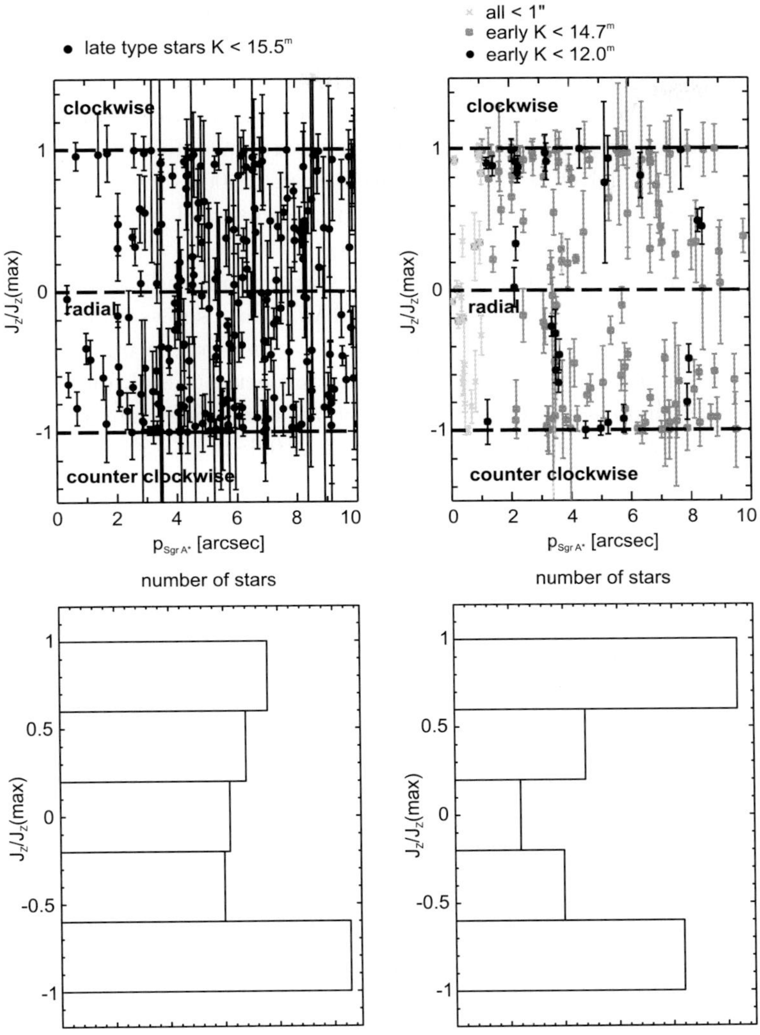

Fig. 3.1 Angular momenta as seen projected on the plane of the sky for late-type (left panel) and early-type (right panel) stars within 10″ of Sgr A* (Genzel *et al.* 2003b). The early-type stars are found to be preferentially on clockwise ($J_z/J_z(\max) \lesssim 1$) or counter-clockwise ($J_z/J_z(\max) \gtrsim -1$) orbits while no such pattern can be found in the orbits of the late-type stars. See also color Fig. A.10. The two histograms at the bottom are integrated (along radius) versions of the top panels.

tified. A convenient anisotropy estimator compares the projected tangential v_T^2 and radial velocities v_R^2 with each other via $\gamma_{TR} = (v_T^2 - v_R^2)/v^2$, where v is the proper motion velocity of a star. A value of $+1$ denotes projected tangential motion, -1 projected radial motion of a star. The properties of the anisotropy parameter γ_{TR} are discussed in detail in Genzel *et al.* (2000) and Schödel *et al.* (2003). Genzel *et al.* (2000) show that an intrinsic three-dimensional radial/tangential anisotropy will be reflected in the two-dimensional anisotropy estimator γ_{TR}.

A comparison of γ_{TR} histograms for stars within radii of 0.6 arcsec to 1.2 arcsec of Sgr A* using the SHARP and VLT data shows that in all cases, the number of stars on radial orbits is $2 - 3\sigma$ above the number of stars on projected tangential orbits. From the projected radial and tangential velocity dispersions one can estimate the anisotropy of the stellar cluster via (Leonard & Merritt 1989; Genzel *et al.* 2003b):

$$\langle\beta\rangle = 1 - \langle\sigma_t^2\rangle/\langle\sigma_r^2\rangle = 3(\langle\sigma_R^2\rangle - \langle\sigma_T^2\rangle)/(3\langle\sigma_R^2\rangle - \langle\sigma_T^2\rangle). \tag{3.1}$$

Here σ_t and σ_r are the radial and tangential three dimensional velocity dispersions. The sky projected radial and tangential velocity dispersions that are measured via proper motions of the stars are σ_T and σ_R. The anisotropy estimator can take on values $-\infty < \langle\beta\rangle \leq +1$, with positive values indicating radial and negative values tangential anisotropy. Averaging the results of the 4 lists in Schödel (2004), we find $\langle\beta\rangle = 0.5 \pm 0.2$. However, only for 3 of the lists the value of the anisotropy is in good agreement with the average value. For one list the authors obtain a value of $\beta \sim 0$. Although it has been shown (Genzel *et al.* 2000) that a measurement of β from a small sample of stars can easily be skewed towards positive values, the simulations also show that the probability of measuring $\beta \geq 0.5$ is as low as 25% for a sample of ≥ 30 stars out of intrinsically isotropic or tangentially anisotropic clusters. This supports the statistical significance of the mean value of β for the central stars.

The kinematic properties of the $K_s \leq 15$ stars can be analyzed further by calculating their angular momentum $J(z)$ along the line of sight. If the total proper motion v_p of a star were perpendicular to its projected radius p the angular momentum along the line of sight would have a maximum value $J_z(\max)$. A convenient way of looking at the true angular momentum $J(z)$ is to calculate it normalized to this maximum value (Genzel *et al.* 2003b):

$$J_z/J_z(\max) = (xv_y - yv_x)/pv_p\,. \tag{3.2}$$

Here v_x, v_y and v_p are the R.A.-, Dec.- and total proper motion velocities of a star at the projected position (x,y) on the sky and at projected radius p. Depending on whether the stellar orbit projected on the sky is mainly counter-clockwise tangential, radial or clockwise tangential with respect to the projected radius vector from the star to Sgr A* we find values for $J_z/J_z(\max)$ of ~ -1, ~ 0 and $\sim +1$. The analysis of this quantity shows no significant projected overall rotation of the Sgr A* cluster in the central arcsecond (Schödel *et al.* 2003; Schödel 2004). As for the cluster at larger distances from Sgr A*, Fig.3.1 shows the projected radial distribution of $J_z/J_z(\max)$ for late and early type stars, as identified from the narrow-band CO-index (Ott *et al.* 2003; Genzel *et al.* 2003b). In general we find that the late type stars have a random distribution of z-angular momenta, whereas the early type stars have the tendency of being on clockwise tangential orbits ($J_z/J_z(\max) \geq 0.6$). Such a coherent motion is also indicated from spectroscopic results. While the late type stars show a random (relaxed) distribution of line-of-sight velocities (Sellgren *et al.* 1990; Haller *et al.* 1996; Genzel *et al.* 1996) the early type stars with spectroscopically measured line-of-sight velocities north of Sgr A* are almost all blue-shifted, while they are red-shifted south of Sgr A* (Genzel *et al.* 1996, 2000; Paumard *et al.* 2001; Genzel *et al.* 2003b). This is indicative of a coherent rotation pattern, with an axis opposite (counter) to that of the overall galactic rotation.

The origin of the coherent motion of the young massive stars within the Galactic Center star cluster presents a yet unsolved problem. Levin & Beloborodov (2003) have shown that 10 out of 13 of these stars for which the full three-dimensional velocity information is available lie in a thin disk (see also Genzel *et al.* 2003b). The half-opening angle of this disk is less than 10°. The authors propose that this peculiar rotation pattern originates from a formerly existing dense gaseous disk around Sgr A*. That disk is no longer present today because the disk material has either been transformed into these young and massive stars that still show the coherent disk rotation or it has been accreted by the central black hole. Levin & Beloborodov (2003) have searched that disk for a common plane within which the three dimensional velocity vectors can come to lie. The 10 stars that fulfill this criterion are IRS 16SW, 16SE1, 16SE2, 16CC, 16C, 33E, 29N, 29NE1, 34W, and 7W. All of these stars move clockwise on the sky. Star IRS 10W, although moving clockwise, has a large velocity (230 km/s) perpendicular to the plane. Two other stars, 16NW and 16NE, both have large velocity offsets perpendicular to the plane and move counterclockwise.

The best-fit plane is uniquely defined by the unit vector **n** normal to it.

In conventional galactic coordinates with sky-projected vector elements in right ascension and declination, and along the line of sight away from us this vector can be described via: $\mathbf{n}$ = (0.486, 0.725, -0.487). Genzel *et al.* (2003b) found evidence for a second possible disk that rotates almost perpendicular to the disk found by Levin & Beloborodov (2003). In addition Genzel *et al.* (2003b) presented intriguing evidence that the stellar populations of the two disks formed at the same time. No satisfactory explanation could be found so far for the possible common origin of the two disks.

3.2 Scenarios for Star Formation at the Galactic Center

The luminosity of the entire central parsec of our Galaxy is dominated by a cluster of massive stars (e.g. Najarro et al. 1997). Already the 7 most luminous ($L>10^{5.75}$ $L_\odot$), moderately hot ($T<10^{4.5}$ K) blue supergiants (mostly referred to as He-stars) contribute half of the ionizing luminosity of that region (Blum *et al.* 1995; Krabbe *et al.* 1995; Najarro *et al.* 1997). One explanation for the presence of these He-stars in the central parsec is that they formed out of a massive cloud that fell into the center less than 10^7 years ago (e.g. Gerhard 2001) and was highly compressed there. Due to the resulting high mass densities the cloud then became gravitationally unstable (e.g. Sanders 1992; Morris & Serabyn 1996). Similarly, Levin & Beloborodov (2003) suggested that the He-stars might have formed out of a dense accretion disk that was present around Sgr A* at their time of formation. However, the exact mechanism of star formation in the GC still remains enigmatic: the extreme tidal fields due to the presence of a supermassive black hole and an extremely dense stellar cluster, as well as strong magnetic fields and stellar winds in this region need to be taken into account in a general discussion of the possibility of star formation in the immediate vicinity of the center (e.g. Morris et al. 1993; see also Figer et al. 2002 on the nature of massive stars in the Arches Cluster). The identification of the hot luminous objects at the Galactic Center with blue supergiants is, however, not undisputed. Alternative scenarios include late-type stars rejuvenated by collisions.

Morris (1993) suggested that the apparent hot stellar objects may be 10 $M_\odot$ black holes that have temporarily acquired optically thick atmospheres in collisions with red giants and appear externally as luminous blue stars with winds analogous to those of normal O or WR stars. In addition, the number of apparently young, massive, and hot He-stars cannot currently be explained via a theory for normal stellar evolution (e.g. Eckart *et al.* (1999b), see also Thornley *et al.* (2000)). We will now discuss in more detail some of the suggested models for the presence of young stars near Sgr A*.

3.2.1 *Infall of Stellar Clusters*

Due to the strong tidal forces, in-situ star formation at the very center of the Galaxy is difficult to explain. Yet we find young and massive stars at that location, which might imply that they were formed at a larger distance and brought into the center later. A plausible mechanism to bring massive stars into the center of stellar clusters is their interaction with individual lower mass members of the cluster. By transferring kinetic energy to them they can sink into the center. However, the time scales associated with such two-body relaxations are long compared to the lifetime of the young and massive He-stars found within the central stellar cluster.

To overcome this problem Gerhard (2001) proposed that a massive and gravitationally bound system of young stars spiraled into the center by losing kinetic energy via dynamical friction (Fig. 3.2). Such a young star cluster could sink into the central $\sim$ 0.5 pc before it is tidally disrupted there. Gerhard (2001), and more recently Portegies Zwart *et al.* (2003) and Kim & Morris (2003), have shown that such a scenario is indeed feasible. It requires that such a cluster has to be initially very massive ($\gg 10^4\ M_\odot$) and dense (or compact) in order to spiral into the center within the few million years' lifetime of its O-stars. It would then arrive in the central region before dissolving completely. The dynamical friction time scale scales inversely with the cluster mass. With an initial mass of 10^5 -10^6 $M_\odot$ a cluster can make it into the center from a distance of several parsecs within a few Myrs. Such a cluster could be similar to the two known (non-nuclear) young star clusters in the central 30 pc, the Quintuplet and the Arches clusters. They have diameters of $\sim$ 0.4 pc and estimated masses of a few $10^4\ M_\odot$ (Figer *et al.* 1999a). In their most recent N-body simulations Portegies Zwart *et al.* (2003) showed that a cluster of 6×10^4 $M_\odot$ and diameter $<$ 0.3 pc could spiral into the central parsec within the available time from an initial radius of 4–5 pc. Kim & Morris (2003) derived similar results, but with cluster masses $\geq 10^5$ M$_\odot$. Rapid mass segregation and core collapse appear to be indispensable if a significant part of such a cluster is indeed to reach the inner parsec of the Milky Way. Recently, Hansen & Milosavljević (2003) added a new ingredient to the infalling cluster scenario: they raised the point that the effectiveness of dynamical friction would be significantly increased if the cluster contained an intermediate mass black hole (IMBH) of 10^3 to 10^4 $M_\odot$ at its center. Hansen & Milosavljević (2003) discuss the orbital characteristics that the gravitational interaction of the IMBH with the central supermassive black hole would imprint onto the motion of the young stars. They also discuss the possibility of detecting the IMBH via its influence on the proper motion of Sgr A*.

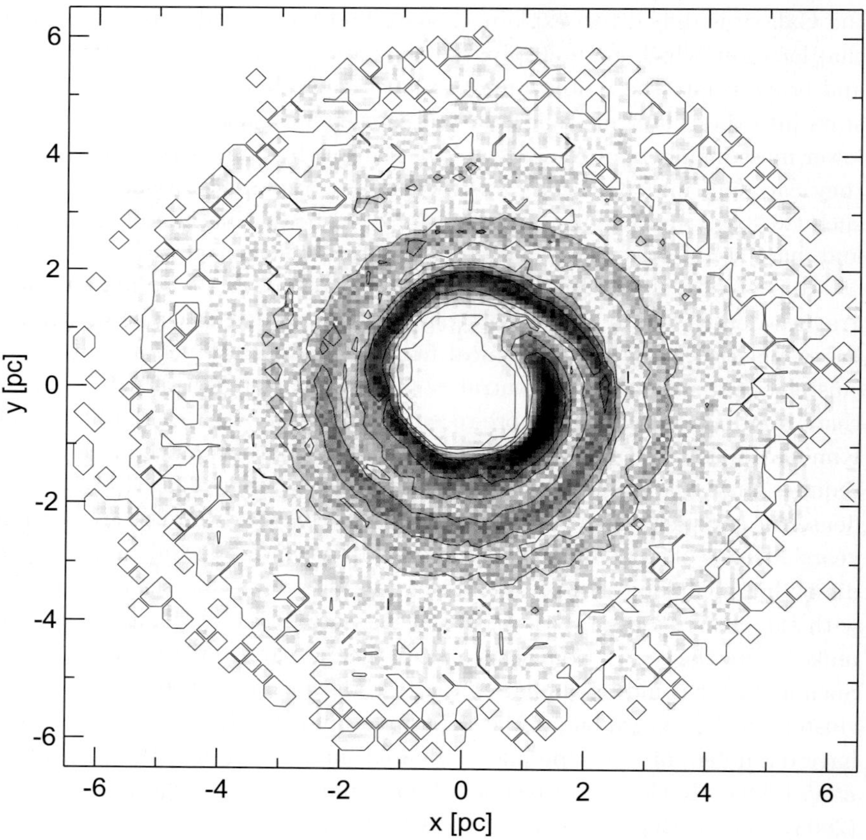

Fig. 3.2 Simulation of a compact stellar cluster spiraling counter-clockwise into the central parsec and dissolving along the way (Portegies Zwart *et al.* 2003). Top view onto the plane of the orbit. Linearly scaled gray shades (black : maximum; white: minimum) indicate the stellar density.

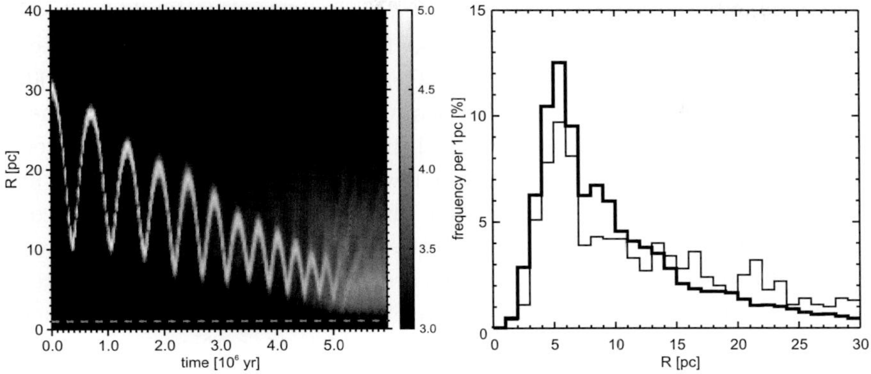

Fig. 3.3 Simulation of the infall of a massive, compact stellar cluster into the central parsec of the Milky Way and its dissolution during this process (Kim & Morris 2003). The grey scale bars give the stellar density in $M_\odot$ pc^{-3}. The right panel shows the distribution of the cluster stars vs. distance from the GC at the end of the simulation.

For possible compact and massive molecular concentrations as candidates for systems that could sink into the central star cluster the situation is different. Until now molecular line surveys did not find such dense ($> 10^6$ cm^{-3}), compact and massive molecular clouds close to the center. Several $> 10^5\ M_\odot$ molecular clouds are located at distances > 10 pc from the inner parsec (Mezger *et al.* 1996). At such a large radius the tidal requirements are less severe and a cluster of young and massive stars could be formed efficiently from such a dense cloud. It will, however, be difficult to let it sink into the central region within the required lifetime of the young stars. The fact that both clockwise and counter-clockwise early-type stars are present within the central parsec (Genzel *et al.* 2003b) may indicate that several of the clusters must have fallen in at about the same time. Alternatively, the infalling system was large and/or fragmented with respect to its impact parameter or the counter-rotating systems may be the result of a collision of two counter-rotating infalling clouds (Genzel *et al.* 2003b). Under these circumstances one would naturally expect that counter rotating orbits would result from the infall.

There are two prominent aggregates of luminous, apparently young stars that could be interpreted as the remnants of young clusters: the IRS 16 and the IRS 13 complexes. Most of the early-type stars in the IRS 16 complex are likely to be spatially confined within the central 0.5 parsecs. They appear to move coherently on clockwise, tangential orbits. In addition, there exists a small (<1.5 arcsec diameter) compact star cluster. This cluster is IRS 13 and is moving counter-clockwise. Both findings are consistent with the "infalling cluster scenario" (Gerhard 2001; Kim & Morris 2003; Portegies Zwart *et al.* 2003). A significant difference, however, is that these calculations predict a severe azimuthal dispersion of the infalling clusters, since they orbit the central region a few thousand times while they are spiraling inward. In contrast to this, the observations show rather concentrated distributions of young and massive stars. Most of the clockwise moving He-stars in the IRS 16 complex are located within the south-eastern quadrant with respect to Sgr A* (Genzel *et al.* 2000, 2003b). This quadrant is three times more densely populated with $K \leq 15$ early type stars than the other quadrants. This asymmetric distribution is, however, not reflected in the distribution of fainter stars in the underlying central cluster. Unless there is a hidden, dark, and probably non-stellar additional mass of the order of $> 10^4\,M_\odot$ it is unlikely that this asymmetric distribution is self gravitating. It therefore is likely that at least the IRS 16 cluster can be explained by a statistical fluctuation. In that case it will be largely dispersed within a few hundred revolutions (i.e. about 10^6 years).

Maillard *et al.* (2003) proposed that the IRS 13 cluster is the remaining core of a massive, young cluster. It is more compact and co-moving with the mini-spiral. It is therefore plausible that IRS 13 is in fact a compact stellar aggregate that is well entrained into the infalling gas-streamers. The overall north-south elongation of the IRS 13N/IRS 13/IRS 2 complex could then be interpreted as being due to the large tidal shear of the central black hole.

3.2.2 *In Situ Formation from Dense Gas*

A second possibility is that stars formed in situ from cloud collapse, following the infall and shock compression of a dense gas cloud (Morris 1993). Numerous molecular and ionized gas clouds are detected in the central parsec, several with large radial velocities, and a few even with the correct angular momentum (e.g. Lacy *et al.* 1991; Jackson *et al.* 1993). In order to let self-gravitation in interstellar clouds overcome the tidal shear and create a large number of massive stars with ages of a few million years, the required gas densities have to be of the order of 10^{10} hydrogen atoms per cm^3. This is several thousand times greater than the density of $n(H, H_2) \leq 10^6$ cm^{-3} currently found in the atomic and molecular gas clouds in the central 1-2 pc (e.g. Jackson *et al.* 1993; Genzel *et al.* 1994, 2003b).

For the high velocity stars bound to the black hole within the central arcsecond the required gas densities are up to 10,000 times higher. High resolution radio maps (e.g. Yusef-Zadeh *et al.* 1998b; Zhao & Goss 1998) and the new VLT adaptive optics images at a wavelength of 3.8 μm provide some evidence for large compression and the action of winds. These maps and images exhibit sharp (diameter $\leq 0.1''$) filaments, possibly caused by the interaction of the mini-spiral gas-streamers with the outflows and winds from the central few arcseconds (see Fig. 0.4 and Clénet *et al.* 2001).

3.2.3 *Formation of Stars in the Mini-Spiral*

If the He-stars are indeed massive objects that were formed $<10^7$ yrs ago, then there may be an even younger population present in the central cluster — and if so — it might still be dust-enshrouded.

Until recently the prime candidates for young stellar objects were the very red, dust embedded sources with featureless K-band spectra — e.g. IRS 21, IRS 1W, IRS 10W, IRS 5 — these objects are strong 10 μm sources and are mostly associated with the mini-spiral streamers (Gezari *et al.* 1985; Genzel *et al.* 1996). The best studied case is IRS 21 which is strongly polarized (17% at 2 μm ; (Eckart *et al.* 1995; Ott *et al.* 1999; Krabbe *et al.* 1995). Initially, Gezari *et al.* (1985) suggested that IRS 21 is an externally heated, high-density dust clump. Given the MIR excess and the featureless NIR spectra several other classifications have been proposed, including an embedded early-type star and a protostar (Krabbe *et al.* 1995; Genzel *et al.* 1996; Clénet *et al.* 2001). Tanner *et al.* (2002) suggested that IRS 21 is an optically thick dust shell surrounding a mass-losing source, such as a dusty WC9 Wolf-Rayet star.

It is very likely that these dust embedded sources are associated with luminous stars that shock and heat the gas and dust of the mini-spiral. Cotera *et al.* (1999a) find that several of these sources are offset from nearby local maxima in the dust emission and temperature distribution. Tanner *et al.* (2002) also indicate that the extended dust emission of IRS 21 is consistent with a bow shock created by the motion of such a hot star through the dust and gas of the mini-spiral. These results suggest that the bow shock scenario may be applicable to most of the luminous dust embedded sources in the central stellar cluster. This is also supported by recent results on several other dusty sources in the mini-spiral by Tanner *et al.* (2003) and the discovery of IRS 8 as a bow shock source (Rigaut *et al.* 2003). Therefore the identification of these sources with young stellar objects is likely to be wrong.

The bulk of the ionized gas in the northern arm, mini-cavity and bar is streaming south/south-west in a counter-clockwise pattern on the sky, with a line-of-sight angular momentum axis parallel to galactic rotation (Lacy *et al.* 1991; Yusef-Zadeh *et al.* 1998a). The gas-streamers are in a plane, which is inclined $\sim 65^o$ with respect to the line of sight (Vollmer & Duschl 2000). A comparison of the proper motion vectors of $K < 15$ stars with significantly reddened colors (K-L$\geq$ 2) due to dust emission shows that the majority of these sources are not comoving with the gas and dust of the mini-spiral (Genzel *et al.* 2003b). IRS 10W, several sources in the northern arm, IRS 1, IRS 21, the IRS 16 sources as well as sources in the bar (IRS 34) have proper motions that are inconsistent with the velocity vectors of the mini-spiral gas streamers. As proposed by Tanner *et al.* (2002), they just happen to move through that area and interact with the mini-spiral's gas and dust. In particular, the proper motion of IRS 21 is fully consistent with the Tanner *et al.* (2002) bow shock model. In fact IRS 1W as well as several other dusty northern arm sources are associated with a bow shock structure on the L-band images. The only possibly convincing case for dusty stars moving along with the gas of the mini-spiral is the IRS 13 complex (see Genzel *et al.* 2003b; Maillard *et al.* 2003). Furthermore, there is the group of compact L-band sources in IRS 13N at the northern tip of the IRS 13 complex with a very large K-L excess. As discussed earlier these may either be low luminosity versions of the brighter bow-shock sources within the central cluster or — alternatively — they may be prime candidates for very young stars deeply embedded in the bar (Eckart *et al.* 2004b).

Forming stars embedded in the mini-spiral gas-streamers also presents a problem for the involved time scales. The dynamical time scale for the motion of the mini-spiral through the central 0.5 pc is $\sim 10^4$ years. This corresponds only to a small fraction of the plausible ages of the massive young stars. In the case of IRS 13 it is unclear how gas and young stars (Eckart *et al.* 2004b) can remain together in the same volume over much longer than the dynamical time scale (Tanner *et al.* 2002). The small luminous IRS 13 cluster contains among the brightest hot stars in the Galactic Center (Najarro *et al.* 1997). Using adaptive optics the central IRS 13 complex can be resolved into individual objects (Paumard *et al.* 2001; Maillard *et al.* 2003; Eckart *et al.* 2004b). Eckart *et al.* (2004b) resolved the object previously known as E3 into 2 components E3N and E3c. The latter one is closest to the 7 mm VLA radio continuum source found at the location of the IRS 13 complex (Zhao & Goss 1998). E3c may in fact be associated with a strong stellar wind or a dusty Wolf-Rayet like star at that location. Given the short time scales for the motion through the central region the young stars must be formed outside the central parsec. However, with initial, shallow or confined surveys (Figer *et al.* 1995; Cotera *et al.* 1999b) only a few such stars have so far been found in the central 10 pc.

In general one can say that with the possible exception of IRS 13, a scenario in which massive stars are formed just outside the dense central region and then pass through the center on highly elliptical orbits does not seem to be very likely. Most of the bright dusty sources associated with the mini-spiral are probably not young and embedded "protostars".

3.2.4 *Capture of Stars by Ejection of Stellar Remnants*

Recently, Alexander & Livio (2004) suggested a dynamical mechanism that explains the presence of young, massive stars, such as S2, via exchanges with compact remnants in the cluster immediately surrounding Sgr A*. In their model, the young stars form at sufficiently large distances from the central massive black hole and are then scattered subsequently into eccentric orbits that bring them close to Sgr A*. There they might interact with a cluster of compact stellar remnants, i.e. stellar mass black holes of a similar mass to the young stars. Such an interaction can lead to the ejection of a stellar mass black hole and to the capture of the young star on a tight orbit around Sgr A*.

3.3 The Central Dark Mass

The three-dimensional radial and transverse velocities that can be derived from the spectroscopic and imaging data indicate the presence of a compact mass at the center of the Milky Way. The lower bound to the mass is of the order of 10^5 $M_{\odot}$and the mass estimates are all consistent with a compact mass of 3-4$\times 10^6$ $M_{\odot}$. In the following we summarize the results obtained for the enclosed masses derived from estimates of increasing complexity.

3.3.1 *First Order Estimates*

One straight forward estimate of the enclosed mass and the mass density can be obtained from the innermost high velocity stars under the assumption that they are just bound, that their radial velocities are not significant, and that their true three-dimensional distance from Sgr A*, r, is close to the projected distance on the sky relative to Sgr A*, p. A lower limit to the mass enclosed within p is

$$M_{min}(p) \geq v_{sky}^2 p/(2G). \quad (3.3)$$

Here v_{sky} is the proper motion velocity of the corresponding high velocity star and G is the gravitational constant. This mass estimate can be evaluated for individual stars or for ensembles of stars. Since v_{sky} and p are the true three dimensional space velocity and the distance from Sgr A* projected onto the plane of the sky, the result is always a lower limit of the enclosed mass. For the measured proper motions of the high velocity stars which are typically of the order of several hundred kilometers per second this results in values of the order of $M_{min} = 1$–2×10^6 $M_\odot$. The corresponding central mass densities can be estimated via

$$\rho_{min} = 3v_{sky}^2/(8\pi G r^2). \quad (3.4)$$

Since the true distance, r, of the corresponding star from Sgr A* is a priori unknown, one may correct the projected distance statistically, assuming a random distribution of radii: $r = 4/\pi \times p$. For typical limits of the enclosed mass, projected proper motions, and separations of the high velocity stars from Sgr A* one typically find mass densities of about $10^{12} M_\odot$ pc^{-3}.

3.3.2 *Mass Estimators*

A more refined estimate of the dark mass can be obtained via the virial theorem (M_{virial}) or by using the projected Bahcall-Tremaine mass estimator (M_{BT}) as given by Bahcall & Tremaine (1981) (see also Genzel *et al.* 1996):

$$M_{virial} = \frac{3\pi}{2\ G} \frac{\sum_i w(i)\ v_i^2}{\sum_i w(i)/p_i} \quad (3.5)$$

and

$$M_{BT} = \frac{16}{\pi G \sum_i w(i)} \sum_i w(i)\ p_i\ v_i^2. \quad (3.6)$$

Here p_i is the projected radius of the star i at radius r_i, which has a weight $w(i)$ and a one-dimensional velocity v_i. Both estimators assume a homogeneous distribution of the stars. The virial theorem assumes additionally an isotropic velocity field. The Bahcall-Tremaine mass estimator assumes that the stars are on isotropic orbits that are dominated by a central mass. Apart from the dominating central mass, these assumptions are usually not fully met in the case of the GC: there are some anisotropies in the velocity field (see sections 3.1 and 3.6.2) and, especially for the innermost stars, where statistics become worse, the available sample with proper motions might not be homogeneously distributed in space.

A very powerful mass estimator that can be applied by using the tangential and radial projections of the two-dimensional proper motion velocity, v_T and v_R, is the Leonard-Merritt (LH) mass estimator (Leonard & Merritt 1989). This mass estimator is less affected by possible traces of anisotropy within the star cluster. For a system with constant mass-to-light ratio the authors find:

$$< M(r) >= \frac{16}{G\pi} < R(\frac{2}{3}v_R^2 + \frac{1}{3}v_T^2) > . \tag{3.7}$$

Applied to the proper motion data obtained within the Sgr A* cluster one finds for both estimators a central enclosed mass of about 2-4×10^6 $M_\odot$ within a mean projected radius of about 10 mpc, depending on the data set and on the methods used for the analysis (Genzel *et al.* 1996; Eckart & Genzel 1997; Ghez *et al.* 1998; Genzel *et al.* 2000; Schödel *et al.* 2003).

3.3.3 *The Jeans Method*

A much better way of deriving the amount of dark mass $M(r)$ that is enclosed within a (three-dimensional) radius r is using the Jeans equation. This method allows the mass calculation without the *a priori* assumption of a dominating central mass or the isotropy of the velocity field. The Jeans equation can be derived from the collisionless Boltzmann equation

$$\frac{\partial f}{\partial t} + v\nabla f + \nabla\Phi\frac{\partial f}{\partial v} = 0 \tag{3.8}$$

with the phase-space density of stars

$$f = f(t, r, v). \tag{3.9}$$

Here v and r are the three-dimensional velocities and separations from the center. For spherical stellar distributions $n(r)$ in a gravitational potential Φ(r) and with velocity dispersions in ϕ and θ directions

$$\sigma_\phi^2(r) = \sigma_\theta^2(r) \tag{3.10}$$

and assuming $\frac{\partial f}{\partial t} = 0$ (no creation or destruction of stars), one finds

$$\Phi r = \frac{GM(r)}{r} = v_{rot}^2(r) - \sigma_r^2(r)(\frac{d\ ln(n(r))}{d\ ln(r)} + \frac{d\ ln(\sigma_r^2(r))}{d\ ln(r)} + 2\beta) \tag{3.11}$$

and can estimate the mass $M(r)$ enclosed within the radius r as a function of v_{rot}, σ_r, n_r, and β:

$$M(r) = M(v_{rot}(r), \sigma_r(r), n_r(r), \beta). \tag{3.12}$$

Here a rotational velocity term v_{rot}(r) has been included to account for a systematical rotation of the cluster. A measure of anisotropy of the velocity field is β, which includes the radial and transverse velocity dispersions:

$$\beta = 1 - \frac{\sigma_\theta^2(r)}{\sigma_r^2(r)}. \tag{3.13}$$

For an isotropic velocity field β=0. If that condition is not fulfilled the degree of anisotropy of the three dimensional velocity field can be determined and introduced into the Jeans equation via the proper motions v_T and v_R (see above). In that case β is given by Eq. (3.1) (see Leonard & Merritt (1989)). The velocity dispersions $\sigma_r(r)$ and densities of stars $n(r)$ as a function of radius r are connected to the observed projected velocity dispersion $\sigma_r(p)$ and the observed surface density $\Sigma(p)$ as a function of projected separation p via Abel integral equations:

$$\Sigma(p) = 2\int_p^\infty \frac{n(r)r}{(r^2-p^2)^{1/2}}dr \tag{3.14}$$

$$\Sigma(p)\sigma_r^2(p) = 2\int_p^\infty \frac{n(r)\sigma_r^2(r)r}{(r^2-p^2)^{1/2}}dr. \tag{3.15}$$

These integral equations can now be used to perform a χ^2 fit of parameterized models for $n(r)$ and $\sigma_r(r)$ to the actually measured projected data $\Sigma(p)$ and $\sigma_r(p)^2$ (see discussions in Kormendy & Richstone 1995; Genzel *et al.* 1996). The used parameterised models are

$$n(r) = \frac{n_o}{1 + (r/r_o)^\alpha} \tag{3.16}$$

$$\sigma_r(r)^2 = \sigma(\infty)^2 \ + \ \sigma(r_{ref})^2 (r/r_{ref})^{-2\beta}. \tag{3.17}$$

Here the central stellar density n_0 is estimated from the projected quantities via $n_0 = \Sigma_0/p_0$, with r_o and the corresponding projected quantity p_o being related to the core radius of the cluster. The modeling parameter β is not to be confused with the anisotropy estimator β, as defined above. The velocity dispersion of the bulge at large distances from the center is $\sigma(\infty)$ and $\sigma(r_{ref})$ the velocity dispersion of the cluster at a reference separation of about 0.4 parsec (corresponding to a projected separation of 10 arcsec) from the center. The exact procedure how the observed data is used to calculate σ(r) and n(r) and the enclosed dark mass M(r) is given in Genzel *et al.* (1996) and Eckart & Genzel (1997). Independent of whether the radial velocities and proper motion velocities are treated as separate or as a combined data set one derives a total enclosed unresolved dark mass of M=3.3×10^6 $M_\odot$ with a combined statistical and systematic uncertainty of $\pm0.35\times10^6$ $M_\odot$. The statistical error is determined by the limited number of stars and therefore the choice of possible annuli around Sgr A* in which they can be grouped. These errors account for about $\pm0.15\times10^6$ $M_\odot$(see detailed discussions in Genzel *et al.* (2000)). The additional systematic error is based on the uncertainties with which the radial and the proper motion velocities can be determined. Since the mass estimates independently derived from the radial velocities and proper motion velocities agree very well with each other the anisotropy in the velocity field must indeed be very small.

Figure 3.4 shows a compilation of various enclosed mass estimates along with three mass models. The blue dash-dotted line is the mass model of the stellar cluster radiating in the near-infrared with a M/L(2 μm) ratio of about 2. In the model shown its core radius is 0.62 pc and its central stellar density $0.5\times10^6 M_\odot$ pc^{-3}. The black straight line is the combination of a $3.0\times10^6 M_\odot$ point mass with the visible (isothermal) stellar cluster and is the best fitting model. The dashed red line represents the combination of the visible stellar cluster with a hypothetical extremely concentrated dark cluster. Here, it was assumed that the central mass is contained in a small compact cluster with a stellar volume density distribution following a model

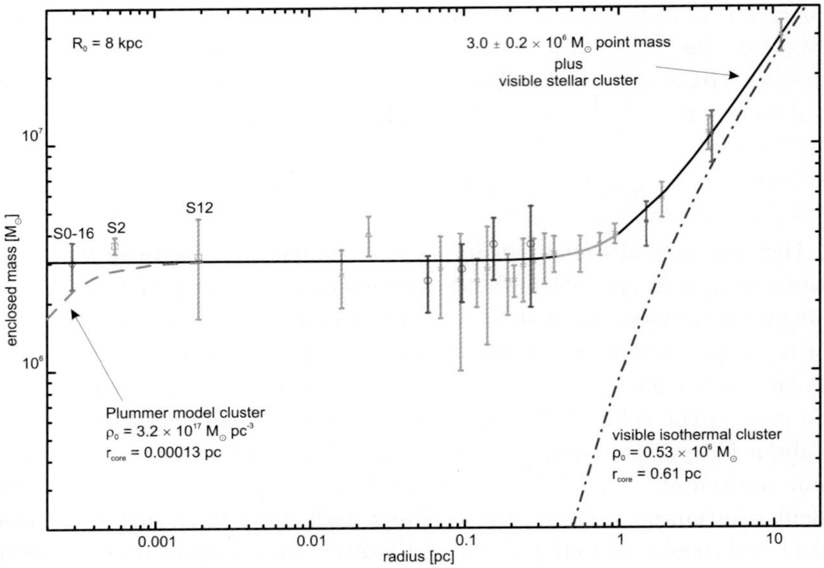

Fig. 3.4 Enclosed mass vs. distance in the GC. Red filled circles are measurements from gas dynamics (Guesten *et al.* 1987). Green crosses are results from Jeans mass modeling of stellar proper motions and line-of-sight velocities by Genzel *et al.* (1997) and Genzel *et al.* (1996). Blue open circles are Leonard-Merritt (LM) mass estimates from Genzel *et al.* (2000). The violet up-pointing triangle is an LM estimate from Schödel (2004). Blue open rectangles are masses from Kepler orbits from Schödel (2004) and Eisenhauer *et al.* (2003b). The dark blue diamond is a Kepler orbit mass estimate from Ghez *et al.* (2003a). Various models for the enclosed are indicated (derived from χ^2-fitting): the blue dash-dotted line is the mass of the large-scale isothermal stellar cluster ($\rho(r) = \rho_0[1 + (\frac{r}{r_{core}})^2]^{\frac{-\alpha}{2}}$, with $\alpha = 2$). The black straight line is this cluster plus a point mass of $3.2 \pm 0.2 \times 10^6$ $M_\odot$; the best fitting model. The red dashed line is the stellar cluster plus a hypothetical compact Plummer model cluster of dark astrophysical objects ($\rho(r) = \rho_0[1 + (\frac{r}{r_{core}})^2]^{\frac{-\alpha}{2}}$, with $\alpha = 5$). The Plummer cluster would be marginally consistent with the mass derived from the innermost orbit (S0-16, see Ghez *et al.* 2003a). Such a cluster would have a lifetime of less than 10^5 years (see section 3.4). See also color Fig. A.11.

of the form

$$\rho(r) = \rho_o[1 + (\frac{r}{r_{core}})^2]^{-\alpha/2}. \tag{3.18}$$

A Plummer model was adopted, which is characterized by an exponential index α=5. A value of α=5 corresponds to the sharpest cluster cutoff that has been observed (Binney & Tremaine 1987). This large value is required since the first order estimates as well as the Keplerian fall-off of the velocity dispersion (as well as the very flat distribution of the velocity dispersion between 0.01 and 0.4 pc) indicate a very compact mass which can only be modeled by a small cluster with a sudden drop off. The core radius and central density of the dark cluster are determined by the innermost mass estimate. In the figure shown this is the orbit of the star S0-16 (Ghez *et al.* 2003a). The cluster can marginally fit the data with a central density of $\rho_0 \geq 3.2 \times 10^{17} M_\odot$ pc^{-3} and a core radius of r_{core}=0.00013 pc. Such a cluster of dark astrophysical bodies would have a lifetime less than 10^5 years (see section 3.4). Further details and similar figures can be found, e.g. in Eckart & Genzel (1996), Genzel *et al.* (1996, 1997, 2000), or Schödel *et al.* (2003).

3.3.4 *The Orbit of the Star S2*

The measurements of proper motion velocities of stars near Sgr A* revealed the presence of a concentrated dark mass of about 3×10^6 $M_\odot$. However, analyses of these data are limited due to various factors. Some of these are the limited amount of stars available for the statistical analyses, insufficient information on possible anisotropy of the stellar cluster, the necessity of binning of the data with radial distance from Sgr A*, or the strongly decreasing number of stars with measured proper motions at distances less than tens of mpc from Sgr A*. Significant progress could be made, however, through the detection of acceleration of individual stars in the immediate vicinity of the dark mass. Ghez *et al.* (2000) reported the first measurements of acceleration for three stars near Sgr A*. This allowed them to pinpoint the position of the dark mass and to derive a lower limit on its density of 8×10^{12} $M_\odot$. Eckart *et al.* (2002a) refined the analysis of stellar accelerations by combining the Keck and NTT proper motion data sets and by taking the influence of geometrical projection effects into account. They found that the amount of the dark mass within the orbits of S1 and S2 must be $3 - 5 \times 10^6$ $M_\odot$, fully consistent with earlier proper motion measurements. They confirmed that S2 was on a bound, highly inclined orbit with large eccentricity, and an orbital period of between 16 and 27 years, consistent with the unique determination of its orbit by Schödel *et al.* (2002), Ghez *et al.* (2003b), Schödel *et al.* (2003), and Eisenhauer *et al.* (2003b) (see also Tab. 2.2 and Fig. 2.17). Spectroscopically obtained line-of-sight velocity measurements of S2 allowed Eisenhauer *et al.* (2003b) a geometric determination of the distance to the Galactic Center, R_0. Eisenhauer *et al.* (2003b) calculated a value of $3.6 \pm 0.6 \times 10^6$ $M_\odot$ and a distance of $R_0 = 7.94 \pm 0.42$ kpc for the dark mass at the focus of the orbit of S2 (the error includes the uncertainty of R_0). This was the first geometrical measurement of the distance to Sgr A* and the Galactic Center, derived from a single stellar orbit. It agreed well with earlier, indirect estimates of that distance (see e.g. Reid 1993).

3.3.5 *Orbits of Other Stars Around Sgr A**

Partial orbits were observed for several other stars besides S2 (Ghez *et al.* 2003a; Schödel *et al.* 2003). The star S0-16 passed Sgr A* at a pericenter distance of just about 60 AU (Ghez *et al.* 2003a). Although currently less well determined than the orbit of S2, they are all consistent with the presence of a compact $3 - 4 \times 10^6$ $M_\odot$ dark object at the position of Sgr A*. Table 2.2 lists the properties of some orbits (Schödel 2004). Figure 2.18 (see chapter 2) illustrates these orbits superposed on a 2003 NACO/VLT image.

3.3.6 *A Lower Limit to the Mass*

As a further constraint on the concentration of the enclosed mass one can estimate a lower limit for the fraction of the dark mass that can be attributed to the radio source Sgr A*. Early speckle spectroscopic observations (Genzel *et al.* 1997; Eckart *et al.* 1999a) and especially most recent high angular and spectral resolution near infrared observations using 8–10 m class telescopes and adaptive optics (Gezari *et al.* 2002; Ghez *et al.* 2003b; Eisenhauer *et al.* 2003b) have resulted in an estimate of the spectral stellar type of some of the sources in the Sgr A* cluster. They are most likely main sequence stars of late O- or early B-type with masses of the order of $m_{star} = 15$ to 20 $M_\odot$. Through several epochs of very long baseline radio interferometry upper limits to the proper motion v_{SgrA*} of Sgr A* itself could be measured (Backer & Sramek 1999; Reid *et al.* 1999). The most recent value is $v_{SgrA*} \leq 8$ km/s (Reid *et al.* 2003b). For the typical proper motions of the high velocity stars within the Sgr A* cluster one finds from the velocity dispersion of the closest, fastest stars $v_{star} \sim \sigma_{SgrA*cluster} = 700$ km/s. The exceptionally low proper motion of Sgr A* indicates that Sgr A* is either moving towards the observer under a very small angle to the line of sight — this can be considered as being very unlikely — or that it is associated with a large mass. Assuming equipartition of kinetic energy between the stars and Sgr A* and physically identifying the dark mass concentration with the radio source Sgr A* one can write:

$$M_{Sgr\ A*} \geq m_{star} \frac{v_{star}^2}{v_{SgrA*}^2}. \tag{3.19}$$

This results in a lower bound to the mass of Sgr A* of the order 10^5 $M_\odot$. Combining this mass estimate with the limit on the size of Sgr A* at 7 mm wavelength of 2.9×10^{13} cm (Bower *et al.* 2004) one obtains a lower limit on the mass density of 3.0×10^{19} $M_\odot$ pc^{-3}. Further details on the subject are given, e.g. in Genzel *et al.* (1997), Reid *et al.* (1999), Schödel *et al.* (2003), and Reid *et al.* (2003b). If all the $\sim3.6\times10^6$ $M_\odot$ as derived from the orbit of S2 are contained in that volume the mass density will be about 35 times times higher: $1.1\times10^{21} M_\odot$ pc^{-3}. These density values have to be compared to the equivalent density of a 3.6×10^6 $M_\odot$ black hole within the volume defined by the Schwarzschild radius of 1.1×10^{12} cm. This density is 2.1×10^{25} $M_\odot$ pc^{-3}.

3.4 Stability of the Enclosed Dark Mass

Knowing the amount and concentration of the central dark mass as well as the velocity dispersion of the central Sgr A* cluster one can examine the stability of hypothetical clusters of dark astrophysical objects under these conditions as a function of mass of the individual cluster members. These arguments are outlined, e.g. in Maoz (1995, 1998) or Genzel *et al.* (1997). For an upper mass range of the cluster constituents of 0.2 to 100 $M_\odot$ the evolution of the cluster is determined by the relaxation time scale for gravitational interactions of the cluster members. This time scale t_{rh} can be calculated within the half core radius r_h. For units of years, solar masses and parsecs (including the gravitational constant G given in these units) and the number of cluster members N, this time scale is given by Spitzer & Hart (1971) (see also Binney & Tremaine 1987) as:

$$t_{rh} = \frac{0.060 M^{0.5} r_h^{1.5}}{m G^{0.5} log(0.4N)}. \tag{3.20}$$

Calculating N from the ratio between the cluster mass and the mean mass m of a cluster member (here $m = 1$ $M_\odot$ for convenience) and choosing

$M_3 = 3.6\times10^6$ M$_\odot$ and $r_{0.01} = 0.01$ pc as suitable units for the Galactic Center one obtains $t_{rh} = 2.8 \times 10^5$ years.

After about 100 $\times$ t$_{rh}$ a significant fraction of the mass of the cluster would get lost through evaporation (Spitzer & Thuan 1972; Binney & Tremaine 1987). Model calculations (Binney & Tremaine 1987) also show that such a cluster undergoes core collapse after

$$t_{collapse} = (10 - 16) \times t_{rh}. \tag{3.21}$$

This is mainly due to binary formation, an increasing collision rate between binaries and single stars (resulting in hardened — i.e. tighter bound — binaries) and energy loss of the binary systems via gravitational wave radiation.

In the case of the Galactic Center, core collapse of a hypothetical dark cluster would thus happen in 5.3×10^5 years for 10 M$_\odot$ black holes, 3.3×10^6 years for 1.4 M$_\odot$ neutron stars and 7.2×10^6 years for 0.6 M$_\odot$ white dwarfs. Evaporation would make such clusters unstable on 10 times these scale estimates, all of which are extremely small compared to the total age of the galactic nucleus. It is likely that during the dense collapse phase a central seed black hole is formed (Rees 1984; Lee 1995; Maoz 1998).

In addition, even if the cluster re-expands after core collapse (due to heating in binary-binary collisions), the stellar number densities during the collapse are so high that the formation of a seed black hole is very likely. However, the Sgr A* cluster cannot be the center of a core collapsing stellar cluster since the mass content of such a center usually is much smaller and its density distribution much shallower than what is found for the Sgr A* cluster.

For masses below 0.2 M$_\odot$ the number of objects that then have to account for the overall measured mass becomes so large, that physical collisions will dominate the cluster dynamics. The collisional time scale can then be written as (Binney & Tremaine 1987):

$$t_{coll} = \frac{1}{16\pi^{0.5}\nu\sigma r^2}. \tag{3.22}$$

Here ν is the number density of stars , σ the velocity dispersion, and r the stellar radius. This scenario will lead to a collapse in less than a few times 10^6 years.

Figure 3.5 is based on the work of (Maoz 1998), who examined the

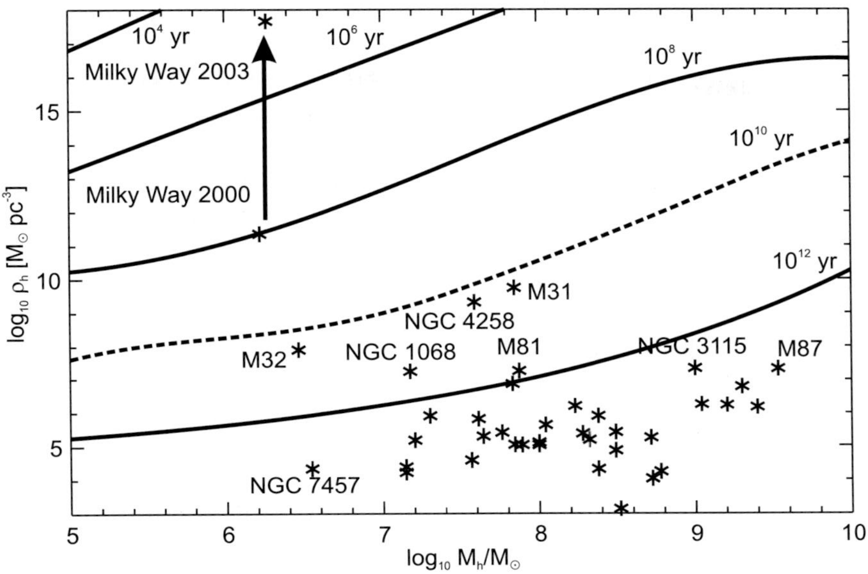

Fig. 3.5 Comparison of masses and densities of observed central dark objects in the nuclei of external galaxies and our Milky Way. Following Maoz (1998), the maximum lifetimes of clusters of dark astrophysical objects (brown dwarfs, white dwarfs, neutron stars, black holes) with a given density and mass are indicated in the diagram. The clusters were assumed to follow a Plummer model density profile, the steepest asymptotic density profile observed in any astrophysical system: $\rho(r) = \rho_0(1 + r^2/r_c^2)^{-5/2}$, with central density ρ_0 and cluster core radius r_c. The horizontal axis represents the half-mass of a cluster, $M_h = M_{total}/2$, and the vertical axis its half-density, which is the density at the half-mass radius (see Maoz 1998). In the case of the Milky Way, the lifetime of any dark cluster with the mass and density corresponding to the observed values at the Galactic Center would have a lifetime considerably lower than the age of our Galaxy. The constraints on the inferred central mass density increased considerably between 2000 (e.g. Genzel *et al.* 2000; Ghez *et al.* 2000) and 2003 (e.g. Schödel *et al.* 2002; Ghez *et al.* 2003b; Schödel *et al.* 2003), reducing the lifetime of any such hypothetical cluster to less than 10^5 years. For comparison, the masses and densities of dark mass concentrations found at the nuclei of other galaxies are indicated as well (the values were taken from Kormendy 2004).

lifetime of clusters of all kinds of astrophysical objects against evaporation and collisional destruction of its constituents. The figure was created in the following way: for each given total mass, mass density, and cluster composition (from stellar mass black holes to sub-brown dwarf objects) the lifetime of a Plummer model cluster is limited by the evaporation and collision time scales, i.e. the lower one of these two values. From all these values the cluster composition with the maximum lifetime was chosen as the limit for a given mass-density point in the diagram. In the case of the Milky Way, the life time of any dark cluster with the mass and density corresponding to the observed values at the Galactic Center would have a life time considerably lower than the age of our galaxy. The constraints on the inferred central mass density increased considerably between 2000 (e.g. Genzel *et al.* 2000; Ghez *et al.* 2000) and 2003 (e.g. Schödel *et al.* 2002; Ghez *et al.* 2003b; Schödel *et al.* 2003), reducing the life time of any such hypothetical cluster from less than 10^8 to less than 10^5 years. For comparison, the masses and densities of dark mass concentrations found at the nuclei of other galaxies are indicated in Fig. 3.5 as well (the values were taken from Kormendy 2004).

A lifetime of 10^5 yr is more than five orders of magnitude shorter than the lifetime of the Milky Way, making the dark cluster hypothesis seem extremely unlikely. Hence, a compact cluster of astrophysical objects, e.g., brown dwarfs, neutron stars, or stellar mass black holes, with the total mass and mass density corresponding to the Galactic Center cannot be stable and does not provide a viable alternative to the supermassive black hole paradigm.

3.5 Agglomerations of Exotic Particles?

In order to explain the extreme mass concentration at the center of the Milky Way two "dark particle matter" models have been under discussion as an alternative to a supermassive black hole. These are the so called "fermion ball" and the "boson star" scenarios.

The fermion ball as an attempt to explain large compact nuclear masses observed at the centers of galaxies was introduced by Viollier *et al.* (1992). They are stabilized by the degeneracy pressure of the corresponding fermion candidates, e.g. neutrinos. The self-gravity of a ball of degenerate fermions can be balanced by the degeneracy pressure of the fermions due to the Pauli principle. In this case, the relation between the mass M and the radius R of a fermion ball, composed of fermions with mass m and degeneracy g, can be described by the non-relativistic Lane-Emden equation:

$$R = \left[\frac{91.869\hbar^6}{m^8 G^3}\left(\frac{2}{g}\right)^2 \frac{1}{M}\right]^{1/3} = r_x \left(\frac{15\text{keV}}{mc^2}\right)^{8/3}\left(\frac{2}{g}\right)^{2/3}\left(\frac{M_\odot}{M}\right)^{1/3}. \tag{3.23}$$

Here the result R is given in lightdays and the used constant r_x equals 3610.66. The maximum mass of a degenerate fermion ball is given by the Oppenheimer-Volkoff limit

$$M_{OV} = 0.38322\frac{M_{Pl}^3}{m^2}\left(\frac{2}{g}\right)^{1/2} = 2.7821 \times 10^9 \left(\frac{15 \quad \text{keV}}{mc^2}\right)^2\left(\frac{2}{g}\right)^{1/2}, \tag{3.24}$$

where the result M_{OV} is given in $M_\odot$ and $M_{Pl} = (\hbar c/G)^{1/2} = 1.2210 \times 10^{19}$ GeV is the Planck mass. For a given fermion mass m all objects heavier than M_{OV} must be black holes.

In case of the GC, the Oppenheimer-Volkoff limit gives us maximum fermion masses of 351 keV for $g = 4$ and 417 keV for $g = 2$. The orbit of S2 tells us $M = 3.6 \times 10^6$ M$_\odot$ and $R = 0.00055$ pc$= 0.655$ ld. Therefore, we obtain as a minimum fermion mass 48 keV for $g = 4$ and 57 keV for $g = 2$. The most massive central dark object currently known is located at the center of M87, with a mass of $> 3 \times 10^9$ M$_\odot$. The Oppenheimer-Volkoff limit would allow a maximum neutrino mass of 14 keV in that case. Comparing this value with the above derived constraints on the properties of a putative neutrino ball at the Galactic Center, one can exclude the possibility that all compact dark objects at the centers of galaxies can be explained by a neutrino ball model.

A motivation for the development of the neutrino ball scenario was that a resolved mass — and therefore a gravitational potential that decreases near the very center — would account for a decreasing radiative efficiency towards its center. This would have helped to explain the low luminosity of Sgr A*. That scenario, however, does not explain what happens to the permanently in-falling (baryonic) matter. It appears plausible that it will be trapped and condense at the bottom of the potential well, where it might eventually form a seed black hole. This scenario defeats the purpose of having a ball of degenerated matter (especially neutrinos; Melia & Falcke 2001). A neutrino ball could not account for the compactness of Sgr A* observed at radio/mm wavelengths. Observations of X-ray and NIR flares from Sgr A* (Baganoff *et al.* 2001, 2003; Eckart *et al.* 2003b; Porquet *et al.* 2003; Goldwurm *et al.* 2003; Ghez *et al.* 2004; Genzel *et al.* 2003a) suggest that the emission comes from structures smaller than about ten Schwarzschild radii of a 3.6×10^6 $M_\odot$ million solar mass black hole. This is more than two orders of magnitude more compact than the radius of a neutrino ball with a neutrino mass of 48 keV.

The only dark particle matter explanation that cannot be ruled out by the present data is a ball of mini-bosons (Maoz 1998), since such a ball could form a very compact configuration that is difficult to distinguish from a black hole. However, it would be hard to understand how the bosons managed to cool sufficiently in order to settle down into such a small volume, and did not form a black hole during that process (Maoz 1998). Boson stars (Kaup 1968) are supposed to be supported by the Heisenberg uncertainty principle. Ruffini & Bonazzola (1969) showed that – e.g. for a boson mass of 1GeV – a stable object of total mass of $10^{-19}M_\odot$ and 1 fm diameter could be formed. If a hypothetical weak repulsive force between bosons is introduced (*ad hoc*) (Colpi *et al.* 1986), it would be possible to form objects with total masses as large as they are found in galactic nuclei (Colpi *et al.* 1986). For a large range of hypothetical boson masses they can have sizes of only several times their Schwarzschild radii. This makes it difficult to clearly distinguish observationally between boson stars and black holes as candidates for supermassive objects at the nuclei of galaxies (see also Torres *et al.* 2000; Mielke & Schunck 2000).

However, even if a boson star had formed at the center, it should eventually have collapsed to a black hole through accretion, during its lifetime, of the abundant gas and dust in the Galactic Center. Therefore we conclude that similar to the fermion ball solution, a supermassive boson star is not an astrophysically attractive explanation for the high mass concentration at the center of the Milky Way. As for the possibility of definitely ruling out the boson star scenario, simultaneous multi-wavelength measurements of the emission from Sgr A* (see Eckart *et al.* 2004a) will allow one to constrain the emission mechanism and therefore the compactness of the emitting region around Sgr A* even further. Probably within the next decade it will be possible to image the "shadow" cast by the putative black hole through deflection of light rays with global radio interferometry at sub-millimeter wavelengths. Such an experiment will involve very long baseline interferometry in the sub-mm regime (Falcke *et al.* 2000; Melia & Falcke 2001).

3.6 The Central Stellar Cusp

The cluster in the Galactic Center is the densest stellar system that can be observed and analyzed in detail at scales from parsecs down to light days. It provides the unique opportunity to study the processes and evolutionary history of a cluster that is dominated by a supermassive black hole. As such it serves as an exemplary case for other quiescent galactic nuclei. Theoretical considerations and numerical modeling predict the formation of a local stellar density excess, a so-called *cusp*, around a SMBH (see e.g. Bahcall & Wolf 1976, 1977; Young 1980; Binney & Tremaine 1987; Quinlan *et al.* 1995). The stellar density is expected to increase approximately with a power-law r^{α} toward the central black hole. The exponents of the power-law density distribution vary between values of $-2.5 \leq \alpha \leq -0.5$ and depend on factors such as the growth rate of the central black hole, the range of stellar masses in the cluster, or the importance of inelastic stellar collisions. Stellar collisions will tend to flatten the slopes of very dense cusps.

3.6.1 *Spatial Distribution of the Stars*

Speckle imaging observations at the NTT and Keck telescopes found the first indications for a cusp around Sgr A* (Eckart *et al.* 1995; Alexander 1999). With the more sensitive and higher spatial resolution adaptive optics observations available at 8–10 m class telescopes since about the beginning of this decade in the near-infrared, spatial scales from light days to a few light years can be probed effectively. The diffraction limited imaging data of NAOS/CONICA at the VLT allowed oneto derive counts of sources with magnitudes $K_s \leq 17$ in annuli centered on the position of Sgr A*. These counts can be corrected for crowding and confusion. The influence of these effects are usually estimated by adding artificial stars to the images and finding the probability of recovering them at the various known locations in the images (see chapter 2.8). Sensitive high-resolution observations of the central stellar cluster and analysis of its surface density were carried out by Genzel *et al.* (2003b). The obtained stellar densities clearly confirm the presence of a stellar cusp. Genzel *et al.* (2003b) found that the cusp has a power-law exponent of $\alpha = -1.4 \pm 0.1$ and is centered on Sgr A* with an uncertainty of just $0.2''$.

In order to estimate the stellar mass of the cusp with a slightly different approach, Mouawad *et al.* (2004) fitted the combined SHARP and CON-

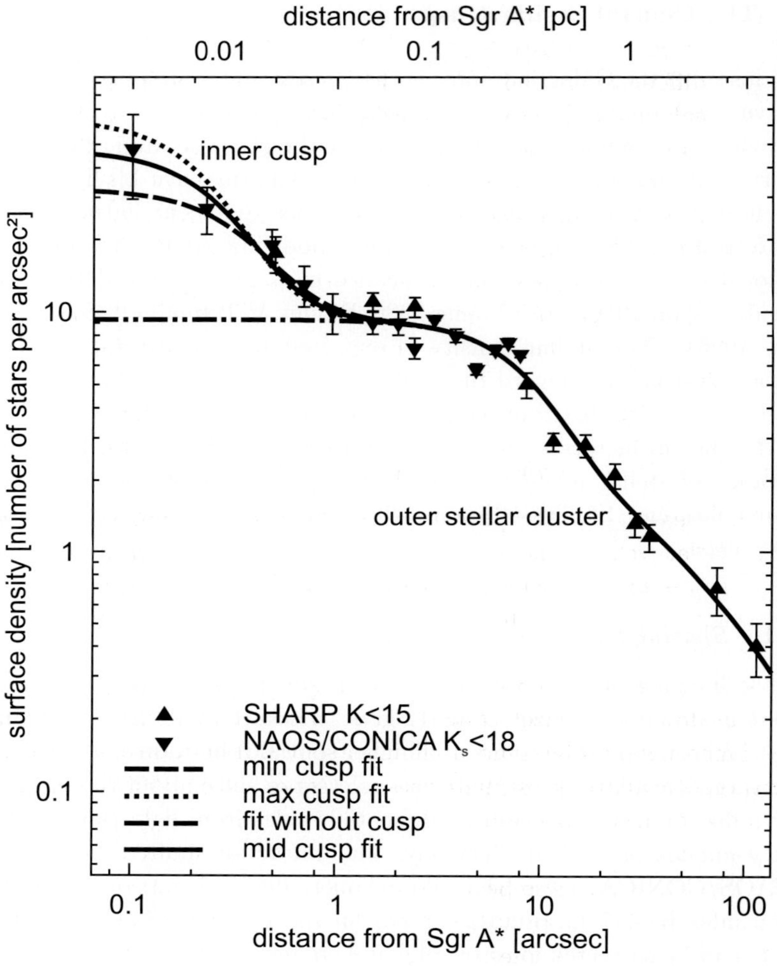

Fig. 3.6 A Plummer model fit to the surface density of stars as a function of distance as given by Mouawad *et al.* (2004). The grey, filled circles represent the CONICA/NAOS data for $K_s \leq 17$. The darker, filled diamonds represent the SHARP/NTT data for stars with $K_s \leq 15$ (Genzel *et al.* 2000), scaled upward by a factor of 5 in order to match the fainter CONICA/NAOS counts. The dashed and dotted curves represent the minimum and maximum fit to the data, respectively. The thin straight curve represents the average fit. The dash-dotted curve shows the stellar density distribution expected for an isothermal stellar cluster. See also color Fig. A.12.

ICA stellar count data with a superposition of several Plummer models of the form: $\rho(r) = \rho(0)[1 + r/r_{core}^2]^{-\alpha/2}$ (α=5), with different densities $\rho(0)$, and different core radii r_{core} (Fig. 3.6). This model emphasizes a distinct break in the density profile at a radial distance of about 1 arcsecond. The cusp is described by a Plummer model with the smallest core radius, r_{core}=0.55 arcsec, and a spatial density $\rho(r)$=4.35×10^7 M$_\odot$ pc^{-3}. The cluster at larger distances is described by the superposition of further Plummer models with larger core radii and lower densities. The Plummer model next to the innermost one (that describes the cusp) has r_{core}=0.135" and $\rho(r)$=6.5×10^8 M$_\odot$ pc^{-3}. The average fit in Fig. 3.6 is outlined by the straight curve. The dashed and the dotted lines represent the uncertainties of the model.

3.6.2 *Velocity Distribution of the Stars*

With the measured stellar proper motions, one can examine the isotropy of the stellar velocity field in the cusp with the anisotropy estimator $\gamma_{TR} = (v_T^2 - v_R^2)/v^2$, where v is the proper motion velocity of a star, with v_T and v_R its projected tangential and radial components. A value of +1 signifies projected tangential motion, −1 projected radial motion of a star. The properties of the anisotropy parameter γ_{TR} are discussed in detail in Genzel *et al.* (2000). They showed that an intrinsic three-dimensional radial/tangential anisotropy will be reflected in the two-dimensional anisotropy estimator γ_{TR}. Genzel *et al.* (2000) found first indications of a possible radial anisotropy in the central arcsecond of the Sgr A* cluster.

Schödel *et al.* (2003) repeated this analysis with a more complete data set and could confirm a prevalence of radial motion in the central arcsecond (see Fig. 3.7). With densities of the order 10^7 to 10^8 M$_\odot$pc^{-3} in the central arc-second, two-body interactions should be comparably frequent and the relaxation time is less than 10^8 yr. In this case we would expect to observe an isotropic velocity field. Therefore, the observed radial anisotropy comes as a surprise. Possibly, only the brighter, potentially young stellar component is un-relaxed and is characterised by radial anisotropy. A larger and deeper proper motion data base or a statistically significant sample of well determined orbits of individual stars in the cusp is needed before one can definitely exclude isotropy of the central cluster. But should the radial anisotropy indeed be proven to be true, theoretical and modelling efforts will be needed to understand this property of the Sgr A* stellar cluster.

The Sgr A* cluster appears to contain young, O/B-type stars the origin of which cannot be explained satisfactorily by current models. A potential anisotropy of the cluster might well provide clues that might help the understanding of its formation. In the scenario proposed by Alexander & Livio (2004), for example, massive stars may become trapped on preferentially eccentric orbits near Sgr A* through interaction with stellar mass black holes, whereby the latter are ejected from the cusp.

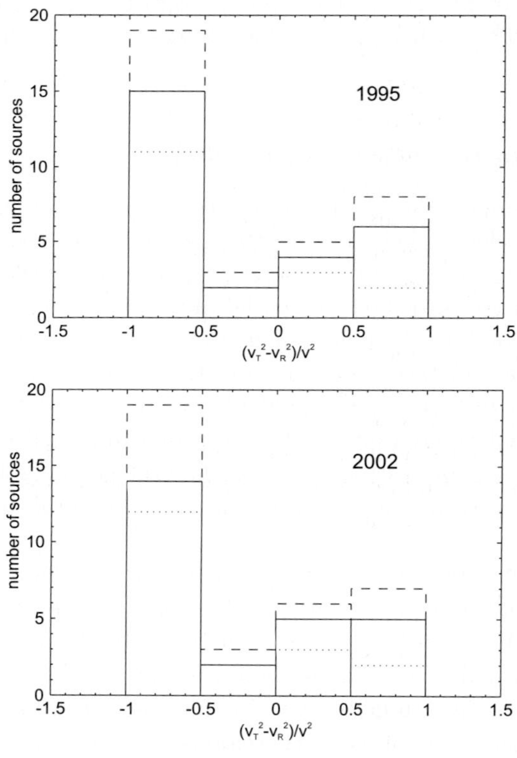

Fig. 3.7 Indications for anisotropic orbital motion of stars within the central cusp of stars around Sgr A* (Schödel *et al.* 2003; Schödel 2004). The anisotropy parameter $\gamma_{TR} = (v_T^2 - v_R^2)/v^2$, evaluated for two epochs separated by 7 years, indicates that the stars are preferentially moving along radial orbits.

3.6.3 *Mass of the Cusp*

Under the assumption that the mass to light ratio of the inner cluster is comparable to that of the outer cluster (M/L$\sim$ 2 μm), Mouawad *et al.* (2004) calculated the enclosed mass as a function of distance from Sgr A*. With the assumed M/L-ratio, the mass present at a distance of 0.55 arcsec from the black hole was found to lie between 8000 $M_\odot$ and 9000 M$\odot$, with about 3100 $M_\odot$ located within the orbit of the star S2.

Using a fourth order Hermite integrator similar to the one used in high-accuracy N-body simulations (Makino & Aarseth 1992; Aarseth 1999; Spurzem 1999), and adapting the best fit value of the mass models in Fig. 3.6, Mouawad *et al.* (2004) computed the trajectory of a star around the black hole for an S2-like orbit (Schödel *et al.* 2002, 2003), taking into account the gravitational potential of extended mass in the cusp. The Hermite scheme allows a fourth order accurate integration based on only two time steps. At this point the use of Plummer model superpositions is very convenient, since the calculation requires the analytic computation of the time derivative of the gravitational force. The resulting retrograde periastron shift amounts to a value of $\sim$1.7 arcmins per orbital revolution which is a few times smaller than the relativistic prograde periastron shift (Rubilar & Eckart 2001). By dropping the assumption of Keplerian orbits, Mouawad *et al.* (2004) showed that not more than 10% of the dark mass present near Sgr A* can be due to an extended component. It is, however, unlikely that — if present — such a high cusp mass is composed out of sub-solar mass constituents, but it could be explained fairly well by a cluster of more massive, high M/L stellar remnants, which Mouawad *et al.* (2004) find to form a stable configuration.

3.6.4 *Young, Massive Stars in the Cusp*

In order to analyse the stellar content of the central cusp the K-band luminosity function (KLF) is of great use (Alexander 1999; Genzel *et al.* 2003b). This function investigates the logarithm of the number of stars per K-band magnitude bin. Genzel *et al.* (2003b) find that the cusp is well pronounced within the central 1.5 acrsec and that the KLF of that region is a featureless power-law. With $d\log N/dK \sim 0.21 \pm 0.03$ it has a similar slope as the integral KLF over the central 9 arcsec. The hump at $K_s \sim 16$ in the KLF of the overall cluster, which is due to horizontal branch/red clump (HB/RC) stars, is absent in the central cusp (see Fig.2.12). In addition the dense cusp also lacks late type giants. This is either due to the real absence of old, low mass stars or, alternatively, red giant branch stars are stripped of their envelopes by collisions or close tidal encounters. Losing their envelope well before their helium core reaches critical mass for helium ignition, these stars would evolve directly to become helium white dwarfs.

In addition to the KLF that allows us to analyse the continuum emission of the cusp stars, near-infrared spectroscopy can help to determine the properties and spectral types of the central cusp population. Early low resolution speckle spectroscopy of the central arcsecond has demonstrated the absence of CO absorption for several individual high velocity stars (Genzel *et al.* 1997). As confirmed for the integral spectra of the Sgr A* cluster (Eckart *et al.* 1999a; Figer *et al.* 2000) this result indicates the presence of main-sequence O8/O9 stars in that region. Gezari *et al.* (2002) and Ghez *et al.* (2003b) have recently reported adaptive optics assisted spectroscopy with the Keck telescope of several high-velocity stars in the cusp near Sgr A* (see also Eisenhauer *et al.* 2003b). Star S2 exhibits HI Brγ absorption corresponding to a hot main-sequence O8/O9 star. Star S1 and the fainter star S0-16 do not exhibit CO absorption features, and thus are likely B0 (S1) and B5 (S0-16) early type stars.

3.6.5 *Dynamics of the Cusp and Presence of Young Stars*

Genzel *et al.* (2003b) found that a large fraction of the early type $K \leq 15$ stars within the central few arcseconds of Sgr A* form a clockwise or counter-clockwise rotation pattern around Sgr A* (see Fig. 3.1). This indicates the presence of a non-relaxed distribution of angular momentum. The most obvious interpretation is that these stars are recently formed young, massive O/B and Wolf-Rayet stars, quite similar to the brighter HeI emission line stars. They must be members of the 2-8 Myr starburst component. This behaviour also requires that their ages are significantly smaller than the two-body relaxation time, which is given by (e.g. Genzel *et al.* 2003b)

$$t_r(m) = 10^{8.2} M_3^{3/2} r_{10}^{-0.13} m_{10}^{-1} (\ln(N_\star/12))^{-1} \quad [yr]\,. \tag{3.25}$$

Here M_3 is the mass enclosed within radius r_{10} (in units of $10''$ or 0.38 pc) in units of $3 \times 10^6\ M_\odot$, m_{10} is the stellar mass in units of $10\ M_\odot$ and $N_\star$ is the number of stars. This must be compared to the (main-sequence) stellar lifetime (e.g. Genzel *et al.* 1994; Cox 2000)

$$t_{ms}(m) = \begin{cases} 10^{7.3} m_{10}^{-2.5} & [yr] \text{ for } m_{10} \leq 0.9 \\ 10^{7.2} m_{10}^{-1.5} & [yr] \text{ for } m_{10} > 0.9 \end{cases} . \tag{3.26}$$

Comparison of the two-body relaxation time and the stellar lifetime shows that the presence of individual young massive stars in the central arcseconds cannot be explained via mass segregation. That means it can be excluded that stars more massive than about 2.5 $M_\odot$ have first formed outside the dense central (≥ 1 pc) cluster region and then sunk via two-body relaxation to the central few arcseconds afterwards.

The effectiveness of dynamical friction is much higher, however, if one considers the infall of entire clusters instead of just individual stars (see section 3.2.1). The idea that the young stars in the central parsec may be the remnants of a dissolved cluster is discussed, e.g. by Gerhard (2001); Portegies Zwart *et al.* (2003) and Kim & Morris (2003). They find that young, massive stars could indeed be transported close to the central black hole within their lifetime if they are transported inwards as part of a massive, dense cluster. However, such a cluster must have a mass of the order $\sim 10^5$ $M_\odot$ or even higher and must not form at distances of much more than about 5–10 pc from the center. Considering these fairly severe constraints, such a scenario does not appear very likely, although there are some indications that the IRS 13 complex of young, luminous stars might be the remnant of a cluster core (Maillard *et al.* 2003).

In the case that a substantial accretion stream around Sgr A* was present at an earlier epoch, the high velocity stars near Sgr A* could have migrated inward as a result of tidal interaction with the dense disk (Ward 1986). However the observed high eccentricities of these stars still remain to be explained.

Gould & Quillen (2003) suggest that S2 was injected into its close orbit by the tidal disruption of a massive binary star, whose primary was more massive than S2 and at least 60$M_\odot$. From numerical integrations they find that 1%-2% of incoming binaries with closest approach equal to 130 AU will leave the secondary in an orbit with an eccentricity close to that of S2.

Another possibility is that the massive, apparently young stars are formed through mergers of lower mass stars. The very high mass density in the central cusp may have resulted in the formation of massive — apparently young — stars through collisions and accompanied rejuvenation of older stars (Morris 1993; Lee 1994; Alexander 1999). The probability of massive stars being formed in this way can be qualitatively assessed by comparing the collisional time scale between stars of different masses with their lifetime. Since the collisional time scale increases with mass, while the respective stellar lifetime decreases, the maximum stellar mass formed in a merger sequence occurs roughly when the two time scales are equal. As shown analytically in Appendix A by Genzel *et al.* (2003b), 10 $M_\odot$ stars can be formed in this way in the innermost cusp, within the central ~0.1 arcsec of Sgr A*, if the merger efficiency in high velocity collisions is large and the angular momentum of the rapidly rotating merger is efficiently removed.

Genzel *et al.* (2003b) conclude that the high velocity stars in the Sgr A* cluster may thus be collisional products — possibly similar to blue stragglers found in globular clusters (Bailyn & Pinsonneault 1995). However, detailed Fokker-Planck calculations by Lee (1994) show that the more massive 'HeI' emission line stars (30–100 $M_\odot$) at projected distances from Sgr A* of the order of 1 to 10 arcsec cannot have formed this way.

Stellar collisions are more frequent in the densest regions of the cusp, at distances $< 1''$ (about 40 mpc) from Sgr A*, where the number density exceeds a few times 10^7 $M_\odot$ pc^{-3}. In order to be effective, these collisions have to occur within the lifetime of the cusp stars. The mass dependence of the main-sequence lifetime is given in Eq. (3.26), and for main-sequence stars one can use the standard radius-mass relationship (Cox 2000). It turns out that collisions are effective in forming 1 $M_\odot$ stars within the central r_{col}=0.6 arcsec. If the merging efficiency is high for grazing collisions (see references in Genzel *et al.* 1996), r_{col} is also the radius within which many 1 $M_\odot$ stars continuously merge into 2 $M_\odot$ stars. However, the high velocity S-sources probably require collisions and merging of more massive stars. For a merger product of 10 $M_\odot$, r_{col} is of the order of 1/10 of an arcsecond i.e. well within the Sgr A* cluster zone.

For the merger product to be stable and the merging efficiency to be high it is also required to quickly remove the angular momentum of the initially rapidly rotating system (e.g. Sills & Bailyn 1999; Sills *et al.* 2001). That could be accomplished via magnetic breaking through a wind/outflow or through a locked circumstellar disk (e.g. Sills *et al.* 2001). If star S2 is such a merger product then this breaking has worked quite efficiently. For the star S2 the rotation velocity of 224 km/s as deduced from NIR spectroscopy is reasonable for the range of likely spectral classifications reaching from O8 to B0 (Gathier *et al.* 1981; Ghez *et al.* 2003b). Thus, in the light of these results and of the theoretical considerations on merging efficiency, S2 appears to be an ordinary main-sequence star rather than a collision product.

Another expected effect of stellar collisions in the cusp region is the destruction of late-type giants (e.g. Alexander & Kumar 2001; Alexander & Livio 2001; Alexander & Morris 2003). Alexander (1999) and Genzel *et al.* (2003b) show that this process becomes very effective for 50 $R_\odot$ giants within the central 1 to 2 arcseconds. If such collisions permanently destroy the giant's envelope, one would expect that the density of late type giants decreases in the central few arcseconds. The drop in the CO-index within the central 5 arcseconds and the replacement of late type stars by the bright and hot stars is discussed, e.g. in Lacy *et al.* (1982), Phinney (1989), Sellgren *et al.* (1990), and Genzel *et al.* (1996). Genzel *et al.* (1996) discuss stellar collisions and external photoionization as two possible mechanisms for removing late-type stars in the core. A more recent analysis of the stellar populations and evolution in the environment of the central stellar cluster is given by Genzel *et al.* (2003b).

Freitag & Benz (2002a) have carried out extensive numerical multi particle Monte Carlo simulations that include the effects of stellar collisions and the presence of a massive black hole at the center of the stellar cluster. They are treated in combination with detailed smoothed particle hydrodynamics calculations of individual collision processes (Freitag & Benz 2002b). These calculation reproduce the standard $r^{-7/4}$ density cusp found by Bahcall & Wolf (1976) in the case in which tidal disruptions are taken into account but collisions of stars are switched off or inefficient. In cases in which stellar collisions are included or very efficient, shallower cusps with exponents around -0.5 are produced. The new calculations show that the role of collisions was probably overestimated in previous works.

3.7 Analysis of Stellar Orbits Near the Central Black Hole

We have seen that the center of the Milky Way contains a massive black hole that is surrounded by a dense stellar cluster with a density cusp centered on the black hole. Due to its proximity it is the only such system that can be studied in detail across the electromagnetic spectrum at high sensitivity and angular resolution. Individual stars can be resolved down to distances of the order one light day with NIR adaptive optics observations at telescopes of the 8-10 m class. Hence, this offers the exiting possibility to study the distribution and dynamics of stars in this exemplary system and test theoretical predictions such as cusp formation, cluster relaxation, or relativistic orbital motion. Stellar orbits probe the gravitational potential at the very heart of the galaxy and can be used as tools for testing the theory of relativity and to infer the distribution of mass in the immediate vicinity of the black hole. They might be useful in detecting dark stellar remnants that might be accumulated at the bottom of the potential well due to mass segregation effects. These studies will become even more important in the very near future due to the increased angular resolution and sensitivity that can be achieved with large aperture interferometry.

3.7.1 *Relativistic Periastron Shift*

With the increased point source sensitivity and resolution due to the combination of large telescope apertures, adaptive optics, and NIR interferometry it is likely that stars with orbital time scales of the order of one year will be detected in the near future. Theses sources, however, will most likely not be on simple Keplerian orbits. The effects of measurable *prograde* relativistic and *retrograde* Newtonian periastron shifts were investigated by Fragile & Mathews (2000), Rubilar & Eckart (2001) and recently by Mouawad *et al.* (2004). Both types of periastron shifts will result in rosetta shaped orbits. A substantial Newtonian periastron rotation can already be expected if only a few percent of the central mass are extended. Measuring the periastron shift of a single star will allow us to determine the inclination of the stellar orbits and to derive inclination corrected shift values. Acquiring the periastron shifts for three stars on orbits with different energy or angular momentum will enable us to solve unambiguously for the compactness, extent and shape of any extended mass contribution (Rubilar & Eckart 2001).

An analysis of the expected periastron shifts has been given by Fragile & Mathews (2000) (see also Munyaneza *et al.* 1998; Jaroszynski 1998, 1999). In those cases in which we consider a central black hole we assume that it is non-rotating (although recent observations indicate a non zero angular momentum; see section 3.8.5) because Jaroszynski (1998) and Fragile & Mathews (2000) have shown that the effects of the black hole angular momentum (as least on the currently observable stars) are negligible over several orders of magnitude of the orbital time scale. We also assume that the stars within the central arcsecond are compact main sequence stars and that their orbits are not influenced significantly by tidal effects on their surfaces.

First, we consider the case where all the mass in the central region is in the form of a single black hole. In this case the relativistic *prograde* periastron advance per revolution is given by (see Rubilar & Eckart 2001)

$$\Delta\varphi = \frac{6\pi G}{c^2}\frac{M}{a(1-e^2)}, \tag{3.27}$$

with a being the semi-major axis and e the eccentricity of the orbit, respectively. The positional shift of the apoastron

$$\Delta s \approx a(1+e)\Delta\varphi = \frac{6\pi G}{c^2}\frac{M}{1-e}, \tag{3.28}$$

is, to first order, independent of the semi-major axis a and only depends on the eccentricity e of the orbit. Therefore, the accuracy with which positions can be measured imposes a lower limit on the eccentricities of orbits for which relativistic periastron shifts can be observed. For an accuracy of 1 mas $=$ 0.04 mpc this corresponds to $e > 0.93$ and for an accuracy of 0.1 mas $=$ 0.004 mpc, $e > 0.35$. Table 2.2 shows that at least three of the stars with observed orbits have eccentricities ≥ 0.8. For these objects the positional shifts will be close to the range of 1 mas. Current high-precision astrometric measurements with the Keck and VLT telescopes achieve positional accuracies of the order 2 mas. This means that next generation instruments such as LBT or VLTI will be needed in order to measure periastron shifts of the currently known stars near Sgr A*. However, with the combined angular resolution and sensitivity of NACO at the VLT (which only became operational at the beginning of 2002) there is hope that faint stars with orbital periods of a few years could be identified in the dense cusp around Sgr A* within the next years. Hence, under optimistic assumptions, progress on the issue could be already made within this decade.

3.7.2 *Lense-Thirring Precession*

The Lense-Thirring effect may have an influence on the motion of stars and gas in the vicinity of the massive black hole at the position of Sgr A* and will result in the precession of orbits. Here we will mainly summarize the effects on the stellar orbits. The Lense-Thirring effect may also be of importance in the analysis of light curves of Sgr A* because it is one of the dynamical modes of the matter being accreted (see section 3.8.6).

As pointed out earlier, most of the young massive He-stars appear to be orbiting the central black hole in a thin stellar disk with a disk opening angle of less than 10°. The three-dimensional orbit of S2 is inclined to this stellar disk by 75°. Levin & Beloborodov (2003) predicted that the orbit should undergo Lense-Thirring precession with the period of (6/a) $\sim 10^6$ yr, where a < 1 is the dimensionless spin of the black hole.

The sense of orbital rotation of S2 is similar to that of the stellar disk, however, its orbital plane is different. S2 is so close to Sgr A* that the black hole spin could make the orbital plane of S2 turn slowly around the spin axis. This effect of general relativity is known as Lense-Thirring precession (Misner *et al.* 1973). Using the measured orbit eccentricity for S2 of $e = 0.87$ and its semi-major axis $r_0 = 4.6 \times 10^{-3}$ pc, Levin & Beloborodov (2003) give as the period of precession for a black hole of three million solar masses

$$P_{LT} = 4.03 \times 10^6 a^{-1} (\frac{3.6 \times 10^6 \mathrm{M}_\odot}{M})^2 \times (\frac{1-e^2}{1-0.87^2})^{3/2} \times (\frac{r_0}{4.6 \times 10^{-3}\mathrm{pc}^3})\mathrm{yr} \tag{3.29}$$

where a is the dimensionless spin of the black hole, with $a = 1$ corresponding to a maximal spin. Thus, even though the S2 orbit is not presently in the plane of the stellar disk, it could have been there a few million years ago. The current angle between the disk and the S2 orbital plane is about 75°, and the disk origin of S2 would require

$$a > 0.21(\frac{3 \times 10^6 M_\odot}{M_{BH}})^2 \frac{5.8 \times 10^6 yr}{t_{S2}} \tag{3.30}$$

where t_{S2} is the time for which S2 has resided in the Sgr A* cluster on a precessing orbit close to Sgr A*.

Lense-Thirring precession of this magnitude could turn the S2 orbit by 75° and it therefore appears to be possible that the S2 orbit originally lay in the stellar disk plane.

Once the orbital elements of other stars within the Sgr A* cluster are known with a similar accuracy as it is now the case for S2, the hypothesis that they originated in the same stellar disk and later precessed out of the disk plane could be tested more thoroughly. Just one additional orbit would be sufficient to determine the direction of the black hole spin, and two additional orbits would allow us to confirm or rule out the precession hypothesis (Levin & Beloborodov 2003). All of this, however, relies on the fact that scattering due to two-body interactions is negligible which may not necessarily be the case given the high cusp mass density (see section 3.7.5).

3.7.3 *Newtonian Effects*

Since the stars are moving on elliptical orbits through the extended mass of the dense stellar cluster their orbits will be subject to Newtonian orbital shifts. In order to study this effect Rubilar & Eckart (2001) considered the simplest case of a spherically symmetric mass distribution. Here it is assumed that a given star can enter the extended mass distribution, and neglect any non-gravitational interactions.

As a consequence of the spherical symmetry of the considered mass distribution, the (Newtonian) gravitational force on a given star depends only on the enclosed mass within the radius corresponding to the position of the star. Therefore, as it moves towards the center of forces, the gravitational force and hence the curvature of the orbit is smaller than in the case in which the whole mass is concentrated within a radius smaller than the periastron radius of the stellar orbit. This leads to orbits with a *retrograde* orbital shift — that is a shift in the *opposite direction* as compared with the relativistic orbital shift.

In principle, one could directly compute the shifts using the Newtonian angular integral for a spherically symmetric system

$$\Delta\varphi = \int_{r_\mathrm{p}}^{r_\mathrm{a}} dr \frac{l}{r^2} \left[2E - \frac{l^2}{r^2} - 2\phi\right]^{-1/2} \tag{3.31}$$

and its post-Newtonian generalization. This is, however, difficult to implement in practice due to the singularities of the integrand at the end points. See the following section or Rubilar & Eckart (2001) for further details.

3.7.4 *Orbits in a Uniform Density Sphere*

Jiang & Lin (1985) presented a simple analytical treatment of the orbits of a test particle which is allowed to enter into the inner region of a sphere with uniform matter distribution. Only the Newtonian gravitational force is considered. In this case, the potential is given by

$$\phi(r) = \begin{cases} \frac{GM}{2R^3}r^2 - \frac{3GM}{2R} & r \leq R, \\ \\ -\frac{GM}{r} & r > R, \end{cases} \tag{3.32}$$

where R is the radius of the sphere of total mass M. They have shown, that for a given M and R the resulting orbit precession is given by

$$\Delta\varphi = 2\arccos\left[\Xi_1(e,a)\right] + \arcsin\left[\Xi_2(e,a)\right] - \frac{\pi}{2}, \tag{3.33}$$

with

$$\Xi_1(e,a) = \frac{1}{e}\left[\frac{a}{R}(1-e^2) - 1\right], \tag{3.34}$$

$$\Xi_2(e,a) = \frac{\frac{2}{R^2} - B}{\sqrt{B^2 + 4A}}, \tag{3.35}$$

and

$$A := -\frac{1}{aR^3(1-e^2)}, \qquad B := \frac{1}{a^2(1-e^2)}\left(\frac{3a}{R} - 1\right). \tag{3.36}$$

Here Rubilar & Eckart (2001) have rewritten the results of Jiang & Lin (1985) in terms of the semi-major axis a and the eccentricity e of the Keplerian approximation valid at larger distances from the focus of the orbit.

Some expected periastron shifts calculated as a pure Newtonian effect are given by Rubilar & Eckart (2001) and recently by Mouawad *et al.* (2004). Depending on the amount of extended enclosed mass the shifts can be substantial even after a single revolution. In the case of the stars in the Sgr A* cusp and assuming a cusp mass of about 7000 $M_\odot$ the expected Newtonian shift would be small and of the order of the relativistic periastron shift ($< 10'$ per revolution).

3.7.5 *Stellar Deflections*

From stellar number counts Genzel *et al.* (2003b) derived a mass density law for the cusp by fitting a broken power-law to the observed number densities:

$$\rho_\star(r) = 1.2 \times 10^6 \left(\frac{r}{0.4\ pc}\right)^{-\alpha} \quad [M_\odot \mathrm{pc}^{-3}], \tag{3.37}$$

with $\alpha = 2.0 \pm 0.1$ at $r \geq 0.4\ pc$, and $\alpha = 1.4 \pm 0.1$ at $r < 0.4\ pc$ (see section 2.8.1). We can calculate the probability of deflection of a star in

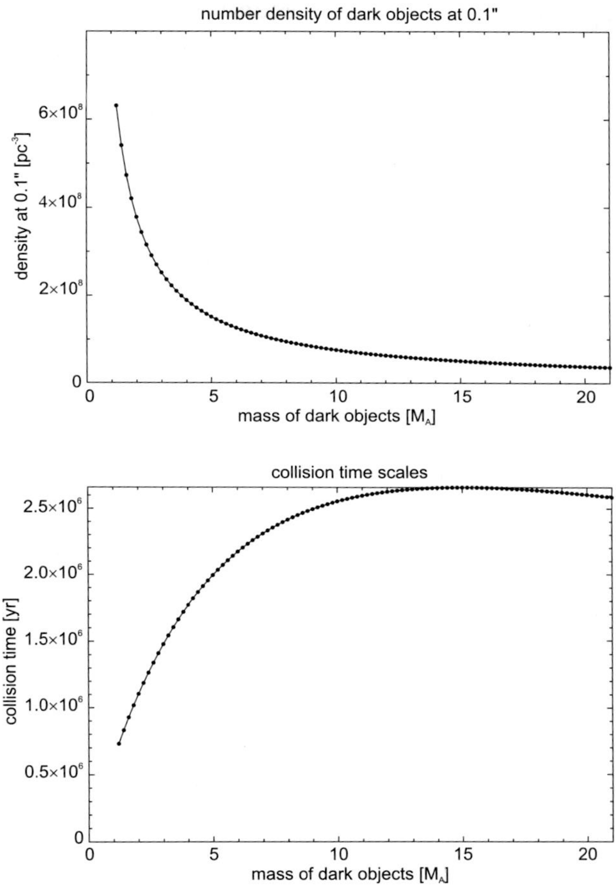

Fig. 3.8 Top panel: number density of objects in the cusp at a distance of $0.1''$ ($\sim$ 3.9 mpc) from Sgr A* versus the mass of these objects. Bottom panel: time after which a star such as S2 experiences a stellar encounter with a deflection $> 10°$ from its path with a probability of one. The time is plotted versus the mass of objects that the cusp might be composed of.

the cusp from its orbit by using this information in combination with the standard scatter law

$$\tan\frac{\theta}{2} = \frac{(m_A + m_B)G}{v^2 s}, \tag{3.38}$$

where θ is the scattering angle, $m_{A,B}$ are the masses of the scattered objects, G the gravitational constant, v the relative velocity at infinity and s the impact parameter (Bowers & Deeming 1984). One might now ask the question of after what length of time the probability for scattering a star by an angle of e.g. $> 10°$ from its path equals one. In order to obtain a rough idea, one can take the orbital parameters of S2 and integrate along its trajectory (neglecting the velocity of the scatter partners). Since one does not know the exact composition of the cusp, this calculation can be repeated for various masses of the particles that constitute the cusp. The upper panel of Fig. 3.8 shows the number density of the cusp at a distance of $0.1''$ (3.9 mpc) from Sgr A* versus the cusp particle mass. The lower panel of Fig. 3.8 shows the time after which the probability reaches one that S2 is scattered by $> 10°$ by an encounter in the cusp, plotted versus the cusp particle mass.

The plot shows that if the cusp consists of objects with masses around 5 $M_\odot$ a star such as S2 will undergo a strong scattering event within a time of roughly two million years. Deflections of the order 1° will even occur on time scales as short as 10^4 years. This means that the nuclear stellar cluster may be dominated by such scattering events. Hence, it is highly improbable that Lense-Thirring precession has a measurable effect on stars such as S2. It also shows that stars in the cusp will lose the characteristics of their original orbits on time scales smaller than the lifetime of stars as massive as S2.

As for the general shape of the probability plots, the mean time between encounters that cause a certain deflection will be very short for a cusp that is composed of low-mass objects. With increasing mass of the cusp particles, the frequency of encounters will decrease. However, the strength of the interactions will increase at the same time. This causes eventually the turnover of the curve at m_{S2} at about 15 $M_\odot$ (see bottom panel in Fig. 3.8).

3.7.6 *Constraints from non-Keplerian Orbits*

In order to determine the central mass and the orbital parameters of the star S2, its positions and radial velocity data have been fitted with Keplerian orbits. The observational data, however, also clearly show that Sgr A* resides in a ~1" diameter stellar cluster of a priori unknown mass (see section 3.6.3). This implies that the proper spherical potential is neither Keplerian nor harmonic. In such a potential the orbits of stars will precess and result in rosetta shaped trajectories in projection on the sky. As Mouawad *et al.* (2004) have pointed out, the assumption of non-Keplerian orbits is a more physical approach in this case. This approach also allows one to derive information on the cusp mass and mass of the central black hole without making preliminary assumptions on the orbits of stars.

Their paper presents the first exemplary modeling efforts in this direction. In addition to the solution of a $3.7\times10^6 M_\odot$ black hole with insignificant cusp mass, Mouawad *et al.* (2004) obtain fits of similarly low χ^2 for a total (MBH plus cusp) mass of $4.1\times10^6 M_\odot$ with a ~10% extended component and $4.8\times10^6 M_\odot$ total mass with a ~25% extended component. Total masses above $\sim5\times10^6 M_\odot$ can be excluded as they are in conflict with enclosed mass estimates at larger radii. It is, however, unlikely that such a large extended mass is composed of sub-solar mass constituents. Such a cluster would very likely be dynamically unstable. A 10% cusp mass could be explained rather well by a cluster of high M/L stellar remnants (e.g. massive stellar black holes etc). Mouawad *et al.* (2004) show that this can lead to a stable stable cusp configuration.

For the detailed investigation presented in Mouawad *et al.* (2004) the authors make use of a realistic description of the potential in which the stars are orbiting both the massive black hole at the very center and the extended stellar cusp. Some of the resulting orbits are shown in Fig. 3.9. Genzel *et al.* (2003b) fitted a broken power-law to their stellar counts, with a distinct break at 10 arcsec. Mouawad *et al.* (2004) show that it is also possible to fit the number density counts with a superposition of several Plummer models (see section 3.6.1). A decomposition using such functions has the advantage of being easily analytically integrable. Plummer models resemble density profiles of actual star clusters with compact cores and an extended outer envelope (Spitzer 1998). In such a decomposition the three-dimensional radial density distribution $\rho(r)$ and the projected mass density $\sigma(r)$ are described by:

$$\rho(r) = \frac{3M}{4\pi r_{core}^3} \frac{1}{(1 + \frac{r^2}{r_{core}^2})^{\frac{5}{2}}} \tag{3.39}$$

and (3.40)

$$\sigma(r) = \frac{4\rho(0) r_{core}}{3} \frac{1}{(1 + \frac{r^2}{r_{core}^2})^2}, \tag{3.41}$$

(3.42)

where r_{core} is the core radius and M the total mass.

The best surface density fit to the data presented by Mouawad *et al.* (2004) consists of a superposition of 6 different Plummer models, which will

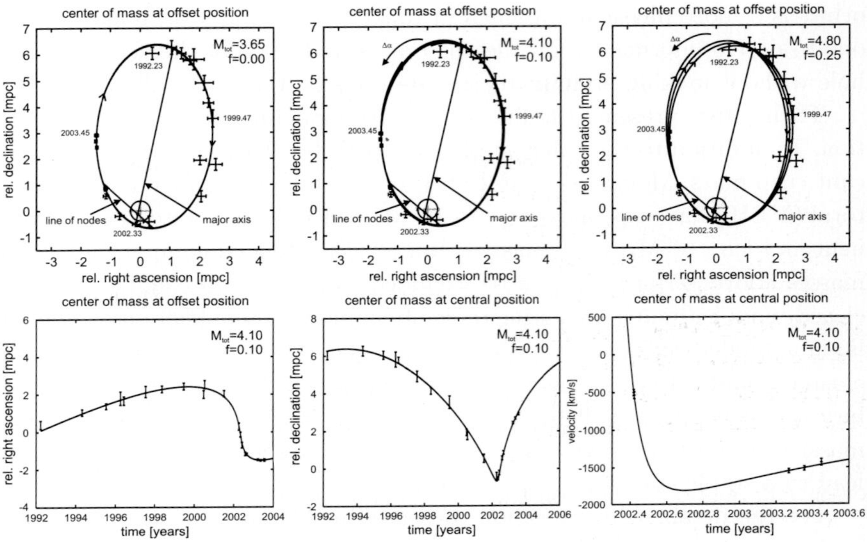

Fig. 3.9 Three exemplary orbits determined by Mouawad *et al.* (2004). Upper left panel: Keplerian orbit with $3.65\times10^6 M_\odot$ point mass. Upper middle panel, non-Keplerian orbits with $4.1\times10^6 M_\odot$ total mass, thereof 10% extended. Upper right panel: non-Keplerian orbits with $4.8\times10^6 M_\odot$ total mass, thereof 25% extended. Here the central mass is at the offset position, 0.082 mpc east and 0.112 mpc south from the nominal radio position of Sgr A* (Schödel *et al.* 2003). The lower panels show the corresponding velocities as a function of relative R.A, the relative Dec., and along the line-of-sight as a function of time for the case of $4.1\times10^6 M_\odot$ total central mass with a 10% extended component. The direction of $\Delta\alpha$, the pericenter-shift, is shown by an arrow. See also color Fig. A.13.

be termed in the following as "composite model":

$$\sigma(r) = \sum_{i=1}^{6} \frac{M_i}{\pi} \frac{r_i^2}{\left(r^2 + r_i^2\right)^2}. \tag{3.43}$$

One can refer to the "inner cusp" as the innermost Plummer model with a core radius $r_1 = 15.5\text{mpc} \cong 0.4''$. The combination of the five outermost components gives a similar configuration as the flattened isothermal sphere of core radius $\sim$0.3 pc derived by Genzel et al. (2003). The mass distribution of the "composite model" is given by

$$M(r) = \sum_{i=1}^{6} \frac{M_i r^3}{(r^2 + r_i^2)^{3/2}}. \tag{3.44}$$

This equation can be scaled and fitted to the dynamical mass estimates as a function of distance from Sgr A* as determined,e.g. by Genzel *et al.* (1996). Such a fit is shown in Fig. 3.6.

The density of the cusp is a few hundred times larger than that of the outer stellar cluster (see Fig. 3.6). The model implies density values as high as 1.68×10^8 $M_\odot pc^{-3}$ at 0.1", and 1.54×10^7 $M_\odot pc^{-3}$ at 1". These values are comparable to the ones derived by Genzel at al. (2003) from a broken power-law density profile (7×10^8 $M_\odot pc^{-3}$ at 0.1"and 3×10^7 $M_\odot pc^{-3}$ at 1"). A future detailed analysis of a larger number of precisely measured stellar orbits (compared to the one available for S2) will allow one to put even stronger constraints on the amount of extended cusp mass.

3.8 The Central Black Hole

The measurement of stellar proper motions in the central cluster, their combination with radial velocities, and especially the detection of orbits in the small Sgr A* cluster has lead to a very reliable measurement of a compact dark mass of $M = 3.6 \times 10^6$ $M_\odot$. From upper limits on the proper motion of Sgr A* (measured with radio interferometry) mass densities of more than 3×10^{19} $M_\odot$ pc^{-3} (e.g. Schödel *et al.* 2003) or even as high as 8×10^{21} $M_\odot$ pc^{-3} can be derived (in the latter estimate all of the enclosed mass was associated with the $<1.2\times10^{12}$ cm Sgr A* radio source Bower *et al.* 2004). Such a compact mass can only exist in the form of a black hole. No stellar cluster of such a mass and mass density can be stable. Also the detection of a variable source at near-infrared and X-ray wavelengths right at the position of Sgr A* is consistent with the presence of a mass accreting black hole. However, the very weak quasi-quiescent and the stronger flare emission from Sgr A* cannot be explained with scaled versions of standard accretion disk models. Current models invoked to explain the luminosity and spectral energy distribution of Sgr A* combine a very low mass accretion rate with a radiatively inefficient accretion flow. In the following sections we will summarize results from recent modeling attempts that are mostly suitable to explain the quasi-quiescent emission of Sgr A*. For an overview of some recent models see, e.g. Melia & Falcke (2001) or

Quataert (2003).

3.8.1 *Standard Accretion Theory*

Interstellar gas in the environment of a black hole will accumulate at the bottom of the gravitational potential well and eventually be accreted onto the central object. Due to the compactness of a black hole, the gas will be strongly compressed during this process. Potential energy of the gas is converted into kinetic energy and dissipation will lead to additional heating of the accreted gas. A certain part of this energy will be radiated away before the material vanishes behind the event horizon. A convenient way to describe accretion and emission of a black hole is the so-called Eddington limit. The Eddington limit on the luminosity L of an object of mass M is set by the balance of gravity and radiation pressure in the object's envelope in the case of spherical accretion. At a distance r from the center of the object the energy flux is

$$f = \frac{L}{4\pi r^2}. \tag{3.45}$$

For each electron this corresponds to a force of $\sigma_t f/c$, where σ_t is the Thomson cross section. If the envelope is bound to the object the radiation pressure is balanced by gravity. For one proton with mass m_p per electron in the fully ionized material of the envelope, the balance condition is

$$\frac{\sigma_t L}{4\pi c r^2} = \frac{GMm_p}{r^2}, \tag{3.46}$$

and the corresponding luminosity is called the Eddington luminosity L_{Edd}

$$L_{Edd} = \frac{4\pi GMm_p c}{\sigma_t} = \frac{4\pi cGM}{\kappa}. \tag{3.47}$$

With σ_t=6.65$\times$10^{-25} cm^2 and m_p=1.67$\times$10^{-24} g the opacity κ due to electron scattering is

$$\kappa = \frac{\sigma_t}{m_p} = 0.4 \ \ cm^2 g^{-1}. \tag{3.48}$$

The rate $\dot{M}$ at which material is accreted at the Eddington luminosity L_{Edd} is $\dot{M}_{Edd}$. L_{Edd} is the maximal luminosity for the case of spherical infall of matter onto a non-rotating black hole. Generally, the accreted gas has a certain angular momentum and will therefore be preferentially accreted in a rotating disk, where it will lose angular momentum due to friction and gradually fall toward the event horizon.

Narayan *et al.* (1998) consider a black hole of m solar masses $M_\odot$

$$M = mM_\odot \tag{3.49}$$

accreting at a rate

$$\dot{M} = \dot{m}\dot{M}_{Edd}. \tag{3.50}$$

$\dot{m}$ is then the accretion rate expressed in Eddington *units* $\dot{M}_{Edd}$, the accretion rate at the Eddington luminosity L_{Edd}:

$$\dot{M}_{Edd} = \frac{L_{Edd}}{c^2 \eta_{eff}} = \frac{4\pi G(M/M_\odot)}{\eta_{eff}\kappa c} \tag{3.51}$$

resulting in an accretion rate of

$$\dot{M}_{Edd} = 1.39 \times 10^{18} g\ s^{-1} = 2.21 \times 10^{-8} (\frac{M}{M_\odot}) \mathrm{M}_\odot \mathrm{yr}^{-1}, \tag{3.52}$$

for $\eta_{eff} = 0.1$, (Narayan & Yi 1995). Since not all the accreted mass is converted into electromagnetic energy, the factor η_{eff} is needed to describe the efficiency of the conversion. Here η_{eff}=0.1 is the accretion efficiency nominally assumed in the case of high density accretion in thin disks (see e.g. Shakura & Sunyaev 1973).

3.8.2 *The Low Luminosity of Sgr A**

There are two ways to estimate the mass accretion rate onto the black hole Sgr A*: from (a) the gas available near the bottom of the potential well and from (b) the measured luminosity of Sgr A* across the electromagnetic spectrum. This comparison is preferentially done in Eddington units.

Stellar winds from surrounding hot, massive He-stars appear to be the primary source of mass available for accretion. The Sgr A* source is located within the central cluster of hot He-stars with winds of the order of 1000 km/s and individual mass loss rates of a few times 10^{-5} $\mathrm{M}_\odot$/yr. These winds supply about 10^{-3} $\mathrm{M}_\odot$ yr^{-1} to the central parsec of the Galactic

Center (e.g. Najarro *et al.* 1997; Melia & Coker 1999; Quataert 2004). The maximum possible accretion rate onto Sgr A* due to this wind was estimated as $\dot{M} \sim 2\times10^{-4}$ M$_\odot$/yr Coker & Melia (see e.g. 1997); Quataert (see e.g. 2003). With Eq. (3.52) this corresponds to an accretion rate in Eddington units of $\dot{m} \approx 1 \times 10^{-3}$ for a $3 - 4 \times 10^6$ M$_\odot$ black hole.

The accretion rate can also be estimated via predicted luminosities. The recent near-infrared identification of Sgr A* results in a de-reddened flux density for the quiescent state of the order 2 mJy at a wavelength of 2.2 μm. This gives a specific luminosity L_ν of 1.5×10^{20} erg s^{-1}Hz^{-1} and a luminosity of $L = \nu L_\nu = 2.1\times10^{34}$ erg s^{-1}. This corresponds to an efficiency for the conversion of mass into electromagnetic energy $\eta \sim 10^{-8} - 10^{-9}$.

Since the luminosity of Sgr A* is wavelength dependent this is a very rough estimate. A more refined analysis with the aid of modeling by Narayan *et al.* (1998), arrives at a bolometric luminosity L_{bol} integrated over all frequencies of only $\sim 2.1\times10^{36}$erg s^{-1}. This corresponds to a radiation efficiency of

$$\eta = \frac{L_{bol}}{\dot{M}_{Edd}c^2} = 5 \times 10^{-6}. \tag{3.53}$$

This has to be compared to the radiation efficiency of 10% that is nominally assumed in the case of high density accretion. For the estimated accretion rate of the Galactic Center this much larger efficiency would imply bolometric luminosities of L$\sim 0.1 \times \dot{M}_{Edd}c^2 \approx 4 \times 10^{43}$erg s^{-1} – a factor about 10^7 higher than what is observed.

3.8.3 *Radiatively Inefficient Accretion Flow Models*

The extremely low luminosity of Sgr A* at all wavelengths was initially a surprise because it was not expected for a black hole of $3 - 4 \times 10^6$ M$_\odot$. However, one has to keep in mind that — in the context of galactic nuclei — initially the theories for accretion onto black holes were developed from observations of extremely energetic AGN, where the mass accretion rate is comparably large. From observations of the linear polarization of the radiation from Sgr A* at sub-millimeter wavelengths, however, it was estimated that it does not accrete more than $\dot{M}10^{-7}$ M$_\odot$ yr^{-1} (Aitken *et al.* 2000; Bower *et al.* 2003).

However, even with this low accretion rate , the luminosity of Sgr A* should be several orders of magnitude higher than what is observed at all wavelengths. It appears that at such low accretion rates, one has to develop a modified theory of accretion that can explain the extremely low radiative efficiency of the gas flow near Sgr A*.

It is the low radiation efficiency of the models of a Radiatively Inefficient Accretion Flow (RIAF, for a brief overview see Quataert 2003) that allows one to fit the observed spectral energy distribution despite the apparently too large mass accretion rate onto the 3-4$\times 10^6$ $M_\odot$ black hole. Current physical models cover a broad range of scenarios (see e.g. Quataert 2003; Yuan *et al.* 2003, for the definitions of abbreviations see following paragraph), such as ADAFs (e.g. Narayan *et al.* 1995; Yuan *et al.* 2002), CDAFs (see Ball *et al.* 2001), ADIOS (Blandford & Begelman 1999), including jet models (Falcke & Markoff 2000), as well as Bondi-Hoyle models (Melia & Falcke 2001), and quasi-monoenergetic electron distributions (Beckert & Duschl 1997). Combinations of models such as a RIAF plus a jet are also discussed (e.g. Yuan *et al.* 2002). Here we summarize some of the essential properties of the different models that are under consideration.

In a purely radial (Bondi-Hoyle) or a low density, non-radial flow most of the rest mass energy of the accretion flow can be advected into the black hole, rather than being radiated away (e.g. Rees *et al.* 1982; Melia 1992; Melia & Falcke 2001). In this case more than 99.9% of the energy in the accretion flow plasma is carried into the black hole by the ions and less than 0.1% is radiated away by the electrons, that — compared to the ions — can cool more efficiently at distances below 100 Schwarzschild radii (see Fig. 2 in Narayan *et al.* 1995).

The Advection Dominated Accretion Flow Model (ADAF) (e.g. Narayan *et al.* 1995, and references therein) was proposed as one of the first RIAFs and was described in an early, qualitative discussion by Rees *et al.* (1982). In an ADAF, the gas accretes on the very short free-fall time scales, and the gravitational energy of the accreting gas is stored as thermal energy and advected into the central black hole. Here the accreted plasma is treated as a two temperature plasma, in which the electrons are Coulomb coupled with the protons. The electrons, which can cool efficiently by radiation carry only a small portion of the energy, since their mass is about 2000 times smaller than the proton mass. The protons radiate less, stay hot, and carry most of their energy beyond the event horizon of the black hole. This advection process works efficiently if the free-fall timescale is short compared to the cooling timescale. In contrast to this scenario, if a "proper" (viscose, luminous) accretion disk is formed the matter that is available for accretion resides a sufficiently long time in the disk for cooling the flow efficiently.

Observations of linear polarization from Sgr A* in the sub-millimeter regime by Aitken *et al.* (2000) and Bower *et al.* (2003) suggest that the rate of mass that effectively accretes onto Sgr A* is lower than 10^{-7} $M_\odot$ yr^{-1}. This is much lower than the gas densities expected from Bondi-Hoyle and ADAF scenarios and makes the latter models unlikely. Also, the observations of the intrinsic size of Sgr A* by Bower *et al.* (2004) agree well with the predictions of a jet model and contradict predictions of thermal, high accretion rate models such as Bondi-Hoyle accretion or ADAFs.

A model in which the gas close to the black hole may react on somewhat longer timescales and thus may be able to explain radiatively inefficient accretion onto a black hole is the convection-dominated accretion flow models (CDAF, Ball *et al.* 2001). In this model angular momentum is transported inward and energy is transported outward by long-wavelength convective fluctuations that determine the structure of the accretion flow.

Theoretical concerns about the assumptions used in the ADAF theory led Blandford & Begelman (1999) to develop their Advection-Dominated Inflow-Outflow Solution (ADIOS). They propose that only a small fraction of the gas that falls toward the black hole is actually accreted by it. Because of the energy released by this process the remaining gas is no longer bound and driven away in the form of a strong wind. In the ADIOS model, the dependency of the gas density on the distance from the black hole scales as $\rho \propto r^{-3/2+p}$, with $p \approx 0.5 - 1$. Hence, the densities near the black hole are lower than in ADAF or Bondi-Hoyle models that predict a dependency $\rho \propto r^{-3/2}$, characteristic for spherical accretion.

This model can explain how the increase in gas density toward Sgr A* can be much shallower than in Bondi-Hoyle or ADAF scenarios, thus delivering an explanation for the very low observed accretion rates.

Falcke & Markoff (2000) propose that the emission from Sgr A* arises primarily in a jet. Yuan *et al.* (2002) developed a jet plus accretion disk model to explain the emission from Sgr A*. In their model the accretion flow is described as an ADAF fed by Bondi-Hoyle accretion of hot plasma. A small fraction of the accretion flow is ejected near the black hole as a short, luminous jet. The electron temperatures obtained in these models are higher than the highest temperature reached in standard ADAF models. The emergent spectrum of Sgr A* is the sum of the emission from jet and underlying ADAF. The strong synchrotron self-Compton radiation from the jet can dominate the Bremsstrahlung from the ADAF. With increasing distance from Sgr A* the plasma in the jet becomes optically thin at ever longer wavelengths. Hence, radiation at different radio wavelengths probes different sections of the jet. Emission at sub-millimeter wavelengths arises at the smallest scales, at the foot of the jet at distances of a few Schwarzschild radii from the black hole. This is a common feature of most models. In the immediate vicinity of the black hole it is hard to distinguish between emission from an accretion flow and from the foot of a jet. It is very difficult to observe a hypothetical jet because the size of the Sgr A* radio source increases quadratically with wavelength because of interstellar scattering. However, recently Bower *et al.* (2004) determined the intrinsic size of Sgr A* using mm-VLBI and closure amplitudes.

The current models predict the spectra from the radio to the γ-ray (Fig. 3.10) domain where the radiation is due to synchrotron radiation (radio to infrared), inverse Compton scattering (infrared to X-ray), bremsstrahlung (X-ray), and possibly pion production above 100 MeV at the highest energies. In most of the models the predicted near-infrared luminosity is produced by the exponential high-energy tail of sub-mm synchrotron radiation or by Compton up-scattered radio synchrotron radiation. Usually, for a given black hole mass M the spectrum of the accretion flow is mainly dependent on a single parameter — namely the mass accretion $\dot{m}$. Changes in the mass accretion rate $\dot{m}$ will most likely also result in a variable luminosity. A large variability is mostly expected in the infrared. For instance changing $\dot{m}$ in the ADAF model by a factor of four changes the infrared luminosity by several orders of magnitude (see Narayan *et al.* 1998, Fig. 3b). Models may also involve shocks in the accretion flow (Melia & Falcke 2001, and references therein) or a quasi monoenergetic synchrotron spectrum (Beckert *et al.* 1996).

The RIAF models are only well suited to explain the quiescent emission of Sgr A*. Recent observational results have shown that the bulk of the quiescent X-ray emission arises in an extended ~0.6 arcsec diameter component, while the NIR quiescent emission is compact and probably unresolved in a ~50 milli-arcsecond beam of a 8-10m class telescope. In summary, current models can explain the sub-luminous radiation from Sgr A* despite the expected fairly large mass accretion rate expected from stellar winds of the surrounding cluster. High angular resolution and time coordinated observations at a wide range of frequencies are required to distinguish between the individual models. Possibly, the emission from Sgr A* arises through a combination of models such as a RIAF plus a jet, similar to the scenario of Yuan *et al.* (2002).

Figure 3.10 summarizes the radio to X-ray spectral energy distribution

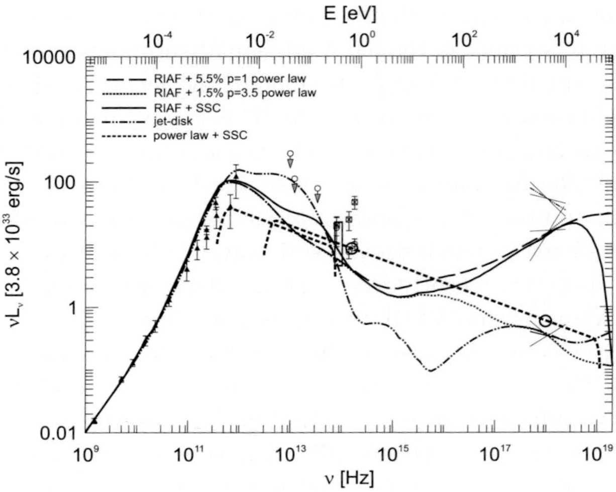

Fig. 3.10 Comparison between model calculations and the luminosity of Sgr A* measured at different wavelengths and in different states. At X-ray and NIR wavelengths: flaring and quasi-quiescent measurements are indicated (see also Genzel *et al.* 2003a). This plot shows the extinction and absorption corrected luminosities νL_ν (energy emitted per logarithmic energy interval), with the observed flux density $S_\nu = L_\nu / 4\pi D^2$, where $D = 7.94$ pc is the GC distance. Black triangles denote the average radio spectrum of Sgr A* and open grey circles mark various infrared upper limits from literature — see Melia & Falcke (2001) and references therein. The three X-ray data ranges are (from bottom to top) the quiescent state as determined with Chandra (black; Baganoff *et al.* 2003), the autumn 2000 Chandra flare (red; Baganoff *et al.* 2001), and the autumn 2002 flare observed by XMM (light blue; Porquet *et al.* 2003). Open red squares with crosses mark the de-reddened peak emission (minus quiescent emission) of the four initially detected NIR flares (Genzel *et al.* 2003a). Open blue circles mark the de-reddened H, K, and L luminosities of the quiescent state, derived from the local background subtracted flux density of the point source at the position at Sgr A*, thus eliminating the contribution from extended, diffuse light due to the stellar cusp around Sgr A*. The thick green solid curve is the jet-starved disk model by Yuan *et al.* (2002). The red long dash-short dash curve is a radiative inefficient accretion flow (RIAF) model of the quiescent emission, where in addition to the thermal electron population of Yuan *et al.* (2002) 1.5% of the electrons have a non-thermal power-law energy spectrum of exponent $p = -3.5$ (Yuan *et al.* 2003). The black thin solid curve is a RIAF model of the flares with 5.5% of the electrons in a power-law of $p = -1$ (Yuan *et al.* 2003) The long-dash thick (grey) curve is a RIAF flare model of the flare with a synchrotron-self Compton component (Yuan *et al.* 2003). The short dashed thick (violet) lines show a typical result for a simple SSC spectrum produced by up-scattering of a power-law distribution of sub-mm wavelength photons into the NIR and X-ray domain given by Eckart *et al.* (2004a), who used the first simultaneous NIR/X-ray measurement of a Sgr A* flare for this purpose. See also the larger color Fig. A.14.

(SED) of the emission from Sgr A* for both quiescent state and flares. The quiescent infrared flux densities lie approximately on the extrapolation of the millimeter/sub-millimeter synchrotron emission to high energies, in accordance with a standard synchrotron SED (e.g. Liu & Melia 2001; Markoff *et al.* 2001; Yuan *et al.* 2002). Current models with only a thermal population of electrons (Liu & Melia 2001; Yuan *et al.* 2002) under-predict the infrared emission. Models with an additional power-law component of energetic ($\gamma_e = E_e/m_e c^2 \geq 10^{2.5}$), non-thermal electrons (Yuan *et al.* 2003) come closer to, but still under-predict the observed NIR flux densities. However, since Sgr A* was only very recently discovered in the NIR, the data from this wavelength regime has not yet been fully taken into account in the modeling efforts and can probably be taken care of by small modifications of the involved parameters. Extensive multi wavelength campaigns are currently monitoring Sgr A* in the 2004 observation season and will also be conducted in the following years. It can be expected that simultaneous measurements and in-band spectral indices (X-ray and NIR) will help to obtain a consistent model of the emission from Sgr A*.

3.8.4 *Sgr A* Flare Models*

The large flares at X-ray and NIR wavelengths (Baganoff *et al.* 2001; Genzel *et al.* 2003a; Porquet *et al.* 2003) added a new and important piece to the puzzle of understanding the nature of gas accretion onto Sgr A*. The discrepancy between the models for the steady-state or quasi-quiescent luminosity from Sgr A* and the luminosity of the flares ranges from factors ≤ 10 in the NIR to more than 100 in the X-ray domain (see Fig. 3.10).

Markoff *et al.* (2001) discuss different models for the flare emission, based on synchrotron or on synchrotron self-Compton emission (SSC). Although they describe these models within the framework of their jet model, they may be regarded as exemplary and applicable to other RIAF models (e.g. Yuan *et al.* 2003; Quataert 2003):

The flare may arise from a short-term overall increase in mass accretion onto Sgr A*. This would result in a significantly enhanced luminosity at all wavelengths. However, Markoff *et al.* (2001) argue that this model is unlikely because of the existing observations at radio wavelengths, where Sgr A* has been monitored regularly for more than two decades and no correspondingly large increase in energy output has ever been observed.

Second, some process in the accretion flow or jet may increase the electron temperature as well as induce a change of the density and/or magnetic field, for example through the conversion of magnetic energy into thermal energy of the electrons by a process such as magnetic reconnection. In combination with the SSC-process, this would result in a large simultaneous increase of emission at mid-infrared and radio wavelengths. This appears unlikely in the light of the observations, which show no large (> 10) variability at these wavelengths (e.g. Hornstein *et al.* 2002; Eckart *et al.* 2003b; Schödel *et al.* 2003).

Markoff *et al.* (2001) call their third third model a "shock flare" or "acceleration flare". It involves processes (e.g. Fermi acceleration or magnetic reconnection) that accelerate a small fraction of the thermal electron population into a power-law tail of the electron distribution function. This model does not cause a large change of emission at infrared and radio/mm wavelengths, but can account for significant changes of the X-ray luminosity. Since the cooling time of synchrotron losses is much shorter than the observed length of the flares, this model requires repeated or continuous acceleration of the electrons.

Flare models such as the ones of Markoff *et al.* (2001) and Yuan *et al.* (2003) raise the idea that the flares from Sgr A* may involve the conversion of magnetic energy into particle energy in the accretion flow. Such events could be similar to solar flares, where magnetic reconnection accelerates the coronal plasma.

Nayakshin *et al.* (2004) proposed that the flaring of Sgr A* may be caused by stars passing through an inactive, fossil, cold accretion disk. However, this scenario appears unlikely because it has problems explaining the lack of a significant offset of the flare emission from the position of Sgr A* and the significant temporal substructure of the flares (see e.g. discussion in Eckart *et al.* 2004a).

In Fig. 3.10 we also show a simple SSC spectrum, based on the first simultaneous NIR/X-ray measurement of a flare from Sgr A*, that was produced by up-scattering of a power-law distribution of sub-mm wavelength photons into the NIR and X-ray domain (Eckart *et al.* 2004a), using the formalism given by Marscher (1983) and Gould (1979). Such a single SSC component model may be too simplistic although it is considered as a possibility in most of the recent modeling approaches. It does not take into account possible deviations of the spectral indices at any specific wavelength from the overall spectral index of the NIR to X-ray spectral index of 1.3. Eckart *et al.* (2004a) find that Lorentz factors γ_e for the emitting electrons of the order of a few thousand are required to produce a sufficient SSC flux in the observed X-ray domain. While relativistic bulk motion, described by the bulk Lorentz factor Γ, is not a necessity to produce sufficient SSC flux density, Eckart *et al.* (2004a) have used modest values for Γ since they will to occur in case of relativistically orbiting gas as well as relativistic outflows — both of which are likely to be relevant in the case of Sgr A*. The size of the synchrotron component is assumed to be of the order of a few times the Schwarzschild radius and a turnover frequency ν_m of a few 100 GHz. For a relativistic bulk motion with Γ=1.2-2 and $\delta=\Gamma^{-1}(1\text{-}\beta\cos\phi)^{-1}$ ranging between 1.3 and 2.0 (i.e. angles ϕ between about 10° and 45°) the corresponding magnetic field strengths are of the order of a few Gauss to about 20 Gauss, which is within the range of magnetic fields expected for RIAF models (e.g. Markoff *et al.* 2001; Yuan *et al.* 2003; Quataert 2003; Narayan *et al.* 2002, 2003).

It can be expected that polarization measurements and simultaneous multi-wavelength observations in the immediate future will provide the necessary constraint for choosing the appropriate model for the Sgr A* flares. It may well be that different models may apply to the weak flares, where the X-ray luminosity changes by factors < 10, and the much rarer strong flares, where the X-ray luminosity increases by factors up to ≥ 100. For example, the first kind of flares could be related to magnetic reconnection events, while the latter type may be related to individual enhanced accretion events. One may also imagine that the true nature of the flares results from a variable contribution from the different proposed mechanisms.

A flaring of Sgr A* could also be caused by a sudden enhancement of accretion as suggested by Liu & Melia (2002). In that case, there might be a significant contribution from bremsstrahlung to the spectrum of Sgr A* at infrared to X-ray wavelengths. A problem of this model is the lack of correspondingly large increases of emission at radio wavelengths (see discussion in Markoff *et al.* 2001).

Although not based on simultaneous observations of the NIR flares at different wavelengths, the data of Genzel *et al.* (2003a) initially suggested that the flares may have a bluer color in the NIR compared to the quiescent emission (Fig. 3.10). A flat to inverted spectrum in the infrared domain would be consistent with a flaring by increased mass accretion as described in the model of Liu & Melia (2002). Such a process is also not excluded by the RIAF/jet models (see above). For the first simultaneous JHK measurements which indicate red flare colors — consistent with theoretical models — see a forthcoming paper of Eisenhauer *et al.*.

Should a blue color of the NIR flares be confirmed in future observations, then the NIR radiation may be thermal bremsstrahlung or blackbody radiation from a component of moderately hot gas (temperatures in excess of a few times 10^3 K). The flares could then be associated to individual accretion events of very dense gas, with a total energy release of the order $\geq 10^{39.5}$ erg. With a standard thin disk radiation efficiency ($\sim 10\%$), this would correspond to a rest mass of a few times 10^{19} g, comparable in mass to that of a comet or a small asteroid. However, what is actually being processed are gaseous filaments out of the turbulent accretion stream due to the high mass loss of the surrounding hot He-stars within the central cluster. Because one would also expect a strong increase at radio/sub-millimeter wavelengths in that case, this radiation would have to come from a jet or else the optical thickness at these wavelengths would be very high.

3.8.5 *A Spin Measurement of the Black Hole?*

The above theories agree amongst each other on the basic physics of the emission mechanisms involved in Sgr A, such as the importance of synchrotron radiation, SSC, or the magnitude of gas densities and magnetic fields. This means that by now we have a basic understanding of the physical processes that create the electromagnetic spectrum of Sgr A*. In this and the following subsections we describe theoretical aspects which are less certain, i.e. are still subject to considerable debate and await further observational confirmation. However, these topics are at the very center of current investigation and will certainly become more important in the near future.*

The light crossing time of the Schwarzschild radius of a 3.6×10^6 $M_\odot$ black hole is just 35 seconds. The short rise-and-decay times of the NIR flares thus are consistent with their origin in a region not larger than 10 R_S, i.e. in the innermost accretion zone. Hence, they may provide information on the conditions in the space immediately surrounding the event horizon of the black hole.

For a creating black hole the most rapid variability of the emission from the in-falling material in its immediate vicinity is expected to occur close to the marginally stable (circular) orbit, $r_{ms}(a)$ (see Bardeen *et al.* 1972), where $a = J/M$ is the black hole's angular momentum parameter. This orbit has a radius $r_{ms}(a = 0) = 3r_S$ for a Schwarzschild black hole, in which the spin parameter $a = 0$. The situation changes, however, if the black hole is rotating (Melia *et al.* 2001; Melia & Falcke 2001). For a maximally rotating object the spin parameter is $a = r_G = GM/c^2$. In that case the location of the innermost stable orbit is strongly influenced by the relative orientation of the spin angular momentum vector of the accretor, i.e. the black hole, and that of the in-falling material. There are three simple cases that can be considered to illustrate the situation. For prograde (+) orbiting material the angular momentum vector on the in-falling matter and the black hole have the same direction and the marginally stable orbit is $r_{ms}(a = r_G, +) = r_G$. For a retrograde (-) marginally stable orbit one finds a radius of $r_{ms}(a = r_G, -) = 9r_G$.

Kepler's third law for circular equatorial orbits in the Kerr metric shows how r_{ms} and the Ω are coupled:

$$\Omega = \frac{2\pi}{P} = \pm \frac{c\, r_G^{1/2}}{r^{1/2} \pm a\, r_G^{1/2}} \quad . \tag{3.54}$$

For these three cases the most rapid fluctuations associated with the 3.6×10^6M$_\odot$ Sgr A* black hole are expected to have the following extreme orbital periods: $P_{min}(a = r_G, +) \sim 3.7$ minutes , $P_{min}(a = r_G, -) \sim 51.8$ minutes, and $P_{min}(a = 0)$ ~27.2 minutes. Therefore, depending on its black hole spin, the emission from Sgr A* is expected to show variations on timescales between 4 and 60 minutes.

Genzel *et al.* (2003a) reported an intriguing 17 min quasi-periodicity that they observed in two flares. This time-scale was not reported in other near-infrared flares because of either the lack of temporal resolution, or an insufficient coverage of the event. Genzel *et al.* (2003a) argued that the fact that the two flares with the 17 min periodicity were observed more than 24 h apart, strongly suggests that the periodicity may be a repeatedly occurring feature.

As for possible frequencies in the accretion zone of the supermassive black hole (see also following section), acoustic waves in a thin disk (Nowak *et al.* 1997), Lense-Thirring or orbital node precession are too slow for explaining the short modulation of the flares (Bardeen *et al.* 1972). The following considerations involve in principle a thin disk. No strong case can be made for the permanent presence of a thin disk near the event horizon. However, one may speculate on the presence of such a structure at least as a transient phenomenon.

For these reasons, Genzel *et al.* (2003a) argued that the 17 min periodicity is the relativistic modulation of the emission of gas orbiting in a prograde disk just outside the last stable orbit (LSO) of a spinning (Kerr) black hole. The period of a non-spinning black hole of 3.6×10^6 M$_\odot$

is 27 min. Therefore, because the LSO has a smaller radius for a spinning (Kerr) black hole, Genzel *et al.* (2003a) determined a spin parameter $a = 0.52(\pm 0.1, \pm 0.08, \pm 0.08)$, where G is the gravitational constant, M_{BH} the mass of the black hole, J its spin, and c the speed of light. The errors in brackets include the influence of the uncertainties of M_{BH}, the flare modulation frequency, and the distance to the GC. The observed spin parameter corresponds to half the maximum value for a Kerr black hole.

Genzel *et al.* (2003a) also reported on a possible — although less firmly established — detection of a 25–30 min quasi-periodicity of the quiescent emission from Sgr A*. Periodicities in the emission from Sgr A* can provide unique insights into the physics of the black hole and the accretion flow as will be further discussed in the following section. Therefore one of the most intriguing questions is whether future observations can confirm the reported modulation frequencies of the emission from Sgr A*.

3.8.6 *Gravitational Disk Modes*

The accretion of matter onto Sgr A* may occur through – perhaps just temporarily present – disks. Different parts of the disk may be luminous at different times and the disk itself may exhibit characteristic oscillations (e.g. Nowak & Lehr 1998). Under these circumstances the overall time dependent energy output of the accreting system may be related to the time varying properties of such a disk structure. Aschenbach *et al.* (2004) have analyzed the two brightest X-ray flares from the X-ray counterpart of Sgr A* one of which was observed by XMM-Newton (Porquet *et al.* 2003) and the other one by Chandra (Baganoff *et al.* 2001) and compared their results to the NIR flare events reported by Genzel *et al.* (2003a). The authors identified quasi-periodic frequencies they discovered in the X-ray flares with those expected characteristic gravitational cyclic modes associated with accretion disks. Predominantly dependent upon fundamental gravitational frequencies there are four cyclic modes. For thin disks they are not strongly affected by hydrodynamic processes. It is still unclear in how far they are responsible for quasi-periodic changes of the energy output at the same frequencies (e.g. Nowak & Lehr 1998).

The circular frequencies of four gravitational cyclic modes associated with black hole accretion disks are given in Nowak & Lehr (1998) and Merloni *et al.* (1999). They are the Kepler frequency (Ω_K), the disk perturbation frequencies in vertical and radial direction called vertical (Ω_V) and radial (Ω_R) epicyclic frequency and the Lense-Thirring precession frequency ($\Omega_{\rm LT}$). Each frequency depends on the central mass M, the angular momentum a and the radial distance r from the center. In the notation by Aschenbach *et al.* (2004) (c=G=1, physical length scales given in units of GM/c^2 and angular frequencies Ω in units of c^3/GM) the corresponding equations can be written as:

$$\Omega_{\rm K} = (r^{3/2} + a)^{-1} \tag{3.55}$$

$$\Omega_V^2 = \Omega_K^2\ (1 - \frac{4a}{r^{3/2}} + \frac{3a^2}{r^2}) \tag{3.56}$$

$$\Omega_R^2 = \Omega_K^2\ (1 - \frac{6}{r} + \frac{8a}{r^{3/2}} - \frac{3a^2}{r^2}) \tag{3.57}$$

$$\Omega_{LT} = \Omega_K - \Omega_V\ \ . \tag{3.58}$$

These equations show how the angular frequencies depend on radius and angular momentum. As shown by Eqs. (3.55) to (3.58) the Kepler frequency $\Omega_{\rm K}$ decreases monotonically with increasing r. For a fixed angular momentum a, both $\Omega_{\rm V}$ and $\Omega_{\rm R}$ initially rise with r, go through a maximum and then decrease with increasing r. The vertical epicyclic frequency $\Omega_{\rm V}$ reaches its maximum at a radial distance denoted by $r = {\rm r_{Vmax}}$ and the radial epicyclic frequency $\Omega_{\rm R}$ takes its maximum value at a distance denoted by $r = {\rm r_{Rmax}}$. For fixed values of radius r and angular momentum a one finds $\Omega_{\rm K} > \Omega_{\rm V} > \Omega_{\rm R}$. This demonstrates that the shortest period is given by the Kepler frequency $\Omega_{\rm K}$ at the smallest possible radius r.

Given the frequencies identified in Sgr A* flare data as they were available until the beginning of 2004, Aschenbach *et al.* (2004) showed that the four relations for $\rm M_{BH}$ versus $(1 - a)$ following from Eqs. (3.55) to (3.58) result in a common value of $\rm M_{BH}$ and a. The best-fit solution is given by $\rm M_{BH} = 2.72^{+0.12}_{-0.19} \times 10^6 M_\odot$ and $a = 0.9939^{+0.0026}_{-0.0074}$. The errors reflect the maximal possible uncertainty given by the maximal uncertainties of the period determination.

In a more recent analysis, Aschenbach (2004) examine 3:2 and 3:1 twin high-frequency quasi-periodic oscillations (QPO) of three microquasars and of Sgr A*. Here it should be mentioned that QPOs have been observed in X-ray binaries for more than two decades (see e.g. review by McClintock & Remillard 2003), and are still subject to highly controversial debate. In the QPO-based interpretation of Sgr A* flares by Aschenbach (2004), there are two orbits at which the vertical and radial epicyclic oscillations are in a 3:2 or 3:1 resonance. Requiring commensurability of these two orbits, i.e. $\Omega_{K,(orbit3:1)} = n \times \Omega_{K,(orbit3:2)}$, with n a natural number, they find that for $n = 3$ there exists a unique solution for the angular momentum, and the radii of the two orbits. With this method, the mass of a black hole that shows QPOs can be calculated conveniently. With the new theory, they find a mass of $3.28 \pm 0.13 \times 10^6$ $M_\odot$for Sgr A*, which is in better agreement with the masses determined through dynamical measurements (see above). If the method suggested by Aschenbach (2004) is confirmed, the independent mass measurement will allow constraining this parameter in orbital fits, thereby reducing significantly the uncertainty of the distance to Sgr A*.

The high value of $a\approx1$ means that the emission from the inner parts of the accretion disk is quite close to the black hole and probably originates on the last marginally stable orbit at a distance from Sgr A* of a few light seconds. Light contributions from further in are very likely to disappear in a very short time and probably do not contribute significantly in this model.

It is currently unclear which value the spin parameter actually has. If the high value of a=0.996 of Aschenbach (2004) is correct then the quasi periodic frequency observed in the previously described NIR flares is well above the Keplerian orbital frequency and must be identified with a lower frequency orbital mode. However, it should be noted that in the case of Kerr black holes with extreme spin, theory indicates that spin-down effects may be important (De Villiers *et al.* 2003). In that case lower spin parameters would be expected. Very high values of a would also be expected with the presence of a strong jet, which is not observed in Sgr A* (e.g. De Villiers *et al.* 2003).

The black hole mass M_{BH} derived from the analysis of the gravitational disk modes is consistent with the Schödel *et al.* (2002) measurement at their 1 σ level ($M_{\rm BH} = 3.7\pm1.5 \times 10^6 M_\odot$) and it also agrees with Ghez *et al.* (2003b) ($M_{\rm BH} = 4.07\pm0.62 \times 10^6 M_\odot$) at their 2σ level, under the assumption of a distance to the center of 8 kpc. Earlier mass estimates derived from statistical ensembles of stars determined $M_{\rm BH}$ to 2.5 - 3.0 $\times$ $10^6 M_\odot$ from measurements of dark matter concentration in the central parsec of the Milky Way (see Fig.3.4, and also, e.g. Genzel *et al.* 1996, 2000). These estimates match the results presented by Aschenbach *et al.* (2004) best.

3.8.7 *Imaging the Event Horizon of Sgr A**

Falcke *et al.* (2000) present a method that may allow us — in the near future — to show the existence of a massive black hole at the center of the Milky Way by direct radio interferometric imaging of the accretion flow. They outline that, to a distant observer, the event horizon will cast a relatively large "shadow" with an apparent diameter of ~10 gravitational radii $r_G = GM/c^2$. That shadow is due to the bending of light by the black hole, and is almost independent of the black hole spin or orientation. The predicted size of that feature of ~30 μas for Sgr A* approaches the resolution of current radio interferometers that operate at short millimeter wavelengths. Falcke *et al.* (2000) show that if the black hole is spinning at maximum and viewed edge-on, then the shadow will be offset by a few μas from the center of mass and will be slightly flattened on one side. Taking into account that the image of the shadow will be broadened by scattering in the interstellar medium and that the achievable resolution of the interferometer is finite, the authors show that the shadow of Sgr A* may be observable with very long baseline interferometry (VLBI) at sub-millimeter wavelengths, which is technically not yet possible. However, this is assuming that the accretion flow is optically thin in this region of the spectrum. Since our current knowledge of the accretion process indicates that this is indeed the fact, there exists a realistic chance of imaging the event horizon of a black hole within the next decade.

3.9 Comparison to Nuclei of Other Galaxies

We can now compare the lower limits on the density of the central dark mass as obtained by different analyses. Observations of high velocity stars and of individual orbits near Sgr A* constrain the central mass density to values $> 10^{18}$ $M_\odot$ pc^{-3}. Even tighter constraints on the mass density come from the proper motion of Sgr A* compared to the velocities in the surrounding star cluster, if one assumes in addition that the size of the dark mass is given by the millimeter radio measurements of Sgr A* (see Rogers *et al.* 1994; Krichbaum *et al.* 1998; Backer & Sramek 1999; Reid *et al.* 1999; Doeleman *et al.* 2001; Reid *et al.* 2003b; Bower *et al.* 2004). These constraints result in a mass density of 10^{19-20} $M_\odot$ pc^{-3}. This limits the lifetime of a hypothetical cluster of dark stellar remnants to less than 10^5 yr (see Fig. 3.5; Maoz 1998). The mass densities have to be compared to the Schwarzschild density, which is the black hole mass divided by the

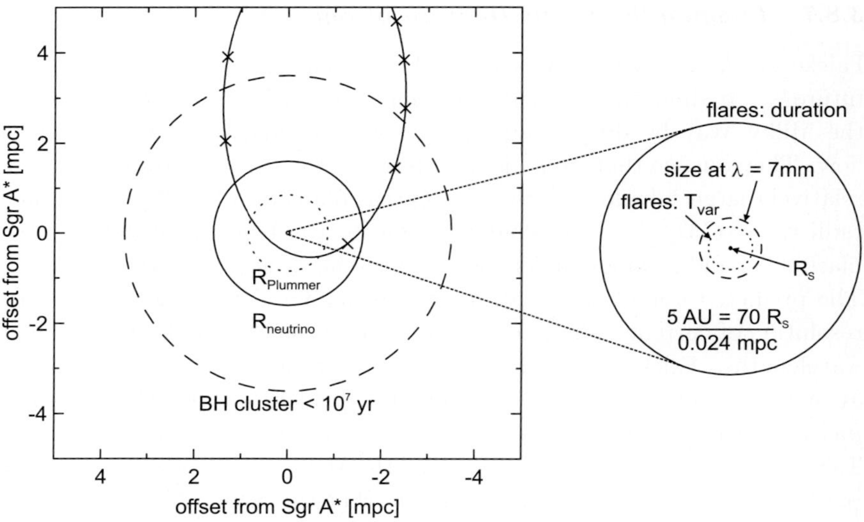

Fig. 3.11 Illustration of the size scales near Sgr A* as probed by different observations. Left panel: Stellar Dynamics, especially the orbits of individual stars, probe the gravitational potential on the scale of several light hours to several light days ($\approx$ 1.2 light days correspond to one mpc). Indicated are the core radii of hypothetical dense clusters (Plummer models): a cluster of 3 $M_\odot$ black holes with a life time of 10^7 yr, marked by a dashed line, and a dark cluster that could marginally fit the gravitational potential as constrained by the orbit of S2. The latter cluster, marked by a dotted line, would have a life time of less than 10^5 yr. The radius of a neutrino ball composed of degenerate 17 keV neutrinos is indicated by a circle with a straight line. All these models are excluded by the observed orbit of S2 and the requirement that a given configuration should have a life time comparable to the life time of the galaxy. Right panel: radio observations and observations of the variability of the X-ray/NIR emission from Sgr A* can probe even smaller scales. Indicated are the size constraints due to the duration of the observed X-ray and NIR flares with a duration of the order 60 min (outer circle, marked by "flares:duration"), the size inferred from 7 mm interferometry (Bower *et al.* 2004), the size limit imposed by the variability (rise-and-fall time, T_{Var}) time of the flares (dotted line), and the Schwarzschild radius, R_S, of a 3.6×10^6 $M_\odot$ black hole.

volume of the Schwarzschild sphere. For a $3{\times}10^6 M_\odot$ black hole we then get a mass density of $9{\times}10^{25} M_\odot$ pc^{-3}. Alternative models like a fermion ball or a boson star (Torres *et al.* 2000) can be largely ruled out and are not very likely (see section 3.5). A tight binary black hole with a separation of less than 10 light hours and a similar total mass can also be largely ruled out (see discussions in Schödel *et al.* 2002, 2003). In summary, current models and measurements suggest very compellingly that the central dark mass concentration at the center of the Milky Way is present in the form of a single supermassive black hole.

In Fig. 3.11 we show the size constraints on the extent of the observed dark mass at the position of Sgr A*: the orbit of the star S2 requires a dark mass of $3.6 \pm 0.3 \times 10^6$ $M_\odot$ to be present in a volume of radius < 0.6 mpc (assuming a distance of 8 kpc; Eisenhauer *et al.* 2003b) and excludes a cluster of dark astrophysical objects as an explanation for the dark mass at the center of the GC. When considering only the size of Sgr A*, radio interferometry and the variability of the NIR and X-ray emission from Sgr A* place even tighter limits, constraining its size to a few astronomical units or a few tens of Schwarzschild radii (for a $3 - 4 \times 10^6$ $M_\odot$ BH).

As summarized in the review by Kormendy & Richstone (1995) the dynamic evidence for large central mass concentrations in galactic nuclei has been growing dramatically over the past decade. The stellar kinematics in elliptical galaxies and central bulges (e.g. Bacon *et al.* 1994; Bender *et al.* 1996; Kormendy *et al.* 1996b,a, 1997; van der Marel *et al.* 1997; van der Marel & van den Bosch 1998), the H_2O maser kinematics in the nearby galaxy NGC 4258 (Greenhill *et al.* 1995b; Miyoshi *et al.* 1995) and the Hα kinematics in M87 (Ford *et al.* 1994; Harms *et al.* 1994) suggest the presence of central dark masses in the range from about 3×10^6 $M_\odot$ in M32 to 3×10^9 $M_\odot$ in M87. Figure 3.11 summarizes the minimum inferred mass densities of the different objects.

From the inferred mass densities and M/L ratios it is clear that almost all cases require unusually dense, and probably unrealistic concentrations of stellar remnants or very low mass stars. M31 (Bacon *et al.* 1994) and M32 (Bender *et al.* 1996; van der Marel *et al.* 1997) are at or above the limits of stable neutron star clusters. The maser disk in NGC 4258 and the Galactic Center stellar dynamics require dark mass concentrations well in excess of 10^9 $M_\odot$ pc^{-3} which can only be explained by a compact cluster of stellar black holes or a single supermassive black hole. The best *extragalactic* case for a central massive black hole is NGC 4258 (Fig. 3.5). For this source the radio interferometrically measured H_2O maser velocities can be fitted by a Kepler rotation curve down to a scale of $\sim$0.1 pc (Herrnstein *et al.* 1998). The implied mass densities are between 10^{11} and 10^{12} $M_\odot$ pc^{-3} (Maoz 1995), close to the estimate for the Galactic Center. As for the Galactic Center, stable dark clusters of neutron stars and stellar black holes at that density can be ruled out. This then leads to the conclusion that most of the dark mass in at least these two objects (i.e. Sgr A* and the nucleus of NGC 4258) must be in form of a (single) massive black hole.

3.10 Massive Black Holes at High Redshifts

An early detailed study of compact dark masses at the centers of external galaxies was given by Kormendy & Richstone (1995). Comparing the derived black hole masses with the luminosities and luminosity based masses revealed a correlation between both quantities (Magorrian *et al.* 1998) which implied that the massive black hole (MBH) mass amounts to 0.6% of the bulge mass.

This value did not remain undisputed. A thorough three-dimensional modeling of HST data by Ho (1999) gave an about three times smaller value of 0.2%. An independent derivation of the black hole masses via reverberation mapping (Ho 1999; Wandel 1999) of the broad line regions of active galaxies indicated that the black hole masses in Seyfert galaxies and PG QSOs may be 5 to 20 times smaller than in more active QSO nuclei.

Most of the discrepancies in the black hole to bulge mass or luminosity relation may have emerged due to the uncertainties in the derivation of the bulge properties. As shown by Ferrarese & Merritt (2000), Gebhardt *et al.* (2000) and others the relation between the black hole mass and the stellar velocity dispersion σ_* of the bulge resulted in a much tighter correlation. This result can probably be explained by the fact that the stellar velocity dispersion traces the stellar mass of the bulge in a more reliable way than the bulge luminosity. A more recent derivation of the relation is given by Tremaine *et al.* (2002):

$$log(MBH/M_\odot) = \alpha + \beta log(\sigma_*/\sigma_0) \tag{3.59}$$

with β=4.02±0.32 and α=8.13±0.06 for σ_0=200 km s^{-1} (see Fig. 3.12). This is in good agreement to earlier results by Ferrarese & Merritt (2000) and Gebhardt *et al.* (2000). The data suggests that most massive galaxies host a massive black hole in their centers. Detailed studies on nearby galaxies like M31, M32 (Bender *et al.* 1996; Corbin *et al.* 2001), M33 (Gebhardt *et al.* 2001), or galaxies with central minima in their stellar luminosity density (Lauer *et al.* 2002) show that the situation may be more complicated, and that each system has to be looked at in more detail.

The determination of the stellar velocity dispersion in the bulge region is very difficult. It relies on the precise determination of stellar absorption line profiles in the bulk spectra along different lines of sight towards the bulges. This requires very high signal to noise spectra. In general it also requires some knowledge or assumptions on the stellar populations for the

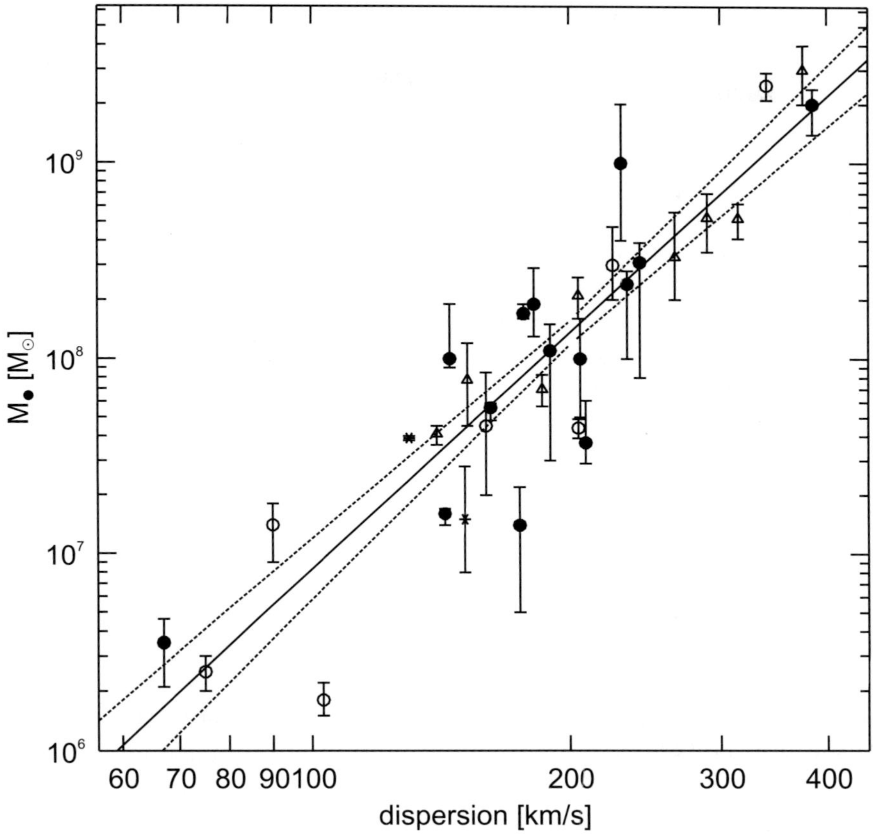

Fig. 3.12 Data presented by Tremaine *et al.* (2002) showing the correlation between black hole mass and velocity dispersion for galactic nuclei, along with the best-fit correlation (see equation in text). Mass measurements based on stellar kinematics are denoted by circles, on gas kinematics by triangles, and on maser kinematics by asterisks. The dashed lines show the 1σ limits on the best-fit correlation. See Tremaine *et al.* (2002) for a detailed discussion.

selection of the appropriate template spectra, which are needed for the cross-correlation analysis of the velocity fields. Especially close to the most luminous active nuclei which can be up to 100 times brighter than the overall host galaxy this is a very challenging task. It is therefore difficult to extend this fairly tight MBH–σ_* relationship to higher redshifts and it is desirable to search for brighter tracers of the nuclear dynamics.

Nelson (2000a) uses the forbidden oxygen line [OIII] at 500.7 nm. He finds no difference between the MBH–$\sigma_{[OIII]}$ masses and black hole masses derived from reverberation mapping methods and proposes to use

$$\sigma_{[OIII]} = FWHM([OIII])/2.35\ km/s \tag{3.60}$$

as a surrogate for the stellar velocity dispersion σ_*. The usage of the [OIII] line, however, raises many problems. It is almost certain that the [OIII] linewidth in many galactic nuclei does not only trace the central mass distribution. In many cases it is affected by nuclear outflows which can give rise to so-called extended narrow line regions (e.g. Schmitt *et al.* 2003; Veilleux *et al.* 2003). The line, however, has the advantage of being a bright (possible) tracer of the nuclear dynamics. It usually originates close to the nuclei and is much easier observable at high redshifts than stellar absorption lines. Future investigations have to show whether this or other lines can in fact be used as reliable surrogates for σ_*. A recent analysis by Shields *et al.* (2003) which is based on the [OIII] line and compares black hole masses that are derived from both the Hβ and the [OIII] line finds that both estimates give comparable results in the redshift range $0<z<3.5$.

It is not clear what the origin of the MBH-bulge mass correlation is. The fact that luminous QSOs are observed at redshifts of $z \geq 6$ suggests that some $10^9 M_\odot$ black holes must have existed already 10^9 years after the big bang (see e.g. Becker *et al.* 2001; Haiman & Loeb 2001). Most of the bright QSOs, however, can be found between z=2 and 3 and probably go through a significant accretion phase at these redshifts. Based on model calculations Yu & Tremaine (2002) find that half the black hole mass is likely to be accreted before $z \sim 1.8$ and only 10% before z=3.0. Shields *et al.* (2003) deduce from their result obtained from the Hβ and the [OIII] lines that black holes typically grow 'contemporaneously' with their host galaxy bulges or else both are already well formed by $z \sim 3$. In summary these results suggest that the MBH to bulge mass relation is obeyed at least by massive QSOs even at times at which much of their accretion lies still in their future.

A recent investigation by Bromm & Loeb (2003) shows how the massive black holes may have formed out of the gas in the bulge environment. The authors present calculations in which at least for metal-free dwarf galaxies with a virial gas temperature of 10^4K and suppressed H_2 formation due to an intergalactic UV background the formation of black hole binaries with masses of a few $10^6 M_\odot$ each may have occurred. In a dense stellar bulge environment these binaries will subsequently result in black hole mergers of large masses (e.g. Portegies Zwart & McMillan 2000). The model calculations by Bromm & Loeb (2003) result in morphologies of the gas concentrations settling into the dark matter potentials at redshifts of z=10, which are very similar to the lumpy appearance of the most distant galaxies known to date (e.g. Conselice *et al.* 2003; Labbé *et al.* 2003).

Appendix A

Color Versions of Selected Figures

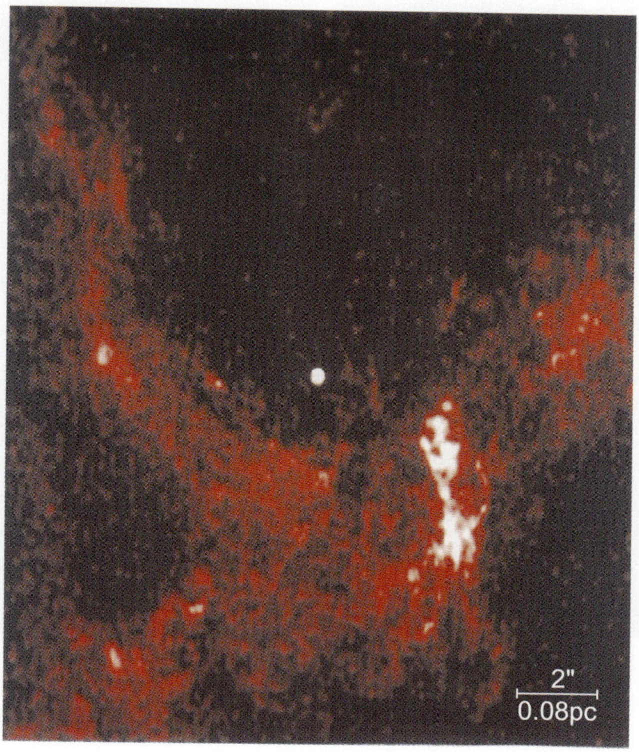

Fig. A.1 VLA image of the mini-spiral at a wavelength of 1.3 cm (Roberts & Goss 1993).

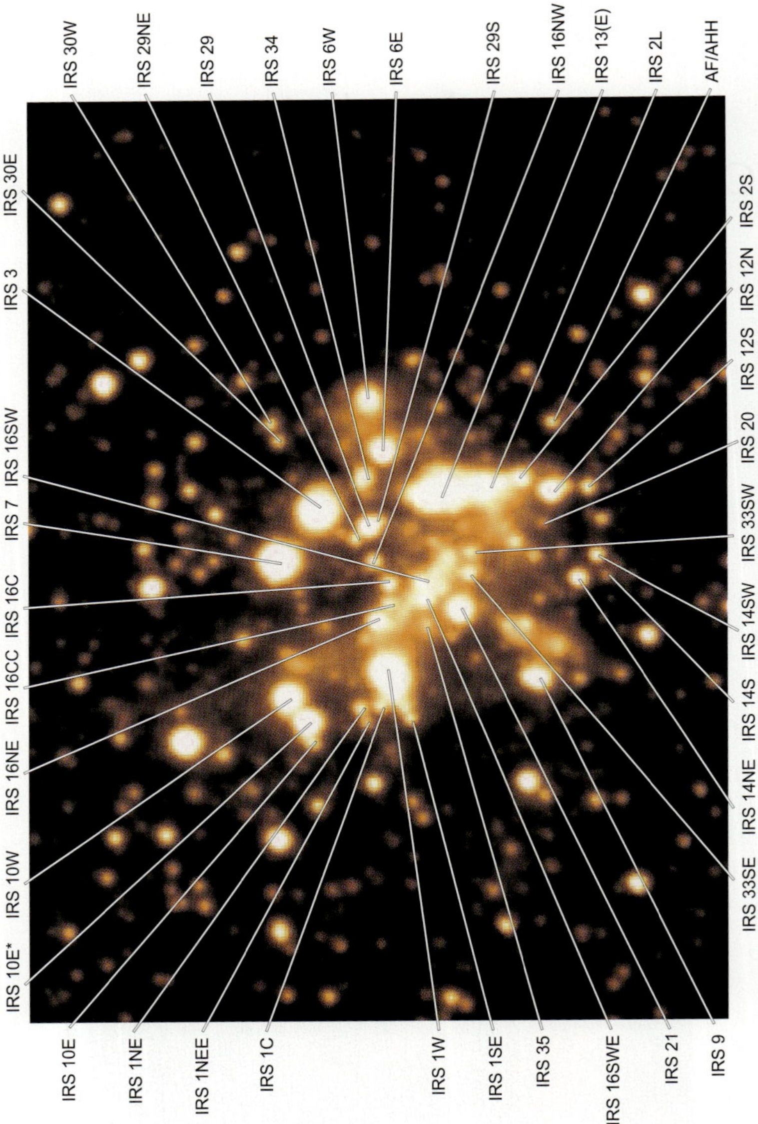

Fig. A.2 M-band image of the central stellar cluster. The names of some individual sources have been indicated (see Viehmann *et al.* 2005) .

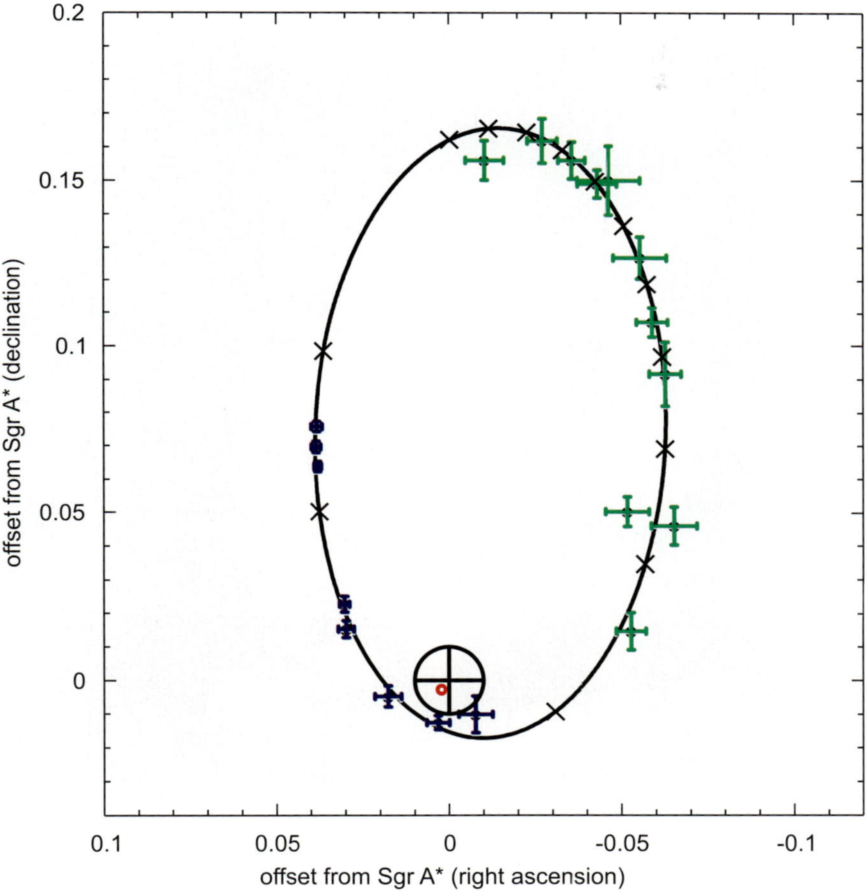

Fig. A.3 Orbit of the star S2. Green: positions measured with SHARP/NTT between 1992 and 2001. Blue: positions measured with NACO/VLT between 2002 and 2003. The black crosses give the position of S2 at January 1 of 1992-2004. The black cross and the 10 mas error circle at (0,0) indicate the nominal radio position of Sgr A*. The small red circle inside the radio position uncertainty indicates the uncertainty of the position of the focus of the orbit (see also Schödel *et al.* 2003; Eisenhauer *et al.* 2003b; Schödel 2004).

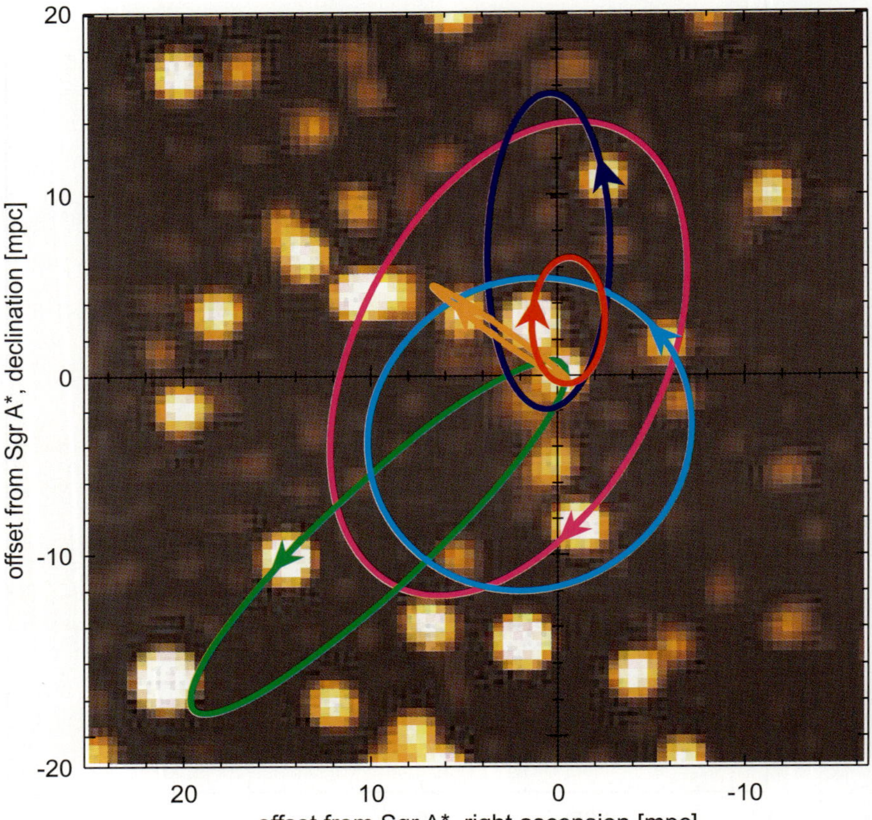

Fig. A.4 Orbits of stars around Sgr A*. Lucy-Richardson deconvolved and beam-restored high-resolution (FWHM∼ 60 milli-arcseconds) near-infrared (2.2 μm) image of the nuclear stellar cluster in the immediate environment of Sgr A*. The image was obtained with CONICA/NAOS at the ESO VLT in June 2003. On this image Sgr A* can be seen in its flaring state (see chapter 2.13.4) as a point source at the origin of the coordinate system. The Keplerian orbits of six stars are over-plotted onto the image (see also Schödel 2004). Arrows indicate the locations of the respective stars and their directions of motion. North is up and east is to the left.

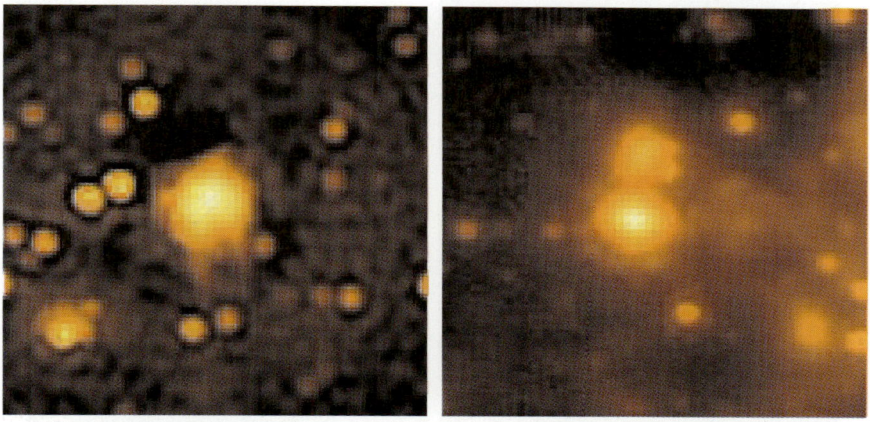

Fig. A.5 Bow shock structures in the GC stellar cluster. Due to the interaction of windy stars with the interstellar medium in the GC, numerous examples of bow shock sources can be found in the central parsec. This Figure shows two exemplary shocks. Left: IRS 1W at 2.2 μm (NACO/VLT May 2003; deconvolved and restored to 60 mas FWHM); the area shown covers 1.6" $\times$ 1.6". Right: bow shock north of IRS 9 (the bright source in the center) at 3.8 μm (NACO/VLT August 2002); the area shown covers 3.2" $\times$ 3.2".

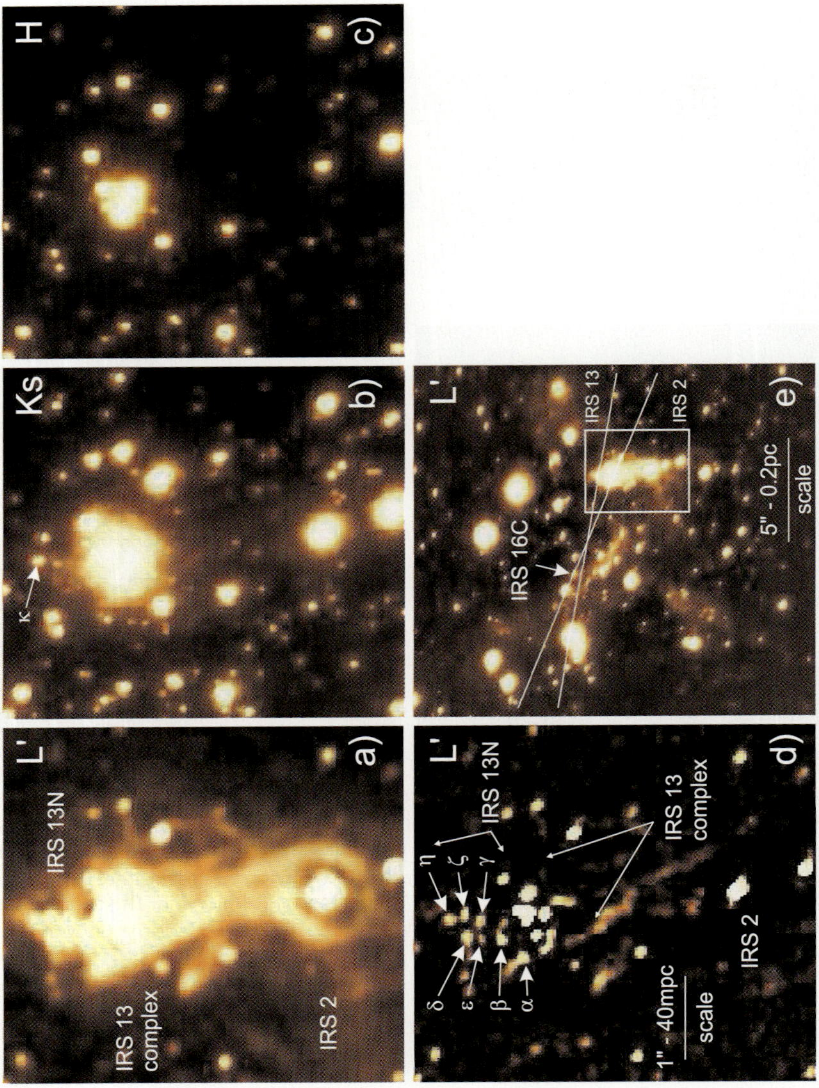

Fig. A.6 H-, K_s-, and L'-band images of the IRS 13/IRS 2 region as shown by Eckart *et al.* (2004b). The image scales of panels (a)-(d) are given in (d). Panel (d) is a high pass filtered L-image which shows the locations of individual stars, including the newly discovered L-band excess sources α through η. Source κ in panel (b) is located between ζ and δ. Panel (e) is an overview of the region; the box indicates the area shown in panels (a)-(d); slit locations for the spectra shown in Eckart *et al.* (2004a) are also indicated.

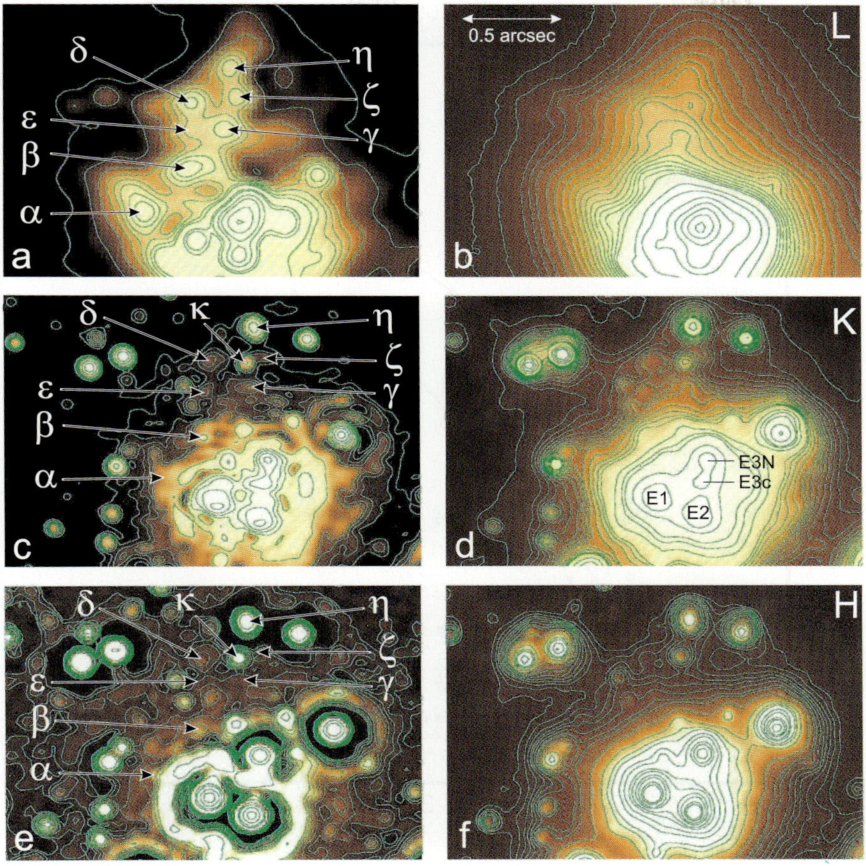

Fig. A.7 A detailed view of the IRS 13 and IRS 13N complexes with NAOS/CONICA as shown by Eckart *et al.* (2004b). They are shown in the L-band, top images (a) and (b), K-band, center images (c) and (d), H-band, bottom images (e) and (f). The images (b), (d), and (f) are raw shift-and-add images of individual adaptive optics exposures. The images (a), (c), and (e) are deconvolved images restored at the nominal angular resolution of the AO system. The ring-like structures around bright sources are artifacts of the deconvolution process. For clarity in the crowded field we show contours overlaid with a color (grey-scale) version of the same image. The positions as well as magnitudes and colors of individual sources are given in Table 2 in Eckart *et al.* (2004a).

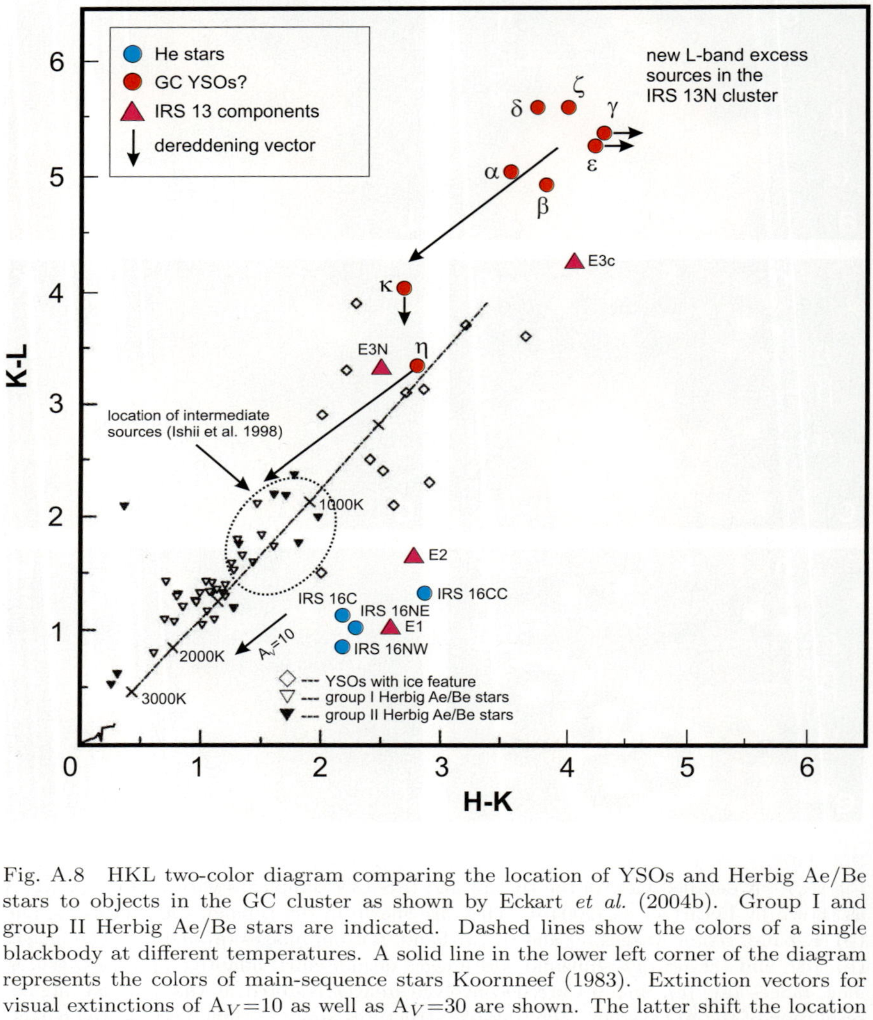

Fig. A.8 HKL two-color diagram comparing the location of YSOs and Herbig Ae/Be stars to objects in the GC cluster as shown by Eckart *et al.* (2004b). Group I and group II Herbig Ae/Be stars are indicated. Dashed lines show the colors of a single blackbody at different temperatures. A solid line in the lower left corner of the diagram represents the colors of main-sequence stars Koornneef (1983). Extinction vectors for visual extinctions of A_V=10 as well as A_V=30 are shown. The latter shift the location of the L-band excess sources to the location of YSOs or the intermediate population "N" and "E" sources introduced by Ishii *et al.* (1998).

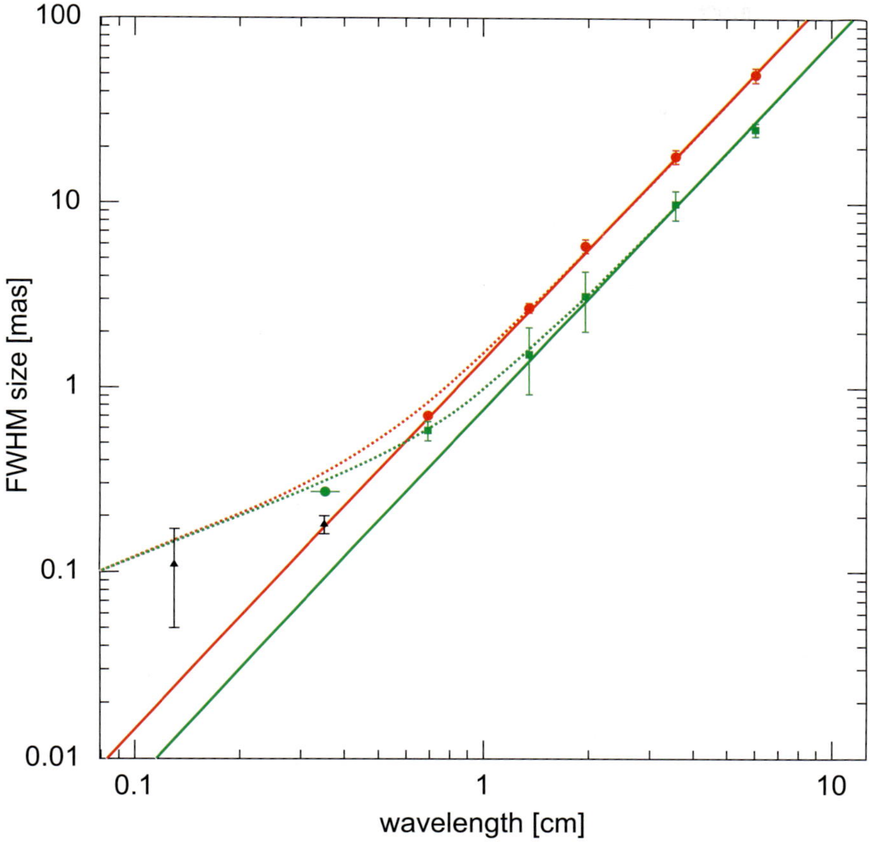

Fig. A.9 Size measurements of Sgr A* from VLBI at multiple wavelengths as shown by Doeleman *et al.* (2001). Green squares are minor-axis sizes and red circles are major-axis sizes of best-fit elliptical models at wavelengths less than 3.5 mm from Lo et al. (1998). Green and red solid lines are λ^2 scattering relations given by Lo et al. (1998). Black triangles show the sizes of circular Gaussian models fit to the 3.5 mm data of Doeleman *et al.* (2001) and to 1.4 mm VLBI data (Krichbaum *et al.* 1998)). Hybrid ADAF model sizes from Özel *et al.* (2000) are added in quadrature to both the minor- and major-axis scattering sizes and are shown as the green and red dotted lines, respectively. The observed sizes predicted by the ADAF model exceed the 3.5 mm upper limit. See also more recent data by Bower *et al.* (2004).

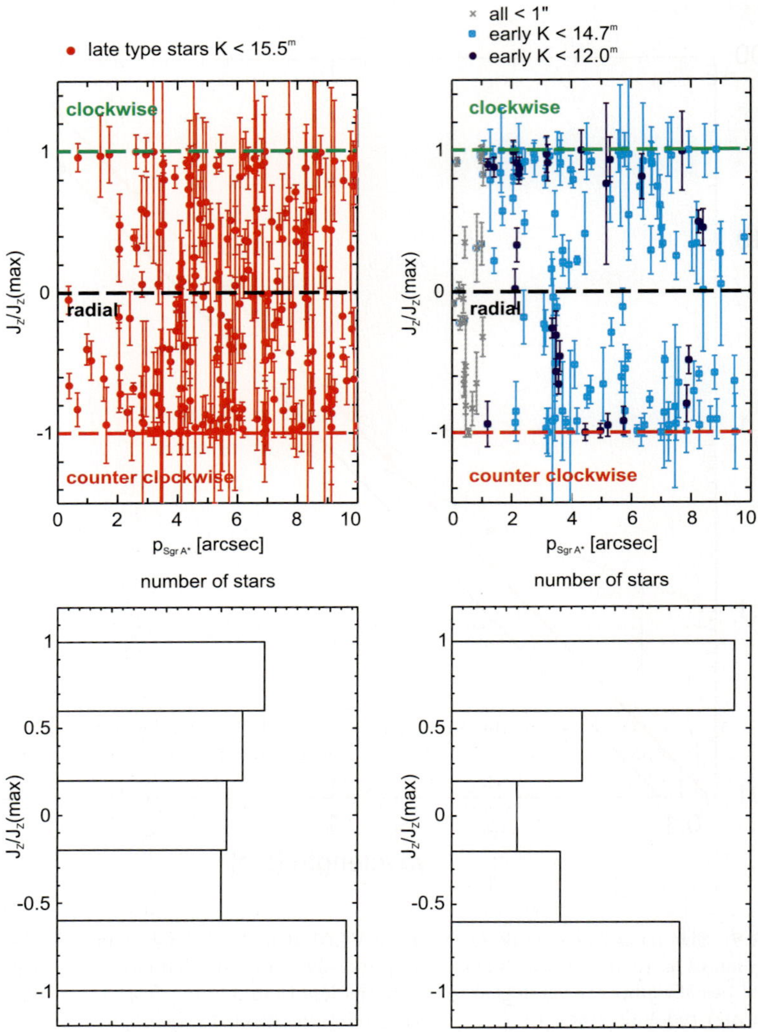

Fig. A.10 Angular momenta as seen projected on the plane of the sky for late-type (left panel) and early-type (right panel) stars within 10″ of Sgr A* (Genzel *et al.* 2003b). The early-type stars are found to be preferentially on clockwise ($J_z/J_z(\mathrm{max}) \lesssim 1$) or counter-clockwise ($J_z/J_z(\mathrm{max}) \gtrsim -1$) orbits while no such pattern can be found in the orbits of the late-type stars. The two histograms at the bottom are integrated (along radius) versions of the top panels.

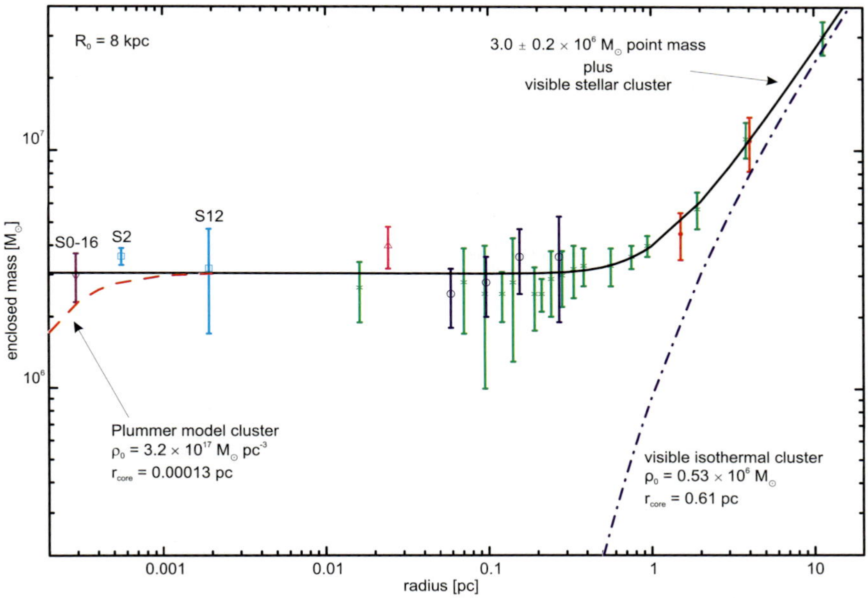

Fig. A.11 Enclosed mass vs. distance in the GC. Red filled circles are measurements from gas dynamics (Guesten *et al.* 1987). Green crosses are results from Jeans mass modeling of stellar proper motions and line-of-sight velocities by Genzel *et al.* (1997) and Genzel *et al.* (1996). Blue open circles are Leonard-Merritt (LM) mass estimates from Genzel *et al.* (2000). The violet up-pointing triangle is an LM estimate from Schödel (2004). Blue open rectangels are masses from Kepler orbits from Schödel (2004) and Eisenhauer *et al.* (2003b). The dark blue diamond is a Kepler orbit mass estiamte from Ghez *et al.* (2003a). Various models for the enclosed are indicated (derived from χ^2-fitting): the blue dash-dotted line is the mass of the large-scale isothermal stellar cluster ($\rho(r) = \rho_0[1 + (\frac{r}{r_{core}})^2]^{\frac{-\alpha}{2}}$, with $\alpha = 1.98$). The black straight line is this cluster plus a point mass of $3.2 \pm 0.2 \times 10^6$ M$_\odot$; the best fitting model. The red dashed line is the stellar cluster plus a hypothetical compact Plummer model cluster of dark astrophysical objects ($\rho(r) = \rho_0[1 + (\frac{r}{r_{core}})^2]^{\frac{-\alpha}{2}}$, with $\alpha = 5$). The Plummer cluster would be marginally consistent with the mass derived from the innermost orbit (S0-16, see Ghez *et al.* 2003a). Such a cluster would have a lifetime of less than 10^5 years (see section 3.4).

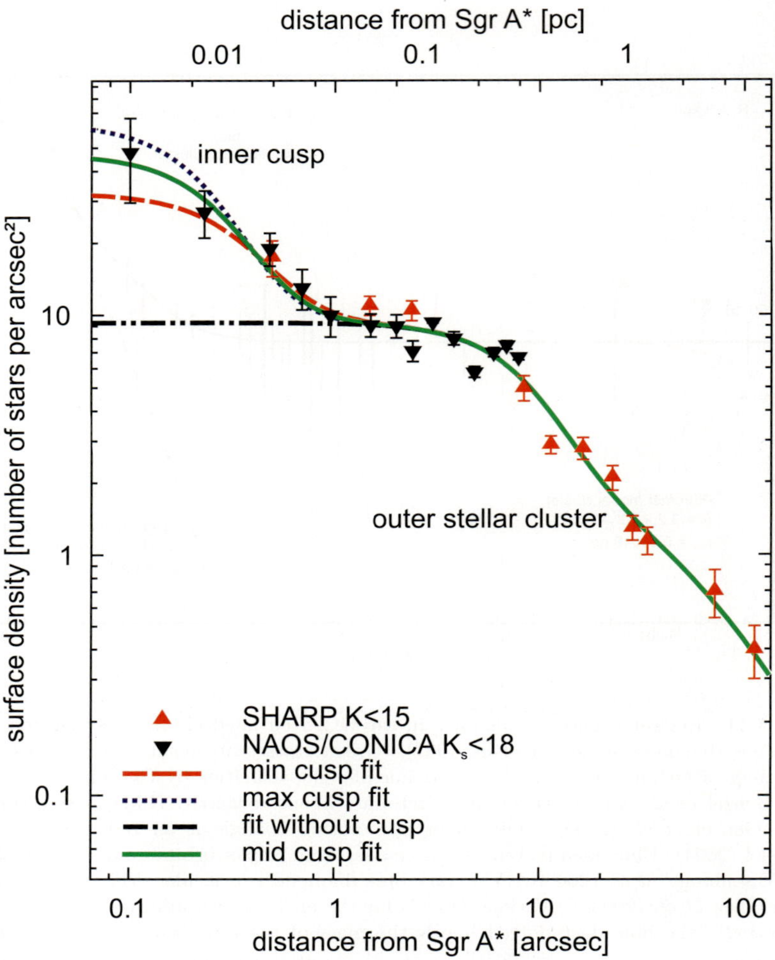

Fig. A.12 A Plummer model fit to the surface density of stars as a function of distance as given by Mouawad *et al.* (2004). The grey, filled circles represent the CONICA/NAOS data for $K_s \leq 17$. The darker, filled diamonds represent the SHARP/NTT data for stars with $K_s \leq 15$ (Genzel *et al.* 2000), scaled upward by a factor of 5 in order to match the fainter CONICA/NAOS counts. The dashed and dotted curves represent the minimum and maximum fit to the data, respectively. The thin straight curve represents the average fit. The dash-dotted curve shows the stellar density distribution expected for an isothermal stellar cluster.

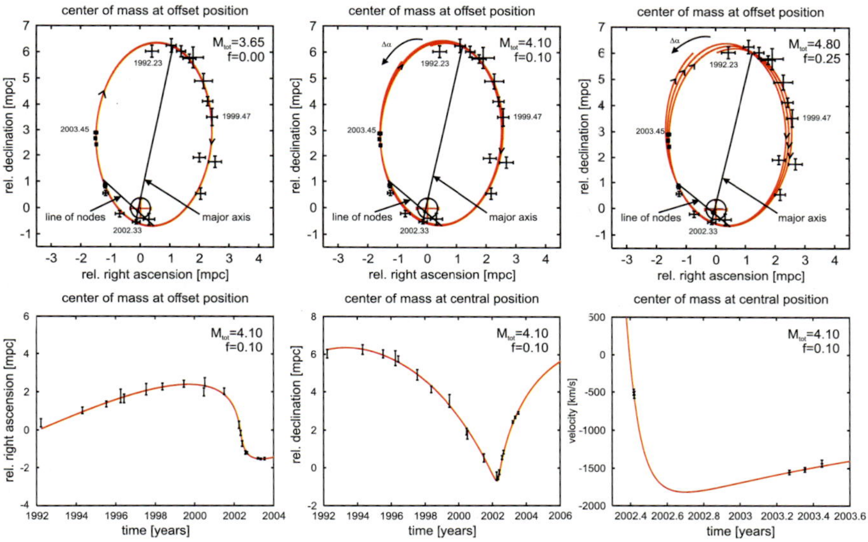

Fig. A.13 Three exemplary orbits determined by Mouawad *et al.* (2004). Upper left panel: Keplerian orbit with $3.65\times10^6 M_\odot$ point mass. Upper middle panel, non-Keplerian orbits with $4.1\times10^6 M_\odot$ total mass, thereof 10% extended. Upper right panel: non-Keplerian orbits with $4.8\times10^6 M_\odot$ total mass, thereof 25% extended. Here the central mass is at the offset position, 0.082 mpc east and 0.112 mpc south from the nominal radio position of Sgr A* (Schödel *et al.* 2003). The lower panels show the corresponding velocities as a function of relative R.A, the relative Dec., and along the line-of-sight as a function of time for the case of $4.1\times10^6 M_\odot$ total central mass with a 10% extended component. The direction of $\Delta\alpha$, the pericenter-shift, is shown by an arrow.

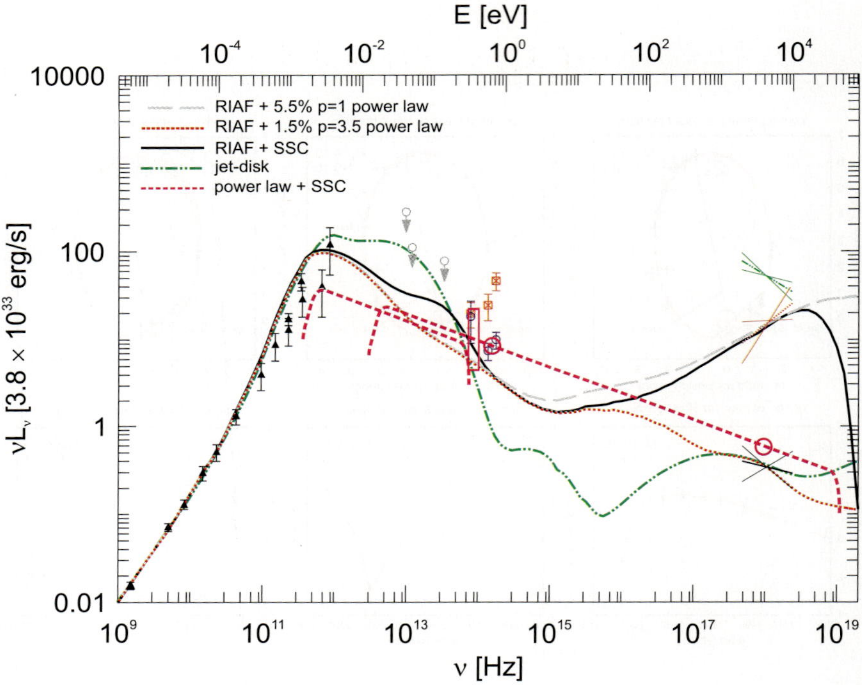

Fig. A.14 Comparison between model calculations and the luminosity of Sgr A* measured at different wavelengths. See caption of Fig. 3.10 for more details. Black triangles denote the average radio spectrum of Sgr A* and open grey circles mark various infrared upper limits from literature — see Melia & Falcke (2001) and references therein. Open red squares with crosses mark the de-reddened peak emission of NIR flares (Genzel *et al.* 2003a). Open blue circles mark the de-reddened H, K, and L luminosities of the quiescent state (Genzel *et al.* 2003a). The thick green solid curve is the jet-starved disk model by Yuan *et al.* (2002). The red long dash-short dash curve, the black thin curve, and the long-dash thick (grey) curve are RIAF models of the quiescent emission (Yuan *et al.* 2002, 2003). The short dashed thick (violet) lines show a typical result for a simple SSC spectrum produced by given by Eckart *et al.* (2004a), who used the first simultaneous NIR/X-ray measurement of a Sgr A* flare for this purpose.

Appendix B

Table of Individual Sources

Table B.1 Bright sources at the Galactic Center.

Name	r(")	$\Delta\alpha$(")	$\Delta\delta$(")	H	K	L	M	$H-K$	$K-L$	$L-M$
IRS 16C	1.12	1.04	-0.42	11.70	9.64	8.07	7.78	2.06	1.57	0.29
IRS 16NW	1.17	-0.03	-1.17	11.66	9.84	8.38	8.24	1.82	1.46	0.14
IRS 16SW	1.43	0.96	1.06	11.53	9.80	7.92	7.53	1.74	1.88	0.39
IRS 16CC	2.01	1.96	-0.42	12.13	10.19	8.50	7.92	1.94	1.69	0.57
IRS 16SWE	2.16	1.75	1.26	13.45	10.67	7.70	7.12	2.78	2.97	0.57
IRS 29S	2.19	-1.99	-0.92	12.33	10.34	9.81	>12.73	2.00	0.53	-3.92
IRS 29	2.19	-1.68	-1.41	14.16	10.36	6.90	6.31	3.81	3.45	0.59
IRS 29NE	2.32	-1.07	-2.06	14.61	11.57	8.91	8.53	3.04	2.66	0.38
IRS 16NE	2.97	2.83	-0.91	10.68	9.18	7.16	6.87	1.49	2.03	0.29
IRS 35	3.19	2.90	1.32	14.55	11.80	8.63	9.00	2.75	3.17	-0.37
IRS 33SE	3.26	0.50	3.22	11.97	9.95	8.17	8.03	2.02	1.78	0.13
IRS 33SW	3.33	-0.52	3.29	12.73	10.69	8.61	8.65	2.05	2.08	-0.04
IRS 21	3.66	2.25	2.89	14.01	10.41	6.29	5.47	3.61	4.12	0.82
IRS 13E	3.70	-3.35	1.58	12.16	9.02	6.18	5.73	3.13	2.84	0.45
IRS 34	4.50	-4.18	-1.68	14.64	11.32	8.23	7.62	3.32	3.09	0.61
IRS 3	4.52	-2.41	-3.82	14.98	10.64	5.08	3.35	4.34	5.56	1.73
IRS 7SE	4.96	2.22	-4.43	13.38	11.20	10.32	10.22	2.18	0.88	0.10
IRS 1W	5.28	5.26	-0.41	11.73	8.90	4.92	4.02	2.83	3.98	0.90

For source identification see Fig. A.2.

Table B.2 Bright sources at the Galactic Center.

Name	r(")	$\Delta\alpha$(")	$\Delta\delta$(")	H	K	L	M	$H-K$	$K-L$	$L-M$
IRS 2L	5.34	-3.64	3.91	14.26	10.60	6.49	5.55	3.67	4.11	0.94
IRS 6E	5.39	-5.33	-0.84	12.57	9.58	6.34	5.71	3.00	3.24	0.62
IRS 20	5.43	-1.09	5.32	12.54	10.57	8.70	8.61	1.97	1.87	0.09
IRS 7	5.54	0.03	-5.54	10.32	8.77	4.48	4.30	1.56	4.28	0.18
IRS 2S	6.12	-3.92	4.70	12.83	10.18	7.92	7.53	2.65	2.26	0.38
IRS 1SW	6.33	6.33	-0.07	12.73	10.43	7.33	6.54	2.30	3.10	0.79
IRS 1C	6.92	6.88	-0.69	12.62	10.42	8.50	8.54	2.20	1.92	-0.04
IRS 1NE	7.24	7.05	-1.64	12.59	10.28	8.52	8.72	2.31	1.76	-0.20
IRS 1SE	7.51	7.48	0.68	12.47	10.48	8.78	>12.73	1.99	1.71	-4.95
IRS 12N	7.81	-3.55	6.96	11.73	9.25	6.63	6.43	2.48	2.63	0.19
IRS 1NEE	7.82	7.70	-1.36	12.76	10.56	9.23	9.57	2.19	1.34	-0.34
IRS 6W	8.06	-7.92	-1.52	12.33	10.19	7.83	6.84	2.14	2.36	0.98
IRS 30E	8.12	-5.76	-5.72	13.12	10.33	8.84	9.12	2.80	1.48	-0.28
IRS 10W	8.25	6.58	-4.98	12.95	9.86	6.25	5.02	3.09	3.61	1.23
IRS 14NE	8.31	0.63	8.28	11.84	9.43	7.71	7.87	2.41	1.72	-0.16
IRS 9	8.59	5.50	6.60	11.78	9.41	6.89	6.90	2.38	2.52	-0.02
IRS 10E*	8.75	7.77	-4.02	13.62	9.76	5.95	5.20	3.86	3.81	0.75
IRS 30W	8.95	-6.57	-6.08	12.86	10.21	8.76	8.87	2.65	1.45	-0.12
IRS 14SW	9.12	-0.50	9.11	11.83	9.56	8.23	8.54	2.27	1.33	-0.31
IRS 12S	9.41	-3.74	8.63	12.07	10.03	8.31	8.91	2.04	1.73	-0.61
IRS 10E	9.53	8.70	-3.88	12.63	10.50	8.84	8.92	2.13	1.65	-0.08
IRS 14S	9.54	0.46	9.53	13.02	10.90	9.71	9.80	2.12	1.18	-0.08
AF/AHH	9.72	-6.85	6.90	12.46	10.43	8.70	8.10	2.03	1.72	0.60

For source identification see Fig. A.2.

Appendix C

Useful Quantities

Table C.1 Important physical constants and units.

Speed of light	c	=	2.998×10^{8} m s^{-1}
Planck's constant	h	=	6.626×10^{-34} J s
Gravitational constant	G	=	6.67×10^{-11} m^{3}s^{-2}kg^{-1}
Boltzmann's constant	k	=	1.38×10^{-23} J K^{-1}
Electron mass	m_e	=	9.11×10^{-31} kg
Proton mass	m_p	=	1.67×10^{-27} kg
Electron charge	e	=	1.60×10^{-19} A s
Stefan-Boltzmann constant	σ	=	5.67×10^{-8} W m^{-2} K^{-4}
Thomson cros section	σ_t	=	6.65×10^{29} m^{2}
Energy flux density (Jansky)	1 Jy	=	10^{-26} W m^{-2} Hz^{-1}
Astronomical unit	1 AU	=	1.5×10^{11} m
Light year	1 ly	=	9.46×10^{15} m
Parsec	1 pc	=	3.09×10^{16} m
Kiloparsec	1 kpc	=	10^{3} pc
Milliparsec	1 mpc	=	10^{-3} pc
Milliarcsecond	1 mas	=	10^{-3} arcseconds
Solar luminosity	$L_\odot$	=	$3.85\ 10^{26}$ W = $3.85\ 10^{33}$ erg s^{-1}
Solar mass	$M_\odot$	=	$1.989\ 10^{30}$ kg
Solar radius	$R_\odot$	=	$6.9599\ 10^{8}$ m

Table C.2 Relations that are specific to the Galactic Center. They are given for a distance to the Center of 8 kpc. *) For a definition of θ see section 2.10.4. **) Uncertainty includes uncertainty in distance R_0.

Distance to the Galactic Center	R_0=(8±0.5) kpc = 26,100 ly = 2.5×10^{17} km
Image scale	1" = 38.78 mpc = 7980 AU
	1 mas = $1'' \times 10^{-3}$
Velocity	1 km/s = 0.0263 mas/yr
	1 mas/yr = 37.92 km/s
Acceleration a	1 mas/yr^2 = 1.22 × 1.22 10^{-6} km/s^2
Mass from projected acceleration*	M $\cos^3\theta$ [10^6 $M_\odot$] = 13 a [mas/yr^2] r^2 ["]
	M $\cos^3\theta$ [10^6 $M_\odot$] = a [mas/yr^2] r^2 [mpc]/116
Mass of Milky Way Black Hole**	M = (3.6 ± 0.6)×10^6 $M_\odot$
Schwarzschild radius	R_S=2 G M c^{-2} = 2.95×$\frac{M}{M_\odot}$ km

Appendix D

Table of Abbreviations

J-band	1.10 - 1.30 μm wavelength range
H-band	1.45 - 1.85 μm
K-band	1.90 - 2.40 μm
L-band	3.20 - 3.80 μm
L'-band	3.48 - 4.08 μm
M-band	4.54 - 5.16 μm
N-band	7.90 - 13.1 μm
3D	MPE imaging spectrometer
A_V	visual extinction
A_K	extinction in K-band
A_λ	extinction at a given wavelength
ADC	Automatic Dispersion Corrector
ADAF	Advection Dominated Accretion Flow
ADIOS	Advection Dominated Inflow Outflow Scenario
ADONIS	ESO AO system operated at the 3.6 m telescope on La Silla

AGB	Asymptotic Giant Branch
AGN	Active Galactic Nucleus
ALADDIN	Santa Barbara Research Center IR array detector
ALMA	Atacama Large Millimeter Array
AMBER	NIR focal instrument of the ESO VLTI
AST/RO	Antarctic Submillimeter Telescope and Remote Observatory
AT	Auxiliary Telescopes (e.g. for ESO VLTI)
BH	Black Hole
BIB	Blocked Impurity Band detector
BLIP	Background LImited Performance
CCD	Charge Coupled Device
CDAF	Convection Dominated Accretion Flow
CELT	Caltech Extremely Large Telescope
CFHT	Canada France Hawaii Telescope
Chandra	US X-ray satellite (CHANDRA)
CONICA	VLT Near-Infrared Imager and Spectrograph
CND	Circum Nuclear Disk (Ring)
Dec	declination
DLR	Deutsche Luft und Raumfahrt Gesellschaft
EGRET	Energetic Gamma Ray Experiment Telescope; instrument flown on the European imaging Compton Telescope (COMPTEL)
ELT	Extremely Large Telescope
ESA	European Space Agency
ESO	European Southern Observatory
Euro50	European 50m diameter Telescope

FASTTRAC	Steward Observatory rapid guiding unit
FET	Field Effect Transistor
FIR	Far-infrared
FOV	Field of View
FWHM	Full Width at Half Maximum
GC	Galactic Center
HII	region of ionized hydrogen
Hawaii 2	Rockwell Infrared array detector
HB	Horizontal Branch region in Hertzsprung-Russel Diagram
HKL	the NIR H-, K-, and L- band
HOKUPA'A	CFHT adaptive optics instrument
HST	Hubble Space Telescope
IBD	Iterative Blind Deconvolution
IMBH	Intermediate Mass Black Hole
IMF	Initial Mass Function
INTEGRAL	International Gamma-Ray Astrophysics Laboratory
ISAAC	VLT Infrared Spectrometer And Array Camera
ISO	Infrared Space Observatory
IR	Infrared
IRTF	Infrared Telescope Facility, University of Hawaii
JCMT	James Clark Maxwell Telescope
JHK	the NIR J-, H-, and K-band
KAO	Kuiper Airborme Observatory
Keck	W.M. Keck Observatory — two 10m diameter telescopes
KLF	K-band Luminosity Function
LBT	Large Binocular Telescope
LGS	LASER Guide Star
LINC/NIRVANA	LBT Interferometric Camera plus MCAO unit
LLAGN	Low Luminosity Active Galactic Nuclei
LR	Lucy Richardson deconvolution

magnitude	apparent brightness proportional to the negative logarithm of the flux density S. m_1-m_2=-2.5 log(S_1/S_2)
m_K	K-band magnitude
MBH	massive Black Hole
MCAO	Multi Conjugated Adaptive Optics
MIDI	MID-infrared Instrument for the VLTI
M/L	mass-to-light ratio
MOS	Metal Oxide Semiconductor
MOSFET	Metal Oxide Semiconductor Field Effect Transistor
MPE	Max-Planck-Institute for Extraterrestrial Physics (Garching)
MPIA	Max-Planck-Institute for Astronomy (Heidelberg)
MPIfR	Max-Planck-Institute for Radioastronomy (Bonn)
NAOS	VLT Nasmyth Adaptive Optics System
NASA	National Aeronautics and Space Administration
NGS	Natural Guide Star
NICMOS	Near Infrared Camera and Multi-Object Spectrometer
NICMOS 3	Rockwell 256^2 pixel HgCdTe array
NIR	near-infrared (1 μm - 2.4 μm wavelength)
NIRC	Keck NIR Camera
NIRC-2	NIR camera for the second Keck telescope
NIRSCEP	Keck NIR spectrometer
NRAO	National Radio Astronomy Observatory
NTF	Non-Thermal filaments
NTT	New Technology Telescope; ESO 3.6 m telescope at La Silla Chile
OVRO	Owens Valley Radio Observatory
OWL	Overwhelmingly Large Telescope
PG QSO	Palomar Green QSO
PRIMA	NIR dual feel facility for the ESO VLTI
PSF	Point Spread Function
PUEO	CFHT adaptive optics instrument

QSO	Quasi Stellar Object
QPO	Quasi-Periodic Oscillation
RA	right ascension
RC	Red Clump stars (region in HRD)
RIAF	Radiatively Inefficient Accretion Flow
seeing	image quality through turbulent atmosphere
SIS	Superconductor-Insulator-Superconductor diode or contact
SKA	Square Kilometer Array
SED	Spectral Energy Distribution
Sgr A	Sagittarius A
Sgr A*	(Sagittarius A star) compact radio source emitting also in the NIR and X-ray domain — location of the $\sim 3.6 \times 10^6 M_\odot$ black hole at the center of the Milky Way
SHARCII	Bolometric camera at the Caltech Submillimeter Observatory
SHARP	System for High Angular Resolution MPE built NIR camera for speckle interferometry
SINFONI	MPE/ESO project: imaging spectrograph SPIFFI plus ESO AO system operated at the VLT
SOFIA	Stratospheric Observatory for Infrared Astronomy
SPIFFI	MPE built imaging spectrograph
SPIFI	Cornell Univ. Submm Fabry-Perot interferometer
SSA	Simple Shift and add Algorithm
SSC	Synchrotron Self-Compton
SSP	Single Stellar Population
SWS	Short Wavelength Spectrometer (ISO)
UCB	University of California Berkeley
UCLA	University of California Los Angeles
UT	Unit Telescope of the VLT

VLA	Very Large Array
VLBA	Very Large Baseline Array
VLBI	Very Large Baseline Interferometry
VLT	Very Large Telescope — 4 ESO telescopes of 8.4m diameter on Paranal, Chile
WFS	Wave Front Sensor used in AO instrumentation
WSA	Weighted Shift and add Algorithm
XMM Newton	European X-ray satellite
YSO	Young Stellar Object

Bibliography

Aarseth, S. J. 1999, Publ. Astron. Soc. Pac., 111, 1333

Adams, D. J., Becklin, E. E., Jameson, R. F., *et al.* 1988, Astrophys. J., Lett., 327, L65

Ageorges, N. 1995, Ph.D. Thesis

Ageorges, N., Eckart, A., Hofmann, R., & Stecklum, B. 1996a, in ASP Conf. Ser. 97: Polarimetry of the Interstellar Medium, 85

Ageorges, N., Fischer, O., Stecklum, B., Eckart, A., & Henning, T. 1996b, Astrophys. J., Lett., 463, L101

Aitken, D. K., Smith, C. H., Gezari, D., McCaughrean, M., & Roche, P. F. 1991, Astrophys. J., 380, 419

Aitken, D. K., Smith, C. H., Moore, T. J. T., & Roche, P. F. 1998, Mon. Not. of the Royal Astron. Soc., 299, 743

Aitken, D. K., Greaves, J., Chrysostomou, A., *et al.* 2000, Astrophys. J., Lett., 534, L173

Alberdi, A., Lara, L., Marcaide, J. M., *et al.* 1993, Astron. & Astroph., 277, L1

Alexander, T. 1999, Astrophys. J., 527, 835

Alexander, T. & Sternberg, A. 1999, Astrophys. J., 520, 137

Alexander, T. & Kumar, P. 2001, Astrophys. J., 549, 948

Alexander, T. & Livio, M. 2001, Astrophys. J., Lett., 560, L143

Alexander, T. & Morris, M. 2003, Astrophys. J., Lett., 590, L25

Alexander, T. & Livio, M. 2004, Astrophys. J., Lett., 606, L21

Allen, D. A. & Sanders, R. H. 1986, Nature, 319, 191

Allen, D. A., Hyland, A. R., & Hillier, D. J. 1990, Mon. Not. of the Royal Astron. Soc., 244, 706

Amico, P., Beletic, J. W., & Beletic, J. E., eds. 2004, Scientific Detectors for Astronomy - The Beginning of a New Era

Anantharamaiah, K. R., Pedlar, A., Ekers, R. D., & Goss, W. M. 1991, Mon. Not. of the Royal Astron. Soc., 249, 262

Andersen, T., Ardeberg, A. L., Beckers, J., *et al.* 2003, in Future Giant Telescopes. Edited by Angel, J. Roger P.; Gilmozzi, Roberto. Proceedings of the SPIE, Volume 4840, pp. 214-225 (2003)., 214–225

Arsenault, R., Salmon, D. A., Kerr, J. M., *et al.* 1994, in Proc. SPIE Vol. 2201, p. 833-842, Adaptive Optics in Astronomy, Mark A. Ealey; Fritz Merkle; Eds., 833–842

Aschenbach, B. 2004, Astron. & Astroph., in press

Aschenbach, B., Grosso, N., Porquet, D., & Predehl, P. 2004, Astron. & Astroph., 417, 71

Backer, D. C. 1978, Astrophys. J., Lett., 222, L9

Backer, D. C. & Sramek, R. A. 1982, Astrophys. J., 260, 512

—. 1999, Astrophys. J., 524, 805

Bacon, R., Emsellem, E., Monnet, G., & Nieto, J. L. 1994, Astron. & Astroph., 281, 691

Baganoff, F. K., Bautz, M. W., Brandt, W. N., *et al.* 2001, Nature, 413, 45

Baganoff, F. K., Maeda, Y., Morris, M., *et al.* 2003, Astrophys. J., 591, 891

Bahcall, J. N. & Wolf, R. A. 1976, Astrophys. J., 209, 214

—. 1977, Astrophys. J., 216, 883

Bahcall, J. N. & Tremaine, S. 1981, Astrophys. J., 244, 805

Bailyn, C. D. & Pinsonneault, M. H. 1995, Astrophys. J., 439, 705

Balick, B. & Brown, R. L. 1974, Astrophys. J., 194, 265

Ball, G. H., Narayan, R., & Quataert, E. 2001, Astrophys. J., 552, 221

Bardeen, J. M., Press, W. H., & Teukolsky, S. A. 1972, Astrophys. J., 178, 347

Bastien, P. & Menard, F. 1990, Astrophys. J., 364, 232

Bates, R. H. T. & Fright, W. R. 1982, Mon. Not. of the Royal Astron. Soc., 198, 1017

Becker, R. H., Fan, X., White, R. L., *et al.* 2001, Astron. J., 122, 2850

Beckert, T., Duschl, W. J., Mezger, P. G., & Zylka, R. 1996, Astron. & Astroph., 307, 450

Beckert, T. & Duschl, W. J. 1997, Astron. & Astroph., 328, 95

Beckert, T. & Falcke, H. 2002, Astron. & Astroph., 388, 1106

Becklin, E. E. & Neugebauer, G. 1968, Astrophys. J., 151, 145

—. 1975, Astrophys. J., Lett., 200, L71

Becklin, E. E., Matthews, K., Neugebauer, G., & Willner, S. P. 1978, Astrophys. J., 219, 121

Becklin, E. E., Gatley, I., & Werner, M. W. 1982, Astrophys. J., 258, 135

Beckwith, S., Dyck, H. M., & Howell, R. R. 1986, Bull. Am. Astron. Soc., 18, 935

Bélanger, G., Goldwurm, A., Goldoni, P., *et al.* 2004, Astrophys. J., Lett., 601, L163

Bender, R., Kormendy, J., & Dehnen, W. 1996, Astrophys. J., Lett., 464, L123+

Benford, G. 1988, Astrophys. J., 333, 735

Beuzit, J.-L., Demailly, L., Gendron, E., *et al.* 1997, Experimental Astronomy, 7, 285

Binney, J. & Tremaine, S. 1987, Galactic dynamics (Princeton, NJ, Princeton University Press, 1987, 747 p.)

Biretta, J. A., Lo, K. Y., & Young, P. J. 1982, Astrophys. J., 262, 578

Biretta, J. A., Lo, K. Y., Boroson, T. A., & Lacy, J. H. 1983, Astron. J., 88, 94

Blandford, R. D. & Begelman, M. C. 1999, Mon. Not. of the Royal Astron. Soc., 303, L1

Blitz, L. 1992, The centre, bulge, and disk of the Milky Way (Kluwer Press, 1992)

Blitz, L., Binney, J., Lo, K. Y., Bally, J., & Ho, P. T. P. 1993, Nature, 361, 417

Blum, R. D., Depoy, D. L., & Sellgren, K. 1995, Astrophys. J., 441, 603

Blum, R. D., Sellgren, K., & Depoy, D. L. 1996, Astrophys. J., 470, 864

Blum, R. D., Ramírez, S. V., Sellgren, K., & Olsen, K. 2003, Astrophys. J., 597, 323

Bonaccini, D., Lacombe, F., Marco, O., Eisenhauer, F., & Hofmann, R. 1996, The Messenger, 82, 16

Bonaccini, D., Prieto, E., Corporon, P., *et al.* 1997, in Conference Paper, SPIE Proceedings, Vol. 3126, 589–594

Bonaccini, D., Hackenberg, W., & Taylor, L. 2003, The Messenger, 114, 15

Bower, G. C., Backer, D. C., Zhao, J., Goss, M., & Falcke, H. 1999a, Astrophys. J., 521, 582

Bower, G. C., Falcke, H., & Backer, D. C. 1999b, Astrophys. J., Lett., 523, L29

Bower, G. C., Wright, M. C. H., Backer, D. C., & Falcke, H. 1999c, Astrophys. J., 527, 851

Bower, G. C., Wright, M. C. H., Falcke, H., & Backer, D. C. 2001, Astrophys. J., Lett., 555, L103

Bower, G. C., Falcke, H., Sault, R. J., & Backer, D. C. 2002, Astrophys. J., 571, 843

Bower, G. C., Wright, M. C. H., Falcke, H., & Backer, D. C. 2003, Astrophys. J., 588, 331

Bower, G. C., Falcke, H., Herrnstein, R. M., *et al.* 2004, Science, 304, 704

Bowers, R. L. & Deeming, T. 1984, Astrophysics. Volume 2 - Interstellar matter and galaxies (Research supported by the University of Texas, Los Alamos National Laboratory, and Digicon Geophysical Corp. Boston, MA, Jones and Bartlett Publishers, Inc., 1984, 283 p.)

Bradford, C. M., Stacey, G. J., & et al. 2002, Appl.Opt., 41, 2561

Brandner, W., Rousset, G., Lenzen, R., *et al.* 2002, The Messenger, 107, 1

Bromm, V. & Loeb, A. 2003, Astrophys. J., 596, 34

Burton, M. & Allen, D. 1992, Proceedings of the Astronomical Society of Australia, 10, 55

Chan, K. 1995, Ph.D. Thesis

Chan, K., Moseley, S. H., Casey, S., *et al.* 1997, Astrophys. J., 483, 798

Chiar, J. E. & et al. 2001, in ASP Conf. Ser. 231: Tetons 4: Galactic Structure, Stars and the Interstellar Medium, 179

Chini, R., Kreysa, E., Mezger, P. G., & Gemuend, H.-P. 1986, Astron. & Astroph., 157, L1

Christou, J. C. 1991, Experimental Astronomy, 2, 27

Clegg, P. E., Ade, P. A. R., Armand, C., *et al.* 1996, Astron. & Astroph., 315, L38

Clénet, Y., Rouan, D., Gendron, E., *et al.* 2001, Astron. & Astroph., 376, 124

Close, L. M., McCarthy, D. W., Christou, J. C., & Melia, F. 1992, Bulletin of the American Astronomical Society, 24, 1178

Close, L. M. & McCarthy, D. W. 1994, Publ. Astron. Soc. Pac., 106, 77

Close, L. M., McCarthy, D. W., & Melia, F. 1995, Astrophys. J., 439, 682

Coker, R. F. & Melia, F. 1997, Astrophys. J., Lett., 488, L149

—. 2000, Astrophys. J., 534, 723

Colavita, M. 2001, Bulletin of the American Astronomical Society, 33, 859

Colpi, M., Shapiro, S. L., & Wasserman, I. 1986, Physical Review Letters, 57, 2485

Conselice, C. J., Bershady, M. A., Dickinson, M., & Papovich, C. 2003, Astron. J., 126, 1183

Corbin, M. R., O'Neil, E., & Rieke, M. J. 2001, Astron. J., 121, 2549

Cotera, A. S., Erickson, E. F., Simpson, J. P., *et al.* 1992, Bulletin of the American Astronomical Society, 24, 1262

Cotera, A., Morris, M., Ghez, A. M., *et al.* 1999a, in ASP Conf. Ser. 186: The Central Parsecs of the Galaxy, 240

Cotera, A. S., Simpson, J. P., Erickson, E. F., *et al.* 1999b, Astrophys. J., 510, 747

Cotera, A., Markoff, S., Geballe, T., & Falcke, H. 2003, The central 300 parsecs of the Milky Way (Astron. Nachr.)

Cox, A. 1999, Allen's Astrophysical Quantities (Edited by Arthur N. Cox, Los Alamos, NM, ISBN: 0-387-98746-0)

Cox, A. N. 2000, Allen's astrophysical quantities (Allen's astrophysical quantities, 4th ed. Publisher: New York: AIP Press; Springer, 2000. Editedy by Arthur N. Cox. ISBN: 0387987460)

Cruzalebes, P., Tessier, E., Lopez, B., Eckart, A., & Tiphene, D. 1996, A&AS, 116, 597

Dainty, J. C. 1975, Laser speckle and related phenomena (Topics in Applied Physics, Berlin: Springer, 1975, edited by Dainty, J.C.)

Dainty, J. C. & Scaddan, R. J. 1975, Mon. Not. of the Royal Astron. Soc., 170, 519

Davidge, T. J., Simons, D. A., Rigaut, F., Doyon, R., & Crampton, D. 1997, Astron. J., 114, 2586

Davies, R. D., Walsh, D., & Booth, R. S. 1976, Mon. Not. of the Royal Astron. Soc., 177, 319

de Graauw, T., Haser, L. N., Beintema, D. A., *et al.* 1996a, Astron. & Astroph., 315, L49

—. 1996b, Astron. & Astroph., 315, L49

De Villiers, J., Hawley, J. F., & Krolik, J. H. 2003, Astrophys. J., 599, 1238

Depoy, D. L. & Sharp, N. A. 1991a, Astron. J., 101, 1324

—. 1991b, Astron. J., 101, 1324

Diamond, P. J., Kemball, A. J., Junor, W., *et al.* 1994, Astrophys. J., Lett., 430, L61

Diolaiti, E., Bendinelli, O., Bonaccini, D., *et al.* 2000, A&AS, 147, 335

Doeleman, S. S., Shen, Z.-Q., Rogers, A. E. E., *et al.* 2001, Astron. J., 121, 2610

Dowell, C. D., Allen, C. A., Babu, R. S., *et al.* 2003, in Millimeter and Submillimeter Detectors for Astronomy. Edited by Phillips, Thomas G.; Zmuidzinas, Jonas. Proceedings of the SPIE, Volume 4855, pp. 73-87 (2003)., 73–87

Duncan, W. D., Sandell, G., Robson, E. I., Ade, P. A. R., & Griffin, M. J. 1990, Mon. Not. of the Royal Astron. Soc., 243, 126

Eckart, A. & Duhoux, P. R. M. 1990, in ASP Conf. Ser. 14: Astrophysics with Infrared Arrays, 336

Eckart, A., Genzel, R., Krabbe, A., *et al.* 1992, Nature, 355, 526

Eckart, A., Genzel, R., Hofmann, R., *et al.* 1994, in NATO ASIC Proc. 445: The Nuclei of Normal Galaxies: Lessons from the Galactic Center, 305

Eckart, A., Genzel, R., Hofmann, R., Sams, B. J., & Tacconi-Garman, L. E. 1995, Astrophys. J., Lett., 445, L23

Eckart, A. & Genzel, R. 1996, Nature, 383, 415

—. 1997, Mon. Not. of the Royal Astron. Soc., 284, 576

Eckart, A., Genzel, R., Hofmann, R., *et al.* 1997, in Science with the VLT Interferometer, 259

Eckart, A., Ott, T., & Genzel, R. 1999a, Astron. & Astroph., 352, L22

Eckart, A., Ott, T., Genzel, R., & Lutz, D. 1999b, in IAU Symp. 193: Wolf-Rayet Phenomena in Massive Stars and Starburst Galaxies, 449

Eckart, A., Hippler, S., Glindemann, A., *et al.* 2000, Experimental Astronomy, 10, 1

Eckart, A., Genzel, R., Ott, T., & Schödel, R. 2002a, Mon. Not. of the Royal Astron. Soc., 331, 917

Eckart, A., Mouawad, N., Krips, M., Straubmeier, C., & Bertram, T. 2002b, in Future Research Direction and Visions for Astronomy. Edited by Dressler, Alan M. Proceedings of the SPIE, Volume 4835, pp. 12-21 (2002)., 12–21

Eckart, A., Bertram, T., Mouawad, N., *et al.* 2003a, Astrophys. Space. Sci., 286, 269

Eckart, A., Moultaka, J., Viehmann, T., *et al.* 2003b, Astronomische Nachrichten Supplement, 324, 557

Eckart, A., Baganoff, F. K., Morris, M., *et al.* 2004a, Astron. & Astroph., 427, 1

Eckart, A., Moultaka, J., Viehmann, T., Straubmeier, C., & Mouawad, N. 2004b, Astrophys. J., 602, 760

Eisenhauer, F., Tecza, M., Mengel, S., *et al.* 2000, in Proc. SPIE Vol. 4008, p. 289-297, Optical and IR Telescope Instrumentation and Detectors, Masanori Iye; Alan F. Moorwood; Eds., 289–297

Eisenhauer, F., Abuter, R., Bickert, K., *et al.* 2003a, in Instrument Design and Performance for Optical/Infrared Ground-based Telescopes. Edited by Iye, Masanori; Moorwood, Alan F. M. Proceedings of the SPIE, Volume 4841, pp. 1548-1561 (2003)., 1548–1561

Eisenhauer, F., Schödel, R., Genzel, R., *et al.* 2003b, Astrophys. J., Lett., 597, L121

Ekers, R. D., van Gorkom, J. H., Schwarz, U. J., & Goss, W. M. 1983, Astron. & Astroph., 122, 143

Erickson, E. F., Colgan, S. W. J., Simpson, J. P., *et al.* 1995, in ASP Conf.

Ser. 73: From Gas to Stars to Dust, 489–498

Falcke, H., Goss, W. M., Matsuo, H., *et al.* 1998, Astrophys. J., 499, 731

Falcke, H., Cotera, A., Duschl, W. J., Melia, F., & Rieke, M. J., eds. 1999, The Central Parsecs of the Galaxy

Falcke, H. & Markoff, S. 2000, Astron. & Astroph., 362, 113

Falcke, H., Melia, F., & Agol, E. 2000, Astrophys. J., Lett., 528, L13

Falcke, H. & Hehl, F. W. 2003, The Galactic black hole (The Galactic Black Hole)

Ferrarese, L. & Merritt, D. 2000, Astrophys. J., Lett., 539, L9

Figer, D. F., McLean, I. S., & Morris, M. 1995, Astrophys. J., Lett., 447, L29

Figer, D. F., Najarro, F., Morris, M., *et al.* 1998, Astrophys. J., 506, 384

Figer, D. F., Kim, S. S., Morris, M., *et al.* 1999a, Astrophys. J., 525, 750

Figer, D. F., McLean, I. S., & Morris, M. 1999b, Astrophys. J., 514, 202

Figer, D. F., Becklin, E. E., McLean, I. S., *et al.* 2000, Astrophys. J., Lett., 533, L49

Figer, D. F., Najarro, F., Gilmore, D., *et al.* 2002, Astrophys. J., 581, 258

Figer, D. F., Rich, R. M., Kim, S. S., Morris, M., & Serabyn, E. 2004, Astrophys. J., 601, 319

Fomalont, E. B. & Wright, M. 1974, Galactic and Extragalactic Radio Astronomy (eds. Verschuur, C.L., Kellermann, K.I., Springer Verlag)

Ford, H. C., Harms, R. J., Tsvetanov, Z. I., *et al.* 1994, Astrophys. J., Lett., 435, L27

Forrest, W. J., Pipher, J. L., & Stein, W. A. 1986, Astrophys. J., Lett., 301, L49

Fragile, P. C. & Mathews, G. J. 2000, Astrophys. J., 542, 328

Freitag, M. & Benz, W. 2002a, Astron. & Astroph., 394, 345

—. 2002b, Astron. & Astroph., 394, 345

Fried, D. L. 1966, Optical Society of America Journal, 56, 1372

Gathier, R., Lamers, H. J. G. L. M., & Snow, T. P. 1981, Astrophys. J., 247, 173

Gatley, I., Beattie, D. H., Lee, T. J., Jones, T. J., & Hyland, A. R. 1984, Mon. Not. of the Royal Astron. Soc., 210, 565

Gatley, I., Jones, T. J., Hyland, A. R., Wade, R., & Geballe, T. R. 1986, Mon. Not. of the Royal Astron. Soc., 222, 299

Gebhardt, K., Kormendy, J., Ho, L. C., *et al.* 2000, Astrophys. J., Lett., 543, L5

Gebhardt, K., Lauer, T. R., Kormendy, J., *et al.* 2001, Astron. J., 122, 2469

Genzel, R. & Townes, C. H. 1987, ARA&A, 25, 377
Genzel, R., Stacey, G. J., Harris, A. I., *et al.* 1990, Astrophys. J., 356, 160
Genzel, R., Hollenbach, D., & Townes, C. H. 1994, Reports of Progress in Physics, 57, 417
Genzel, R., Thatte, N., Krabbe, A., Kroker, H., & Tacconi-Garman, L. E. 1996, Astrophys. J., 472, 153
Genzel, R., Eckart, A., Ott, T., & Eisenhauer, F. 1997, Mon. Not. of the Royal Astron. Soc., 291, 219
Genzel, R., Pichon, C., Eckart, A., Gerhard, O. E., & Ott, T. 2000, Mon. Not. of the Royal Astron. Soc., 317, 348
Genzel, R., Schödel, R., Ott, T., *et al.* 2003a, Nature, 425, 934
—. 2003b, Astrophys. J., 594, 812
Gerhard, O. 2001, Astrophys. J., Lett., 546, L39
Gezari, D. Y., Shu, P., Lamb, G., *et al.* 1985, Astrophys. J., 299, 1007
Gezari, D., Dwek, E., & Varosi, F. 1994, in NATO ASIC Proc. 445: The Nuclei of Normal Galaxies: Lessons from the Galactic Center, 343
Gezari, S., Ghez, A. M., Becklin, E. E., *et al.* 2002, Astrophys. J., 576, 790
Ghez, A. M., Klein, B. L., Morris, M., & Becklin, E. E. 1998, Astrophys. J., 509, 678
Ghez, A. M., Morris, M., Becklin, E. E., Tanner, A., & Kremenek, T. 2000, Nature, 407, 349
Ghez, A. M., Becklin, E., Duchjne, G., *et al.* 2003a, Astronomische Nachrichten Supplement, 324, 527
Ghez, A. M., Duchêne, G., Matthews, K., *et al.* 2003b, Astrophys. J., Lett., 586, L127
Ghez, A. M., Wright, S. A., Matthews, K., *et al.* 2004, Astrophys. J., Lett., 601, L159
Gilmozzi, R., Dierickx, P., & Monnet, G. 2001, in Quasars, AGNs and Related Research Across 2000. Conference on the occasion of L. Woltjer's 70th birthday, 184
Glass, I. S. 1999, Handbook of infrared astronomy / Cambridge U. Press, 1999 (Cambridge: University Press, 1999)
Glasse, A. C. H., Aitken, D. K., & Roche, P. F. 2003, Astronomische Nachrichten Supplement, 324, 563
Glindemann, A. & Lévêque, S. 2000, in From Extrasolar Planets to Cosmology: The VLT Opening Symposium, Proceedings of the ESO Symposium held at Antofagasta, Chile, 1-4 March 1999. Edited by Jacqueline Bergeron and Alvio Renzini. Berlin: Springer-Verlag, 2000. p. 468., 468
Glindemann, A., Algomedo, J., Amestica, R., *et al.* 2003, EAS Publications

Series, Volume 6, 2003, Observing with the VLTI, Proceedings of the conference held 3-8 February, 2002 in Les Houches, France. Edited by G. Perrin and F. Malbet, pp.91., 6, 91

Goldwurm, A., Brion, E., Goldoni, P., *et al.* 2003, Astrophys. J., 584, 751

Goss, W. M. & McGee, R. X. 1996, in The Galactic Center, Astronomical Society of the Pacific Conference Series, Volume 102, Manuscripts presented at the 4th international meeting jointly organized by the European Southern Observatory (ESO) and Cerro Tololo Inter-American Observatory (CTIO), held March 10-15, 1996 in La Serena, Chile, San Francisco: Astronomical Society of the Pacific (ASP), —c1996, edited by Roland Gredel, p.369, 369

Goss, W. M., Brown, R. L., & Lo, K. Y. 2003, Astronomische Nachrichten Supplement, 324, 497

Gould, R. J. 1979, Astron. & Astroph., 76, 306

Gould, A. & Quillen, A. C. 2003, Astrophys. J., 592, 935

Graves, J. E. & McKenna, D. L. 1991, in Active and adaptive optical systems; Proceedings of the Meeting, San Diego, CA, July 22-24, 1991 (A93-39451 15-74), p. 262-272., 262–272

Graves, J. E., Roddier, F. J., Northcott, M. J., Anuskiewicz, J., & Monnet, G. 1994, in Proc. SPIE Vol. 2201, p. 502-507, Adaptive Optics in Astronomy, Mark A. Ealey; Fritz Merkle; Eds., 502–507

Graves, J. E., Northcott, M. J., Roddier, F. J., Roddier, C. A., & Close, L. M. 1998, in Proc. SPIE Vol. 3353, p. 34-43, Adaptive Optical System Technologies, Domenico Bonaccini; Robert K. Tyson; Eds., 34–43

Graves, J. E., Northcott, M. J., Roddier, F. J., *et al.* 2000, in Proc. SPIE Vol. 4007, p. 26-30, Adaptive Optical Systems Technology, Peter L. Wizinowich; Ed., 26–30

Gray, A. D., Nicholls, J., Ekers, R. D., & Cram, L. E. 1995, Astrophys. J., 448, 164

Greenhill, L. J., Colomer, F., Moran, J. M., *et al.* 1995a, Astrophys. J., 449, 365

Greenhill, L. J., Jiang, D. R., Moran, J. M., *et al.* 1995b, Astrophys. J., 440, 619

Griffin, M. J., Ade, P. A. R., Orton, G. S., *et al.* 1986, Icarus, 65, 244

Grindlay, J. E. & Liller, W. 1978, in COSPAR, Plenary Meeting, 83

Guesten, R., Genzel, R., Wright, M. C. H., *et al.* 1987, Astrophys. J., 318, 124

Haiman, Z. & Loeb, A. 2001, Astrophys. J., 552, 459

Haller, J. W., Rieke, M. J., Rieke, G. H., *et al.* 1996, Astrophys. J., 456,

194
Hansen, B. M. S. & Milosavljević, M. 2003, Astrophys. J., Lett., 593, L77
Harms, R. J., Ford, H. C., Tsvetanov, Z. I., *et al.* 1994, Astrophys. J., Lett., 435, L35
Hehl, F. W., Kiefer, C., & Metzler, R. J. K. 1998, LNP Vol. 514: Black Holes: Theory and Observation
Henry, J. P., Depoy, D. L., & Becklin, E. E. 1984, Astrophys. J., Lett., 285, L27
Herbst, T. M., Beckwith, S. V. W., & Shure, M. 1993, Astrophys. J., Lett., 411, L21
Herbst, T. M., Rix, H., Bizenberger, P., & Ollivier, M. 2000, in Proc. SPIE Vol. 4006, p. 673-680, Interferometry in Optical Astronomy, Pierre J. Lena; Andreas Quirrenbach; Eds., 673–680
Herrnstein, J. R., Moran, J. M., Greenhill, L. J., Blackman, E. G., & Diamond, P. J. 1998, Astrophys. J., 508, 243
Herrnstein, R. M. & Ho, P. T. P. 2002, Astrophys. J., Lett., 579, L83
Heyvaerts, J., Norman, C., & Pudritz, R. E. 1988, Astrophys. J., 330, 718
Hildebrand, R. H., Gonatas, D. P., Platt, S. R., *et al.* 1990, Astrophys. J., 362, 114
Hildebrand, R. H., Davidson, J. A., Dotson, J., *et al.* 1993, Astrophys. J., 417, 565
Hill, J. M. 1997, in Proc. SPIE Vol. 2871, p. 57-68, Optical Telescopes of Today and Tomorrow, Arne L. Ardeberg; Ed., 57–68
Ho, L. 1999, in ASSL Vol. 234: Observational Evidence for the Black Holes in the Universe, 157
Hofmann, R., Eckart, A., Genzel, R., & Drapatz, S. 1993, Astrophys. Space. Sci., 205, 1
Hofmann, R., Brandl, B., Eckart, A., Eisenhauer, F., & Tacconi-Garman, L. E. 1995, in Proc. SPIE Vol. 2475, p. 192-202, Infrared Detectors and Instrumentation for Astronomy, Albert M. Fowler; Ed., 192–202
Holland, W. S., Robson, E. I., Gear, W. K., *et al.* 1999, Mon. Not. of the Royal Astron. Soc., 303, 659
Honingh, C. E., Haas, S., Hottgenroth, D., Jacobs, K., & Stutzki, J. 1997, in 8th Int. Symp. ed. R. Blundell and E. Tong (Cambridge: Harvard Univ.)
Hornstein, S. D., Ghez, A. M., Tanner, A., *et al.* 2002, Astrophys. J., Lett., 577, L9
Horrobin, M., Eisenhauer, F., Tecza, M., *et al.* 2004, Astronomische Nachrichten, 325, 88

Hubin, N. 1997, in Proc. SPIE Vol. 2871, p. 827-836, Optical Telescopes of Today and Tomorrow, Arne L. Ardeberg; Ed., 827–836

Ishii, M., Nagata, T., Sato, S., *et al.* 1998, Astron. J., 116, 868

Jackson, J. M., Geis, N., Genzel, R., *et al.* 1993, Astrophys. J., 402, 173

Jaroszynski, M. 1998, Acta Astronomica, 48, 653

—. 1999, Acta Astronomica, 49, 117

Jauncey, D. L., Tzioumis, A. K., Preston, R. A., *et al.* 1989, Astron. J., 98, 44

Jefferies, S. M. & Christou, J. C. 1993, Astrophys. J., 415, 862

Jiang, H. X. & Lin, J. Y. 1985, American Journal of Physics, 53, 694

Kaup, D. J. 1968, Physical Review, 172, 1331

Khokhlov, A. & Melia, F. 1996, Astrophys. J., Lett., 457, L61

Kim, S. S. & Morris, M. 2003, Astrophys. J., 597, 312

Klein, B. L., Ghez, A. M., Morris, M., & Becklin, E. E. 1996, in The Galactic Center, Astronomical Society of the Pacific Conference Series, Volume 102, Manuscripts presented at the 4th international meeting jointly organized by the European Southern Observatory (ESO) and Cerro Tololo Inter-American Observatory (CTIO), held March 10-15, 1996 in La Serena, Chile, San Francisco: Astronomical Society of the Pacific (ASP), —c1996, edited by Roland Gredel, p.228, 228

Knacke, R. F. & Capps, R. W. 1977, Astrophys. J., 216, 271

Knox, K. T. & Thompson, B. J. 1974, Astrophys. J., Lett., 193, L45

Knox, K. T. 1976, Optical Society of America Journal, 66, 1236

Koornneef, J. 1983, Astron. & Astroph., 128, 84

Korff, D., Dryden, G., & Miller, M. G. 1972, Opt. Commun., 5, 187

Kormendy, J. & Richstone, D. 1995, ARA&A, 33, 581

Kormendy, J., Bender, R., Ajhar, E. A., *et al.* 1996a, Astrophys. J., Lett., 473, L91

Kormendy, J., Bender, R., Richstone, D., *et al.* 1996b, Astrophys. J., Lett., 459, L57

Kormendy, J., Bender, R., Magorrian, J., *et al.* 1997, Astrophys. J., Lett., 482, L139

Kormendy, J. 2001, in Revista Mexicana de Astronomia y Astrofisica Conference Series, 69–78

Kormendy, J. 2004, in Coevolution of Black Holes and Galaxies

Krabbe, A., Genzel, R., Drapatz, S., & Rotaciuc, V. 1991, Astrophys. J., Lett., 382, L19

Krabbe, A., Genzel, R., Eckart, A., *et al.* 1995, Astrophys. J., Lett., 447, L95

Krichbaum, T. P., Zensus, J. A., Witzel, A., *et al.* 1993, Astron. & Astroph., 274, L37

Krichbaum, T. P., Graham, D. A., Witzel, A., *et al.* 1998, Astron. & Astroph., 335, L106

Labbé, I., Rudnick, G., Franx, M., *et al.* 2003, Astrophys. J., Lett., 591, L95

Lacombe, F., Marco, O., Geoffray, H., *et al.* 1998, Publ. Astron. Soc. Pac., 110, 1087

Lacy, J. H., Townes, C. H., & Hollenbach, D. J. 1982, Astrophys. J., 262, 120

Lacy, J. H., Achtermann, J. M., & Serabyn, E. 1991, Astrophys. J., Lett., 380, L71

LaRosa, T. N., Kassim, N. E., Lazio, T. J. W., & Hyman, S. D. 2000, Astron. J., 119, 207

Latvakoski, H., Stacey, G., Hayward, T., & Gull, G. 1996, Bulletin of the American Astronomical Society, 28, 1350

Latvakoski, H. M., Stacey, G. J., Gull, G. E., & Hayward, T. L. 1999, Astrophys. J., 511, 761

Lauer, T. R., Gebhardt, K., Richstone, D., *et al.* 2002, Astron. J., 124, 1975

Lebofsky, M. J., Rieke, G. H., Deshpande, M. R., & Kemp, J. C. 1982, Astrophys. J., 263, 672

Lebofsky, M. J. & Rieke, G. H. 1987, in AIP Conf. Proc. 155: The Galactic Center, 79

Lebrun, F., Leray, J. P., Lavocat, P., *et al.* 2003, Astron. & Astroph., 411, L141

Lee, H. M. 1994, in NATO ASIC Proc. 445: The Nuclei of Normal Galaxies: Lessons from the Galactic Center, 335

Lee, H. M. 1995, Mon. Not. of the Royal Astron. Soc., 272, 605

Leinert, C. & Graser, U. 1998, in Proc. SPIE Vol. 3350, p. 389-393, Astronomical Interferometry, Robert D. Reasenberg; Ed., 389–393

Lenzen, R., Hofmann, R., Bizenberger, P., & Tusche, A. 1998, in Proc. SPIE Vol. 3354, p. 606-614, Infrared Astronomical Instrumentation, Albert M. Fowler; Ed., 606–614

Leonard, P. J. T. & Merritt, D. 1989, Astrophys. J., 339, 195

Lesch, H. & Reich, W. 1992, Astron. & Astroph., 264, 493

Levin, Y. & Beloborodov, A. M. 2003, Astrophys. J., Lett., 590, L33

Levine, D. A. 1995, Ph.D. Thesis

Lindqvist, M., Habing, H. J., & Winnberg, A. 1992, Astron. & Astroph.,

259, 118
Lis, D. C. & Menten, K. M. 1998, Astrophys. J., 507, 794
Liszt, H. S. & Spiker, R. W. 1995, ApJS, 98, 259
Liu, T., Becklin, E. E., Henry, J. P., & Simons, D. 1993, Astron. J., 106, 1484
Liu, S. & Melia, F. 2001, Astrophys. J., Lett., 561, L77
—. 2002, Astrophys. J., Lett., 566, L77
Lo, K. Y., Schilizzi, R. T., Cohen, M. H., & Ross, H. N. 1975, Astrophys. J., Lett., 202, L63
Lo, K. Y. & Claussen, M. J. 1983, Nature, 306, 647
Lo, K. Y., Backer, D. C., Kellermann, K. I., *et al.* 1993, Nature, 362, 38
Lo, K. Y., Shen, Z., Zhao, J., & Ho, P. T. P. 1998, Astrophys. J., Lett., 508, L61
Lohmann, A. W., Weigelt, G., & Wirnitzer, B. 1983, Appl. Opt., 22, 4028
Looney, L. W., Raab, W., Poglitsch, A., *et al.* 2003, in Airborne Telescope Systems II. Edited by Ramsey K. Melugin, Hans-Peter Roeser . Proceedings of the SPIE, Volume 4857, pp. 47-55 (2003)., 47–55
Lucy, L. B. 1974, Astron. J., 79, 745
Lugten, J. B. 1987, Ph.D. Thesis
Lund, N., Budtz-Jørgensen, C., Westergaard, N. J., *et al.* 2003, Astron. & Astroph., 411, L231
Lutz, D., Feuchtgruber, H., Genzel, R., *et al.* 1996, Astron. & Astroph., 315, L269
Lynden-Bell, D. 1969, Nature, 223, 690
Lynden-Bell, D. & Rees, M. J. 1971, Mon. Not. of the Royal Astron. Soc., 152, 461
Maeda, Y., Baganoff, F. K., Feigelson, E. D., *et al.* 2002, Astrophys. J., 570, 671
Magorrian, J., Tremaine, S., Richstone, D., *et al.* 1998, Astron. J., 115, 2285
Maillard, J.-P., Paumard, T., Stolovy, S., & Rigaut, F. 2003, in Galactic Center Workshop 2002: The Central 300 Parsecs of the Milky Way, 263–269
Makino, J. & Aarseth, S. J. 1992, Publ. Astron. Soc. Jpn., 44, 141
Maoz, E. 1995, Astrophys. J., Lett., 447, L91
—. 1998, Astrophys. J., Lett., 494, L181
Marcaide, J. M., Alberdi, A., Bartel, N., *et al.* 1992, Astron. & Astroph., 258, 295
Marchetti, E., Hubin, N. N., Fedrigo, E., *et al.* 2003, in Adaptive Opti-

cal System Technologies II. Edited by Wizinowich, Peter L.; Bonaccini, Domenico. Proceedings of the SPIE, Volume 4839, pp. 317-328 (2003)., 317–328

Mariotti, J. M., Chelli, A., Sibille, F., *et al.* 1983, Astron. & Astroph., 120, 237

Mariotti, J.-M. 1989, in NATO ASIC Proc. 274: Diffraction-Limited Imaging with Very Large Telescopes, 3–23

Markoff, S., Falcke, H., Yuan, F., & Biermann, P. L. 2001, Astron. & Astroph., 379, L13

Marscher, A. P. 1983, Astrophys. J., 264, 296

Martinez, A., Bujarrabal, V., & Alcolea, J. 1988, A&AS, 74, 273

Mas-Hesse, J. M., Giménez, A., Culhane, J. L., *et al.* 2003, Astron. & Astroph., 411, L261

Matthews, K. & Soifer, B. T. 1994, Experimental Astronomy, 3, 77

Matthews, K., Ghez, A. M., Weinberger, A. J., & Neugebauer, G. 1996, Publ. Astron. Soc. Pac., 108, 615

Mayer-Hasselwander, H. A., Bertsch, D. L., Dingus, B. L., *et al.* 1998, Astron. & Astroph., 335, 161

McClintock, J. E. & Remillard, R. A. 2003, astro-ph/0306213

McGinn, M. T., Sellgren, K., Becklin, E. E., & Hall, D. N. B. 1989, Astrophys. J., 338, 824

McLean, I. S., Becklin, E. E., Bendiksen, O., *et al.* 1998, in Proc. SPIE Vol. 3354, p. 566-578, Infrared Astronomical Instrumentation, Albert M. Fowler; Ed., 566–578

McLean, I. S., Wilcox, M. K., Becklin, E. E., *et al.* 2000, Astrophys. J., Lett., 533, L45

McLean, I. S. & Sprayberry, D. 2003a, in Instrument Design and Performance for Optical/Infrared Ground-based Telescopes. Edited by Iye, Masanori; Moorwood, Alan F. M. Proceedings of the SPIE, Volume 4841, pp. 1-6 (2003)., 1–6

McLean, I. S. & Sprayberry, D. 2003b, in Instrument Design and Performance for Optical/Infrared Ground-based Telescopes. Edited by Iye, Masanori; Moorwood, Alan F. M. Proceedings of the SPIE, Volume 4841, pp. 1-6 (2003)., 1–6

Meeks, M. L. 1976, New York, Academic Press, Inc. (Methods of Experimental Physics. Volume 12, Pt. C), 1976. 367 p, 12

Melia, F. 1992, Astrophys. J., Lett., 387, L25

Melia, F., Jokipii, J. R., & Narayanan, A. 1992, Astrophys. J., Lett., 395, L87

Melia, F. 1994, Astrophys. J., 426, 577
Melia, F. & Coker, R. 1999, Astrophys. J., 511, 750
Melia, F., Bromley, B. C., Liu, S., & Walker, C. K. 2001, Astrophys. J., Lett., 554, L37
Melia, F. & Falcke, H. 2001, ARA&A, 39, 309
Menten, K. M., Reid, M. J., Eckart, A., & Genzel, R. 1997, Astrophys. J., Lett., 475, L111
Merloni, A., Vietri, M., Stella, L., & Bini, D. 1999, Mon. Not. of the Royal Astron. Soc., 304, 155
Mezger, P. G., Zylka, R., Salter, C. J., *et al.* 1989, Astron. & Astroph., 209, 337
Mezger, P. G., Duschl, W. J., & Zylka, R. 1996, A&AR, 7, 289
Mielke, E. & Schunck, F. 2000, Nucl.Phys. B, 594, 1985
Miralda-Escudé, J. & Gould, A. 2000, Astrophys. J., 545, 847
Misner, C. W., Thorne, K. S., & Wheeler, J. A. 1973, Gravitation (San Francisco: W.H. Freeman and Co., 1973)
Miyazaki, A., Tsutsumi, T., & Tsuboi, M. 2003, in Galactic Center Workshop 2002: The Central 300 Parsecs of the Milky Way, 363–369
Miyoshi, M., Matsumoto, K., Kameno, S., Takaba, H., & Iwata, T. 1994, Nature, 371, 395
Miyoshi, M., Moran, J., Herrnstein, J., *et al.* 1995, Nature, 373, 127
Moneti, A., Cernicharo, J., & Pardo, J. R. 2001, Astrophys. J., Lett., 549, L203
Moran, J. M. 1998, in Proc. SPIE Vol. 3357, p. 208-219, Advanced Technology MMW, Radio, and Terahertz Telescopes, Thomas G. Phillips; Ed., 208–219
Moroz, V. I. 1961, Soviet Astronomy, 5, 361
Morris, M. & Yusef-Zadeh, F. 1985, Astron. J., 90, 2511
—. 1989, Astrophys. J., 343, 703
Morris, M., Davidson, J. A., Werner, M., *et al.* 1992, Astrophys. J., Lett., 399, L63
Morris, M. 1993, Astrophys. J., 408, 496
Morris, M. 1994, in NATO ASIC Proc. 445: The Nuclei of Normal Galaxies: Lessons from the Galactic Center, 185
Morris, M. 1996, in IAU Symp. 169: Unsolved Problems of the Milky Way, 247
Morris, M. & Serabyn, E. 1996, ARA&A, 34, 645
Morris, M., Ghez, A. M., & Becklin, E. E. 1999, Advances in Space Research, 23, 959

Mouawad, N., Eckart, A., & Pfalzner, S. 2003a, LNP Vol. 626: Galaxies and Chaos, 302

Mouawad, N., Eckart, A., Pfalzner, S., *et al.* 2003b, Astronomische Nachrichten Supplement, 324, 315

—. 2004, AN submitted

Moultaka, J., Eckart, A., Viehmann, T., *et al.* 2004, Astron. & Astroph., 425, 529

Muno, M. P., Baganoff, F. K., Bautz, M. W., *et al.* 2003, Astrophys. J., 589, 225

Munyaneza, F., Tsiklauri, D., & Viollier, R. D. 1998, Astrophys. J., Lett., 509, L105

Murphy, B. W., Cohn, H. N., & Durisen, R. H. 1991, Astrophys. J., 370, 60

Nagata, T., Woodward, C. E., Shure, M., Pipher, J. L., & Okuda, H. 1990, Astrophys. J., 351, 83

Nagata, T., Woodward, C. E., Shure, M., & Kobayashi, N. 1995, Astron. J., 109, 1676

Najarro, F., Hillier, D. J., Kudritzki, R. P., *et al.* 1994, Astron. & Astroph., 285, 573

Najarro, F., Krabbe, A., Genzel, R., *et al.* 1997, Astron. & Astroph., 325, 700

Narayan, R. & Yi, I. 1995, Astrophys. J., 452, 710

Narayan, R., Yi, I., & Mahadevan, R. 1995, Nature, 374, 623

Narayan, R., Mahadevan, R., Grindlay, J. E., Popham, R. G., & Gammie, C. 1998, Astrophys. J., 492, 554

Narayan, R., Quataert, E., Igumenshchev, I. V., & Abramowicz, M. A. 2002, Astrophys. J., 577, 295

Narayan, R., Igumenshchev, I. V., & Abramowicz, M. A. 2003, Publ. Astron. Soc. Jpn., 55, L69

Nayakshin, S., Cuadra, J., & Sunyaev, R. 2004, Astron. & Astroph., 413, 173

Nelson, C. H. 2000a, Astrophys. J., Lett., 544, L91

Nelson, J. E. 2000b, in Proc. SPIE Vol. 4004, p. 282-289, Telescope Structures, Enclosures, Controls, Assembly/Integration/Validation, and Commissioning, Thomas A. Sebring; Torben Andersen; Eds., 282–289

Nelson, J. E. 2003, in Future Giant Telescopes. Edited by Angel, J. Roger P.; Gilmozzi, Roberto. Proceedings of the SPIE, Volume 4840, pp. 47-59 (2003)., 47–59

Nord, M. E., Lazio, T. J. W., Kassim, N. E., Goss, W. M., & Duric, N.

2004, Astrophys. J., Lett., 601, L51
Norton, P. 1991, Optical Engineering, 30, 1649
Nowak, M. A., Wagoner, R. V., Begelman, M. C., & Lehr, D. E. 1997, Astrophys. J., Lett., 477, L91
Nowak, M. & Lehr, D. 1998, in Theory of Black Hole Accretion Disks, 233
Özel, F., Psaltis, D., & Narayan, R. 2000, Astrophys. J., 541, 234
Okuda, H., Shibai, H., Nakagawa, T., *et al.* 1990, Astrophys. J., 351, 89
Ott, T., Eckart, A., & Genzel, R. 1999, Astrophys. J., 523, 248
Ott, T., Genzel, R., Eckart, E., & Schödel, R. 2003, in Galactic Center Workshop 2002: The Central 300 Parsecs of the Milky Way, 543–549
Ott, T. 2004, Ph.D. Thesis
Paresce, F., Delplancke, F., Derie, F., *et al.* 2003, in Interferometry for Optical Astronomy II. Edited by Wesley A. Traub. Proceedings of the SPIE, Volume 4838, pp. 486-495 (2003)., 486–495
Park, S., Muno, M. P., Baganoff, F. K., *et al.* 2004, Astrophys. J., 603, 548
Paumard, T., Maillard, J. P., Morris, M., & Rigaut, F. 2001, Astron. & Astroph., 366, 466
Paumard, T., Maillard, J., & Morris, M. 2003, Astronomische Nachrichten Supplement, 324, 605
Pedlar, A., Anantharamaiah, K. R., Ekers, R. D., *et al.* 1989, Astrophys. J., 342, 769
Perrier, C. 1989, in NATO ASIC Proc. 274: Diffraction-Limited Imaging with Very Large Telescopes, 99
Phinney, E. S. 1989, in IAU Symp. 136: The Center of the Galaxy, 543
Pierce-Price, D., Richer, J. S., Greaves, J. S., *et al.* 2000, Astrophys. J., Lett., 545, L121
Poglitsch, A., Stacey, G. J., Geis, N., *et al.* 1991, Astrophys. J., Lett., 374, L33
Porquet, D., Predehl, P., Aschenbach, B., *et al.* 2003, Astron. & Astroph., 407, L17
Portegies Zwart, S. F. & McMillan, S. L. W. 2000, Astrophys. J., Lett., 528, L17
Portegies Zwart, S. F., McMillan, S. L. W., & Gerhard, O. 2003, Astrophys. J., 593, 352
Predehl, P. & Truemper, J. 1994, Astron. & Astroph., 290, L29
Predehl, P. 1995, Advances in Space Research, 16, 159
Press, W. H., Flannery, B. P., Teukolsky, S. A., & Vetterling, W. T. 1989, Numerical recipes in C. The art of scientific computing (Cambridge: University Press, 1989)

Quataert, E. 2003, Astronomische Nachrichten Supplement, 324, 435
—. 2004, Astrophys. J., 613, 322
Quinlan, G. D., Hernquist, L., & Sigurdsson, S. 1995, Astrophys. J., 440, 554
Quirrenbach, A., Coude Du Foresto, V., Daigne, G., *et al.* 1998, in Proc. SPIE Vol. 3350, p. 807-817, Astronomical Interferometry, Robert D. Reasenberg; Ed., 807–817
Ragazzoni, R., Soci, R., Arcidiacono, C., *et al.* 2003, in Astronomical Adaptive Optics Systems and Applications. Edited by Tyson, Robert K.; Lloyd-Hart, Michael. Proceedings of the SPIE, Volume 5169, pp. 181-189 (2003)., 181–189
Rees, M. J., Phinney, E. S., Begelman, M. C., & Blandford, R. D. 1982, Nature, 295, 17
Rees, M. J. 1984, ARA&A, 22, 471
Reich, W. 1994, in NATO ASIC Proc. 445: The Nuclei of Normal Galaxies: Lessons from the Galactic Center, 55
Reid, M. J. 1993, ARA&A, 31, 345
Reid, M. J., Readhead, A. C. S., Vermeulen, R. C., & Treuhaft, R. N. 1999, Astrophys. J., 524, 816
Reid, M. J., Menten, K. M., Genzel, R., *et al.* 2003a, Astrophys. J., 587, 208
Reid, M. J., Menten, K. M., Genzel, R., Ott, T., & Schödel, R. 2003b, in Galactic Center Workshop 2002: The Central 300 Parsecs of the Milky Way, 505–511
Richardson, W. H. 1992, J. Opt. Soc. Am., 62, 55
Richardson, E. H. 1994, in NATO ASIC Proc. 423: Adaptive Optics for Astronomy, 227
Richichi, A., Malbet, F., Hofmann, K.-H., & Petrov, R. 2000, in ESA SP-451: Darwin and Astronomy : the Infrared Space Interferometer, 51
Rieke, G. H., Telesco, C. M., & Harper, D. A. 1978, Astrophys. J., 220, 556
Rieke, G. H. & Rieke, M. J. 1988, Astrophys. J., Lett., 330, L33
Rieke, G. H., Rieke, M. J., & Paul, A. E. 1989, Astrophys. J., 336, 752
Rieke, G. H. 1994, Detection of light: from the ultraviolet to the submillimeter (Cambridge, UK: Cambridge University Press, edited by Visnovsky, Karen)
Rieke, M. J. 2003, in A Decade of Hubble Space Telescope Science, 119
Rigaut, F., Salmon, D., Arsenault, R., *et al.* 1998, Publ. Astron. Soc. Pac., 110, 152
Rigaut, F., Geballe, T., Roy, J.-R., & Draine, B. 2003, in Galactic Center

Workshop 2002: The Central 300 Parsecs of the Milky Way, 597–603
Roberts, D. A. & Goss, W. M. 1993, ApJS, 86, 133
Roddier, F. 1989, in NATO ASIC Proc. 274: Diffraction-Limited Imaging with Very Large Telescopes, 33
Roddier, N. 1991, in ACTIVE AND ADAPTIVE OPTICAL SYSTEMS SPIE 1542 P. 120, 1991, 120
Rodríguez-Fernández, N. J., Martín-Pintado, J., & de Vicente, P. 2001, Astron. & Astroph., 377, 631
Rogers, A. E. E., Doeleman, S., Wright, M. C. H., *et al.* 1994, Astrophys. J., Lett., 434, L59
Rosa, M. R., Zinnecker, H., Moneti, A., & Melnick, J. 1992, Astron. & Astroph., 257, 515
Rosner, R. & Bodo, G. 1996, Astrophys. J., Lett., 470, L49
Rousset, G., Lacombe, F., Puget, P., *et al.* 1998, in Proc. SPIE Vol. 3353, p. 508-516, Adaptive Optical System Technologies, Domenico Bonaccini; Robert K. Tyson; Eds., 508–516
Roy, S. & Pramesh Rao, A. 2003, Astronomische Nachrichten Supplement, 324, 391
—. 2004, Mon. Not. of the Royal Astron. Soc., 349, L25
Rubilar, G. F. & Eckart, A. 2001, Astron. & Astroph., 374, 95
Ruffini, R. & Bonazzola, S. 1969, Physical Review, 187, 1767
Ruszkowski, M. & Begelman, M. C. 2002, Astrophys. J., 573, 485
Sakano, M., Warwick, R. S., Decourchelle, A., & Predehl, P. 2004, Mon. Not. of the Royal Astron. Soc., 350, 129
Sanders, R. H. 1992, Nature, 359, 131
Sault, R. J. & Macquart, J.-P. 1999, Astrophys. J., Lett., 526, L85
Schmitt, H. R., Donley, J. L., Antonucci, R. R. J., *et al.* 2003, Astrophys. J., 597, 768
Schödel, R., Ott, T., Genzel, R., *et al.* 2002, Nature, 419, 694
Schödel, R., Ott, T., Genzel, R., *et al.* 2003, Astrophys. J., 596, 1015
Schödel, R. 2004, Ph.D. Thesis
Schoeller, M., Gitton, P. B., Argomedo, J., *et al.* 2003, in Interferometry for Optical Astronomy II. Edited by Wesley A. Traub . Proceedings of the SPIE, Volume 4838, pp. 870-880 (2003)., 870–880
Scoville, N. Z., Stolovy, S. R., Rieke, M., Christopher, M., & Yusef-Zadeh, F. 2003, Astrophys. J., 594, 294
Scribner, D., Kruer, M., & Killiany, J. 1991, Proceedings of the IEEE, 79, 66-85
Sellgren, K., McGinn, M. T., Becklin, E. E., & Hall, D. N. 1990, Astrophys.

J., 359, 112
Serabyn, E. & Guesten, R. 1991, Astron. & Astroph., 242, 376
Serabyn, E. & Morris, M. 1994, Astrophys. J., Lett., 424, L91
Serabyn, E., Shupe, D., & Figer, D. F. 1998, Nature, 394, 448
Shakura, N. I. & Sunyaev, R. A. 1973, Astron. & Astroph., 24, 337
Shields, G. A., Gebhardt, K., Salviander, S., *et al.* 2003, Astrophys. J., 583, 124
Sills, A. & Bailyn, C. D. 1999, Astrophys. J., 513, 428
Sills, A., Faber, J. A., Lombardi, J. C., Rasio, F. A., & Warren, A. R. 2001, Astrophys. J., 548, 323
Simon, M., Chen, W.-P., Forrest, W. J., *et al.* 1990, Astrophys. J., 360, 95
Simons, D. A., Hodapp, K.-W., & Becklin, E. E. 1990, Astrophys. J., 360, 106
Simons, D. A. & Becklin, E. E. 1996, Astron. J., 111, 1908
Sofue, Y. & Handa, T. 1984, Nature, 310, 568
Sofue, Y. 1985, Publ. Astron. Soc. Jpn., 37, 697
Spanoudakis, P., Zago, L., Chetelat, O., Gentsch, R., & Mato Mira, F. 2000, in Proc. SPIE Vol. 4007, p. 408-415, Adaptive Optical Systems Technology, Peter L. Wizinowich; Ed., 408–415
Spitzer, L. J. & Hart, M. H. 1971, Astrophys. J., 164, 399
Spitzer, L. J. & Thuan, T. X. 1972, Astrophys. J., 175, 31
Spitzer, L. J. 1998, Physical Processes in the Interstellar Medium (John Wiley & Sons, inc.: ISBN 0-471-29335-0)
Spurzem, R. 1999, Journal of Computational and Applied Mathematics, 109, 407
Stacey, G. J., Nikola, T., Bradford, C. M., *et al.* 2002, EAS Publications Series, Volume 4, Proceedings of Infrared and Submillimeter Space Astronomy", held 11-13 June, 2001. Edited by M. Giard, J.P. Bernanrd, A. Klotz, and I. Ristorcelli. EDP Sciences, 2002, pp.419-419, 4, 419
Stacey, G. J., Nikola, T., Bradford, C. M., *et al.* 2004, in The Dense Interstellar Medium in Galaxies, 273–276
Staguhn, J., Stutzki, J., Uchida, K. I., & Yusef-Zadeh, F. 1998, Astron. & Astroph., 336, 290
Staguhn, J. G., Benford, D. J., & Dowell, C. D. 2003, American Astronomical Society Meeting, 202
Stark, A. A., Bolatto, A. D., Chamberlin, R. A., *et al.* 1997, Astrophys. J., Lett., 480, L59
Stark, A. A., Bally, J., Balm, S. P., *et al.* 2001, Publ. Astron. Soc. Pac., 113, 567

Stebbins, J. & Whitford, A. E. 1947, Astron. J., 52, 130

Stolovy, S. R., Herter, T., Gull, G. E., Pirger, B., & Vogt, N. P. 1995, in ASP Conf. Ser. 73: From Gas to Stars to Dust, 469–476

Stolovy, S. R., Yusef-Zadeh, F., Burton, M., *et al.* 1999, Bulletin of the American Astronomical Society, 31, 970

Straubmeier, C., Eckart, A., Bertram, T., Zealouk, L., & Wang, Y. 2003, in Interferometry for Optical Astronomy II. Edited by Wesley A. Traub . Proceedings of the SPIE, Volume 4838, pp. 1271-1281 (2003)., 1271–1281

Swinyard, B. M., Clegg, P. E., Ade, P. A. R., *et al.* 1996, Astron. & Astroph., 315, L43

Tamblyn, P., Rieke, G. H., Hanson, M. M., *et al.* 1996, Astrophys. J., 456, 206

Tanner, A., Ghez, A. M., Morris, M., *et al.* 2002, Astrophys. J., 575, 860

Tanner, A., Ghez, A., Morris, M., & Becklin, E. 2003, in Galactic Center Workshop 2002: The Central 300 Parsecs of the Milky Way, 597–603

Tecza, M., Thatte, N. A., Krabbe, A., & Tacconi-Garman, L. E. 1998, in Proc. SPIE Vol. 3354, p. 394-403, Infrared Astronomical Instrumentation, Albert M. Fowler; Ed., 394–403

Telesco, C. M., Davidson, J. A., & Werner, M. W. 1995, in ASP Conf. Ser. 73: From Gas to Stars to Dust, 503–504

Thompson, A. R., Moran, J. M., & Swenson, G. W. 1986, Interferometry and synthesis in radio astronomy (New York, Wiley-Interscience, 1986, 554 p.)

Thornley, M. D., Schreiber, N. M. F., Lutz, D., *et al.* 2000, Astrophys. J., 539, 641

Tiede, G. P., Frogel, J. A., & Terndrup, D. M. 1995, Astron. J., 110, 2788

Timmermann, R., Genzel, R., Poglitsch, A., *et al.* 1996, Astrophys. J., 466, 242

Torres, D. F., Capozziello, S., & Lambiase, G. 2000, Phys. Rev. D, 62, 104012

Tremaine, S., Gebhardt, K., Bender, R., *et al.* 2002, Astrophys. J., 574, 740

Tsuboi, M., Inoue, M., Handa, T., *et al.* 1986, Astron. J. 92, 818

Tsuboi, M., Kawabata, T., Kasuga, T., Handa, T., & Kato, T. 1995, Publ. Astron. Soc. Jpn., 47, 829

Tyson, R. K. 1998, Principles of adaptive optics (Principles of adaptive optics, Edition: 2nd ed., Publisher: Boston, MA: Academic Press, 1998, ISBN: 0127059024)

Ubertini, P., Lebrun, F., Di Cocco, G., *et al.* 2003, Astron. & Astroph., 411, L131

Uchida, K. I., Morris, M., Serabyn, E., & Guesten, R. 1996, Astrophys. J., 462, 768

van der Marel, R. P., de Zeeuw, P., Rix, H.-W., & Quinlan, G. D. 1997, Nature, 385, 610

van der Marel, R. P. & van den Bosch, F. C. 1998, Astron. J., 116, 2220

Vedrenne, G., Roques, J.-P., Schönfelder, V., *et al.* 2003, Astron. & Astroph., 411, L63

Veilleux, S., Shopbell, P. L., Rupke, D. S., Bland-Hawthorn, J., & Cecil, G. 2003, Astron. J., 126, 2185

Viehmann, T., Eckart, A., Schödel, R., *et al.* 2005, Astron. & Astroph. in press

Viollier, R. D., Leimgruber, F. R., & Trautmann, D. 1992, Physics Letters B, 297, 132

Vollmer, B. & Duschl, W. J. 2000, New Astronomy, 4, 581

—. 2002, Astron. & Astroph., 388, 128

Vollmer, B., Beckert, T., & Duschl, W. J. 2004, Astron. & Astroph., 413, 949

Walker, C. K., Kooi, J. W., Chan, M., *et al.* 1992, Int. J. Infrared Millimeter Waves, 13, 785

Wandel, A. 1999, Astrophys. J., Lett., 519, L39

Ward, W. R. 1986, Icarus, 67, 164

Wardle, M. & Konigl, A. 1990, Astrophys. J., 362, 120

Weitzel, L., Krabbe, A., Kroker, H., *et al.* 1996, A&AS, 119, 531

Werner, M. W., Davidson, J. A., Morris, M., *et al.* 1988, Astrophys. J., 333, 729

Winkler, C., Courvoisier, T. J.-L., Di Cocco, G., *et al.* 2003, Astron. & Astroph., 411, L1

Winnberg, A., Baud, B., Matthews, H. E., Habing, H. J., & Olnon, F. M. 1985, Astrophys. J., Lett., 291, L45

Wizinowich, P., Acton, D. S., Shelton, C., *et al.* 2000a, Publ. Astron. Soc. Pac., 112, 315

Wizinowich, P. L., Acton, D. S., Lai, O., *et al.* 2000b, in Proc. SPIE Vol. 4007, p. 2-13, Adaptive Optical Systems Technology, Peter L. Wizinowich; Ed., 2–13

Wright, M. C. H., Coil, A. L., McGary, R. S., Ho, P. T. P., & Harris, A. I. 2001, Astrophys. J., 551, 254

Young, P. 1980, Astrophys. J., 242, 1232

Yu, Q. & Tremaine, S. 2002, Mon. Not. of the Royal Astron. Soc., 335, 965

Yuan, F., Markoff, S., & Falcke, H. 2002, Astron. & Astroph., 383, 854

Yuan, F., Quataert, E., & Narayan, R. 2003, Astrophys. J., 598, 301

Yusef-Zadeh, F., Morris, M., & Chance, D. 1984, Nature, 310, 557

Yusef-Zadeh, F. & Morris, M. 1987, Astrophys. J., 322, 721

Yusef-Zadeh, F., Wardle, M., & Parastaran, P. 1997, Astrophys. J., Lett., 475, L119

Yusef-Zadeh, F., Roberts, D. A., & Biretta, J. 1998a, Astrophys. J., Lett., 499, L159

—. 1998b, Astrophys. J., Lett., 503, L191

Yusef-Zadeh, F., Stolovy, S. R., Burton, M., Wardle, M., & Ashley, M. C. B. 2001, Astrophys. J., 560, 749

Zhao, J. & Goss, W. M. 1998, Astrophys. J., Lett., 499, L163

Zhao, J., Bower, G. C., & Goss, W. M. 2001, Astrophys. J., Lett., 547, L29

Zhao, J., Young, K. H., Herrnstein, R. M., *et al.* 2003, Astrophys. J., Lett., 586, L29

Zoccali, M., Renzini, A., Ortolani, S., *et al.* 2003, Astron. & Astroph., 399, 931

Zylka, R., Mezger, P. G., & Lesch, H. 1992, Astron. & Astroph., 261, 119

Zylka, R., Mezger, P. G., Ward-Thompson, D., Duschl, W. J., & Lesch, H. 1995, Astron. & Astroph., 297, 83

Index